VOLUME THREE HUNDRED AND THIRTY THREE

INTERNATIONAL REVIEW OF CELL AND MOLECULAR BIOLOGY

MiRNAs in Differentiation and Development

INTERNATIONAL REVIEW OF CELL AND MOLECULAR BIOLOGY

VOLUME THREE HUNDRED AND THIRTY THREE

INTERNATIONAL REVIEW OF CELL AND MOLECULAR BIOLOGY

MiRNAs in Differentiation and Development

Edited by

LORENZO GALLUZZI

Department of Radiation Oncology
Weill Cornell Medical College
New York, NY

ILIO VITALE

Department of Biology
University of Rome "Tor Vergata"
Rome, Italy

ELSEVIER

ACADEMIC PRESS

An imprint of Elsevier

Academic Press is an imprint of Elsevier
50 Hampshire Street, 5th Floor, Cambridge, MA 02139, United States
525 B Street, Suite 1800, San Diego, CA 92101-4495, United States
The Boulevard, Langford Lane, Kidlington, Oxford OX5 1GB, United Kingdom
125 London Wall, London, EC2Y 5AS, United Kingdom

First edition 2017

ISBN: 978-0-12-811870-2
ISSN: 1937-6448

Publisher: Zoe Kruze
Acquisition Editor: Alex White
Editorial Project Manager: Fenton Coulthurst
Production Project Manager: James Selvam
Cover Designer: Mark Rogers

Typeset by SPi Global, India

CONTENTS

CONTRIBUTORS

Simona Barbato
Laboratory of Biochemistry and Mitochondrial Pathophysiology, University of Bologna, Bologna, Italy

Thomas Bertero
Université Côte d'Azur, CNRS, INSERM, IRCAN, FHU-OncoAge, Nice, France

Giovanni Blandino
Oncogenomic and Epigenetic Unit, Regina Elena National Cancer Institute, Rome, Italy

George A. Calin
Center for RNA Interference and Non-Coding RNAs, The University of Texas MD Anderson Cancer Center, Houston, TX, United States

Silvia Anna Ciafrè
Department of Biomedicine and Prevention, University of Rome "Tor Vergata", Rome, Italy

Muller Fabbri
Children's Center for Cancer and Blood Diseases and The Saban Research Institute, Norris Comprehensive Cancer Center, Keck School of Medicine, University of Southern California, Children's Hospital Los Angeles, Los Angeles, CA, United States

Linda Fabris
The University of Texas MD Anderson Cancer Center, Houston, TX, United States

Yuanzhang Fang
The University of Texas MD Anderson Cancer Center, Houston, TX, United States

Silvia Galardi
Department of Biomedicine and Prevention, University of Rome "Tor Vergata", Rome, Italy

Frauke Goeman
Oncogenomic and Epigenetic Unit, Regina Elena National Cancer Institute, Rome, Italy

Eleonora Grisard
CRO-Aviano, National Cancer Institute, Aviano, Italy

Cheng Huang
School of Pharmacy, Shanghai University of Traditional Chinese Medicine, Shanghai, China

Yujing Li
The University of Texas MD Anderson Cancer Center, Houston, TX, United States

Zhenghu Li
The University of Texas MD Anderson Cancer Center, Houston, TX, United States

Xiongbin Lu
The University of Texas MD Anderson Cancer Center, Houston, TX, United States

Bernard Mari
Université Côte d'Azur, CNRS, IPMC, FHU-OncoAge, Sophia-Antipolis, France

Neri Mercatelli
Unit of Biology, Genetics and Biochemistry, Department of Movement, Human and Health Sciences, University of Rome "Foro Italico"; Laboratory of Cellular and Molecular Neurobiology, CERC Fondazione Santa Lucia, Rome, Italy

Milena S. Nicoloso
CRO-Aviano, National Cancer Institute, Aviano, Italy

Nicolas Pottier
Faculté de Médecine de Lille, Pole Recherche, Lille, France

Roger Rezzonico
Université Côte d'Azur, CNRS, IPMC, FHU-OncoAge, Sophia-Antipolis, France

Giancarlo Solaini
Laboratory of Biochemistry and Mitochondrial Pathophysiology, University of Bologna, Bologna, Italy

Sabrina Strano
Oncogenomic and Epigenetic Unit, Regina Elena National Cancer Institute, Rome, Italy

Lu Zhang
The University of Texas MD Anderson Cancer Center, Houston, TX, United States; School of Pharmacy, Shanghai University of Traditional Chinese Medicine, Shanghai, China

CHAPTER ONE

MicroRNAs in DNA Damage Response, Carcinogenesis, and Chemoresistance

Yuanzhang Fang*[,1], Lu Zhang*[,†,1], Zhenghu Li*[,1], Yujing Li*, Cheng Huang[†,2], Xiongbin Lu*[,2]

*The University of Texas MD Anderson Cancer Center, Houston, TX, United States
[†]School of Pharmacy, Shanghai University of Traditional Chinese Medicine, Shanghai, China
[2]Corresponding authors: e-mail address: chuang_shutcm@163.com; xlu2@mdanderson.org

Contents

[1] These authors contributed equally to this work.

International Review of Cell and Molecular Biology, Volume 333
ISSN 1937-6448
http://dx.doi.org/10.1016/bs.ircmb.2017.03.001

Abstract

In the classical model of tumorigenesis, cancer develops via slowly accumulating mutations that facilitate uncontrolled cell growth and allow cells to escape apoptosis. Oncogenes and tumor suppressor genes regulate the key signaling pathways involved in tumorigenesis and cancer progression. Moreover, studies indicate that MicroRNAs (MiRNAs) are also key parts of these processes. MiRNAs are short, noncoding RNAs that regulate the expression of target genes at the posttranscriptional level. By regulating the expression of numerous cytokines, MiRNAs play crucial roles in cell growth, apoptosis, and stemness maintenance. Abundant evidence demonstrates that MiRNAs can function as both oncogenes and tumor suppressors in accommodating the proliferation and invasion of cancer cells in solid tumors. Targeting these MiRNAs may significantly alter oncogenic signaling pathways and slow or halt progression of many different types of tumors. A better understanding of the functions of MiRNAs in cancer will enable the development of new treatment strategies for chemoresistant malignancies. This review discusses recent findings about the connections between MiRNAs and carcinogenesis and provides insight into the role of MiRNAs in generating chemoresistance.

1. INTRODUCTION

MicroRNAs (MiRNAs) are small (about 22 nucleotides [nt]), highly conserved, noncoding RNAs that bind to target sequences in the 3′ untranslated region (UTR) of mRNA. MiRNAs mediate inhibition of mRNA translation or promote its degradation, leading to reduced protein expression (Ha, 2011; Ha and Kim, 2014). To date, more than 1000 human MiRNAs have been identified, and the number is increasing (Kozomara and Griffiths-Jones, 2011; Mogilyansky and Rigoutsos, 2013). About one-third of protein-coding genes are controlled by MiRNAs (Jin et al., 2013). MiRNAs are differentially expressed in cases of many diseases, including cancer, in which they may act as tumor suppressors, oncogenes, or both (Croce, 2008; Jansson and Lund, 2012).

MicroRNAs are localized throughout the whole-animal genome. Most are interconnected in genomic clusters and transcribed from a single polycistronic transcription unit (Rodriguez et al., 2004). DNA-dependent RNA polymerase II transcribes encoded genes to generate a primary MiRNA, which contains a region of imperfect double-stranded RNA (Czech and Hannon, 2011). Primary MiRNA transcripts are similar to those of most protein-coding genes (Marsico et al., 2013). They have 5′ cap structures and poly(A) tails and may contain introns. The ribonuclease III enzyme Drosha processes primary MiRNAs in the nucleus and releases an approximately 65- to 70-nt hairpin-shaped precursor MiRNA (pre-miRNA) (Wahid et al., 2010). Some other MiRNAs, called mirtrons, reside within short (about 400-nt) introns of protein-coding genes (Berezikov et al., 2007; Okamura et al., 2007; Ruby et al., 2007). Mirtrons bypass the Drosha processing step and instead use a splicing mechanism to generate pre-miRNAs (Ruby et al., 2007). They are then restructured and refolded into short stem–loop structures that mimic the pre-miRNAs. The nuclear export protein exportin 5 recognizes the 2-nt single-stranded 3′ overhang of pre-miRNAs (characteristic of ribonuclease III-mediated cleavage) and actively transports them in a RanGTP-dependent manner to the cytoplasm (Czech and Hannon, 2011). Once in the cytoplasm, the pre-miRNAs are cleaved into approximately 22-nt MiRNA:MiRNA duplexes by the enzyme Dicer. Mature MiRNAs are then incorporated into the RNA-induced silencing complex, where they mediate the silencing of target genes (Macfarlane and Murphy, 2010; Ota et al., 2013).

Although MiRNAs bind to their target mRNAs through imperfect base pairing, perfect complementary is required between the seed region of the MiRNA (nucleotides 2–7 of the mature sequence) and the target mRNA (Felekkis et al., 2010; Lewis et al., 2005). MiRNA–mRNA binding leads to degradation, destabilization, or translational inhibition of the mRNA and, consequently, gene expression silencing. MiRNAs inhibit the initiation of translation by repressing m7G-cap recognition or the joining of the ribosomal subunit 60S to the translation–initiation complex. Furthermore, MiRNAs block translation elongation by slowing elongation or ribosome drop-off (Filipowicz et al., 2008; Gu and Kay, 2010). Degradation of the mRNA appears to occur when the sequence beyond the seed region of the MiRNA is perfectly complementary to the remaining region in the 3′ UTR of the mRNA (Thomson et al., 2011). Degradation probably involves deadenylation, decapping, and exonucleolytic cleavage of the target mRNA transcript (Behm-Ansmant et al., 2006; Wu et al., 2006).

2. MiRNA REGULATION IN RESPONSE TO DNA DAMAGE

MiRNAs and DNA damage are bidirectional processes. Evidence suggests that MiRNAs can be activated by different treatments with DNA-damaging agents (Bouwman and Jonkers, 2012; Wouters et al., 2011) and that both unique and common activation sets exist for DNA damage (Han et al., 2012). Regulating DNA damage based on the natural and intensity way, MiRNAs are activated on unique as well as common sets by varying levels of exposure to DNA-damaging agents (Simone et al., 2009). Functional targets and their associated processes have been identified for some MiRNAs activated by DNA damage. In the case of DNA damage from ultraviolet radiation, a large subset of MiRNAs can be rapidly modulated, though MiRNA expression returns to original levels after 24 h (Pothof et al., 2009). In other words, MiRNAs modulate DNA damage response in the time between the transcriptional and posttranscriptional responses (Majidinia and Yousefi, 2016; Wang and Taniguchi, 2013; Wouters et al., 2011). DNA damage response targets of many MiRNAs have yet to be uncovered. However, accumulating evidence suggests that DNA damage response regulates MiRNA expression at both the transcriptional and posttranscriptional levels.

2.1 Transcriptional Regulation of MiRNAs

DNA damage can regulate MiRNA at the transcriptional level. Transcription factors like p53, E2F, and MYC control MiRNA transcription by binding to MiRNA promoters or modifying the expression of MiRNA processing-machinery components. P53 is a well-known transcription factor and tumor suppressor protein in animals. Activated p53 directs a transcriptional program that prevents the proliferation of genetically unstable cells. P53 directly binds to MiRNA promoters and thereby modulates transcription or processing-component expression. In this way, p53 regulates MiRNA expression (Fig. 1). A mutation or deletion of *TP53* is found in more than half of human tumors (Lee et al., 2012; Madka et al., 2016).

The MiR-34 family of direct transcriptional targets was the first group of MiRNAs found to interact with p53. Expression of the MiR-34 family is induced by DNA damage and oncogenic stress (Cha et al., 2012; Wong et al., 2011). For example, in $Cav^{-/-}$ breast tumor stromal cells, MiR-34 is upregulated upon oxidative stress and activation of autophagy via HIF1α (Martinez-Outschoorn et al., 2010; Pavlides et al., 2010). MiR-34a is ubiquitous in cell lines and tissues and is transcriptionally regulated individually. It exhibits the highest expression change that induces apoptosis after adriamycin administration in $p53^{+/+}$ cells. MiR-34a expression is decreased in some cells with loss of p53 function, for instance in pancreatic cancer cells (Xiong et al., 2015). MiR-34a is involved in cell-cycle regulation, cell proliferation, apoptosis, angiogenesis, and DNA repair by altering the mRNA transcripts of several genes, such as *BCL2*, *CCND1*, *CCNE2*, *MYC*, *CDK4*, *CDK6*, *RAD51AP1*, *CHK1*, and *MDC1* (Hermeking, 2012; Krause et al., 2016; Surova and Zhivotovsky, 2013; Wang and Martin, 2014). Moreover, MiRNA-34a participates in a positive feedback loop with SIRT1. MiR-34a also increases the acetylation of p53, which in turn

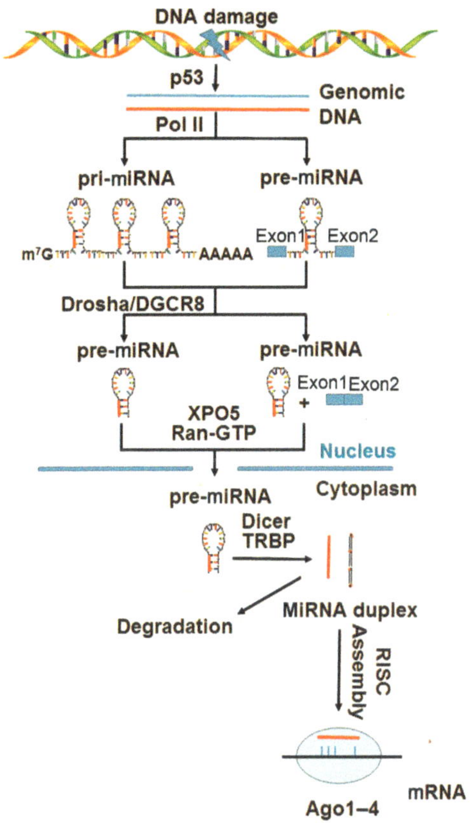

Fig. 1 DNA damage affects MiRNA biogenesis via regulation of transcription. P53 is a representative transcriptional factor that mediates specific MiRNAs through transcription. DNA damage also affects MiRNAs via posttranscriptional regulation.

increases expression of transcriptional targets like P21 and PUMA. In this way, MiR-34a regulates the cell cycle and apoptosis (Park et al., 2014).

MiR-34b and MiR-34c are mainly found in lung tissue and have a common primary transcript. MiR-34c expression is upregulated by p53 upon DNA damage. However, MiR-34c expression can also increase in the absence of p53. Cannell et al. (2010) suggested that this may involve an alternative signaling pathway such as the p38/MAPK pathway. An increase in MiR-34c expression also decreases expression of MYC, which increases DNA synthesis and prevents DNA damage-induced cell-cycle arrest at S phase (Cannell et al., 2010). Because of its diverse functions, MiR-34 is the first MiRNA to reach clinical trials for cancer therapy (Bader, 2012).

Besides the MiR-34 family, expression of several other MiRNAs, including MiR-215, MiR-192, MiR-194, the MiR-17-92 cluster, and MiR-145, is known to be regulated by p53. DNA damage promotes p53-dependent upregulation of MiR-215 expression, which induces cell-cycle arrest and targets several transcripts involved in cell-cycle checkpoints (Georges et al., 2008; Hermeking, 2012). Braun et al. (2008) identified a putative p53-binding element surrounding the MiR-215 cluster, demonstrating that MiRNA-215 is transcriptionally activated by p53 (Braun et al., 2008). MiR-192 has a similar function. MiR-192 expression may participate in a positive feedback loop with p53 activity. It increased p21 levels in a $p53^{+/+}$ colorectal carcinoma cell line (Georges et al., 2008; Song et al., 2008). The finding that both MiR-192 and MiR-215 are downregulated in many cancer cells suggests that these MiRNAs act as tumor suppressors. However, researchers reported that MiR-215 is regulated mainly by *RB1*, a retinoblastoma tumor suppressor gene, in gastric cancer (Deng et al., 2014).

MiR-194 is also transcriptionally regulated by p53. By targeting thrombospondin 1, MiR-194 promotes angiogenesis in colon cancer (Sundaram et al., 2011). In gastric cancer cells, MiR-194 decreases cell proliferation and migration and targets the E3 ubiquitin-protein ligase RBX1 (Chen et al., 2015). In addition, the MiR-17-92 cluster is sensitive to hypoxia-induced apoptosis via a p53-dependent mechanism (Molitoris et al., 2011; Yan et al., 2009). The combination of the TATA-binding protein transcriptional factor and the TATA box containing the p53-binding site within the MiRNA-17-92 promoter is prevented. It further triggers p53-mediated MiR-17-92 repression (Yan et al., 2009). MiR-145 is also regulated by p53. P53 directly binds to the promoter of MiR-145, which in turn suppresses the expression of MYC and CDK6, leading to inhibition of cell proliferation and induction of apoptosis (Khalili et al., 2015; Sachdeva et al., 2009).

Other members of the p53 family are also involved in MiRNA expression. For example, PF63 suppresses metastasis formation. It regulates the expression levels of Dicer, which are very low in p63-knockout mice and p63-deficient human cells. MiR-130b can also be activated by p63, which directly binds to the Dicer promoter and activates MiR-130b transcription (Su et al., 2010).

MYC and E2F, which are involved in DNA damage-induced cell-cycle checkpoints, also regulate the expression of several MiRNAs. Both induce transcription of the MiR-17-92 cluster, which in turn inhibits E2F expression via a feedback loop (Aguda et al., 2008; Li et al., 2014; Psathas et al.,

2013). Other MiRNAs repressed by E2F transcription factors include MiR-106a-92, MiR-106b-25, MiR-210, MiR-128, MiR-34, and MiR-20 (Wan et al., 2011; Zhang et al., 2015).

2.2 Posttranscriptional Regulation of MiRNAs

Posttranscriptional maturation is an irreplaceable step in biogenesis; it is of equal importance to transcriptional regulation of MiRNAs. DNA damage can modulate the maturation process of MiRNAs and thereby regulate MiRNA expression (Ren and Yu, 2012) (Fig. 2). Several MiRNAs with growth-suppressive functions, including MiR-16-1, MiR-143, and MiR-145, have been found to be posttranscriptionally upregulated by DNA damage (Suzuki et al., 2009). Loss of *TP53* function affects MiRNA processing

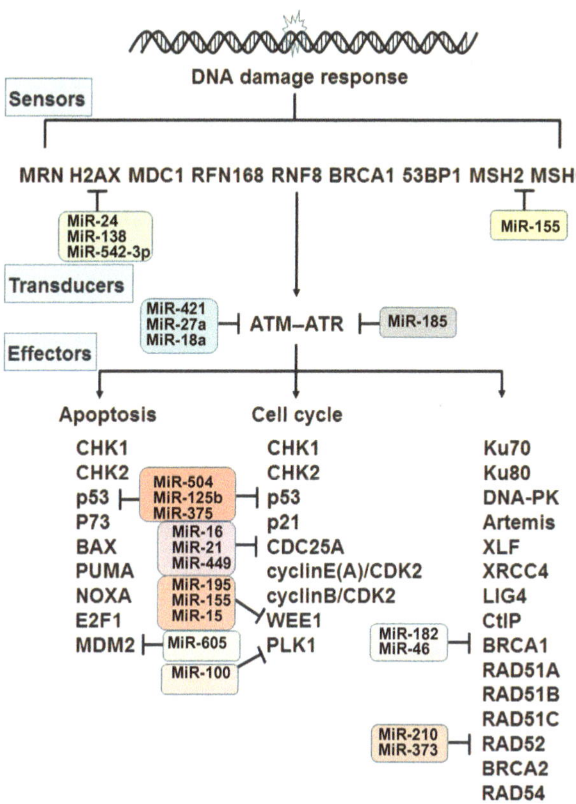

Fig. 2 Schematic diagram of the sensor, transducer, and effector proteins involved in the DNA damage response. The effectors mediate apoptosis, the cell cycle, and DNA repair.

and transcriptional activity and may contribute to cancer progression. P53 associates with the RNA helicase p68 and interacts with the Drosha/DGCR8 complex, which facilitates the transformation of primary MiRNAs into pre-miRNAs. P53 mutants interfere with the functional assembly between the Drosha complex and p68 and thereby diminish MiRNA-processing activity. DNA damage also promotes the expression of MiR-16, which is triggered by cell-cycle-dependent relocalization of the MiRNA-binding protein Ago2 (Krell et al., 2016).

The roles of p63, p73, and p53 in MiRNA maturation have been analyzed using comparative and computational genomic approaches. Through a complex network of transcriptional and posttranscriptional regulation activities, these genes function as both positive and negative regulators of MiRNA processing and maturation. Boominathan (2010) determined that the MiRNAs let-7, MiR-21, MiR-16, MiR-107, MiR-134, MiR-143, MiRNA-145, MiR-200c, MiR-449a, and MiR-503 are involved in p53/p63/p73-axis regulation (Boominathan, 2010). Boominathan's study detailed how p53, p63, and p73 regulate the components of MiRNA processing. In addition, Hermeking (2012) identified other regulating MiRNA venues involving the p53/p63/p73 complex. For example, p53-regulated MiRNAs can target the Drosha–DGCR8, Dicer–TRBP2, and Argonaute MiRNA-processing complexes (Hermeking, 2012).

Regulatory networks functioning independently of the DNA damage-signaling protein kinase ATM are also involved in DNA damage signaling inmiRNA pathways (Han et al., 2012). For instance, ERK-mediated phosphorylation of TRBP in the Dicer complex increases the expression of mature MiRNAs. TRBP may upregulate MiRNAs such as MiR-17, MiR-20a, and MiR-92a and downregulate the let-7 tumor suppressor family of MiRNAs, suggesting that TRBP activates MiRNA expression (Dent et al., 2003). That is to say, DNA damage can induce the expression of specific MiRNAs and thereby regulate specific cellular functions or influence the expression of a subset of MiRNAs.

3. DNA DAMAGE RESPONSE REGULATION BY MiRNAs

In their role as endogenous gene regulators (Fig. 3), MiRNAs affect protein stability and offer another mode of regulation of the DNA damage response. Researchers have reported that MiRNAs regulate DNA damage-responsive and their predicted target-binding sites on DNA damage-response genes (Hu and Gatti, 2011; Wan et al., 2011).

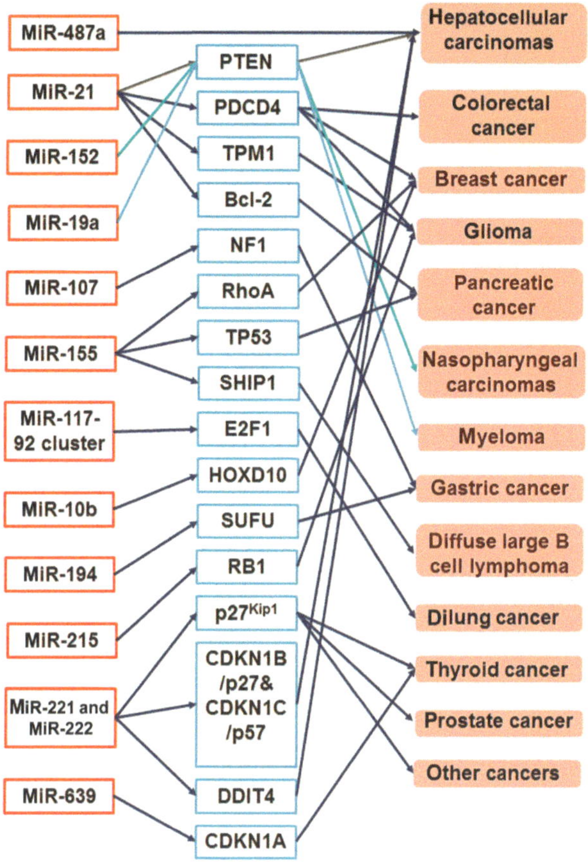

Fig. 3 MiRNAs act as oncogenes in various cancers through multiple signaling pathways.

In this section, we describe the principal avenues of cross talk between MiRNAs and DNA damage-response pathways.

3.1 Sensors of DNA Damage and Mediators of the DNA Damage Response

The histone variant H2AX is immediately and extensively phosphorylated by the ATM and ATR kinases upon DNA damage. Phosphorylation of H2AX is an early signal of DNA damage (Huen and Chen, 2010). ATM is recruited by the MRE11/RAD50/NSB1 (MRN) complex, which acts as a sensor of DNA double-strand breaks (DSBs). Terminally differentiated blood cells exhibit a reduced capacity to repair DSBs. Both H2AX and the

MRN complex are bound by MDC1. It then acts to recruit RNF8 to accumulate the expression of RNF168, BRCA1, and 53BP1 around the DNA break sites (van Attikum and Gasser, 2009). Upregulation of MiR-24 inhibits H2AX expression and DNA repair (Lal et al., 2009b). Upregulation of MiR-24 also increases the sensitivity to DNA-damaging γ-irradiation and mimetic agents. However, MiR-24 expression is decreased by MYC and completely lost in several types of cancer cells (Chang et al., 2008).

The cell-proliferation inhibitors E2F1, CDK4, and CDK6 are also targets of MiR-24 (Lal et al., 2009a). In addition, both MiR-138 and MiR-542-3p reduce the phosphorylation of H2AX at serine 139 after DNA damage in U2OS human osteosarcoma cells (Wang et al., 2011). Overexpression of MiR-138 inhibits homologous recombination and increases cellular sensitivity to multiple DNA-damaging agents. MiR-155, which functions on the DNA mismatch repair proteins MSH2, MSH6, and MLH1, induces microsatellite instability in human colorectal cancer cells, suggesting that MiR-155 modulates the mismatch repair mechanism in carcinogenesis (Valeri et al., 2010; Yang et al., 2015).

3.2 Transducers of the DNA Damage Response

In response to DSBs, ATM and ATR transduce DNA damage signals to downstream proteins. ATM expression is primarily activated by autophosphorylation on serine residues and mediated by the phosphatase WIP1 (Lu et al., 2005). Overexpression of MiR-421, a regulator of ATM, results in a defective cell-cycle checkpoint at S phase and increased radiosensitivity in HeLa cells (Hu et al., 2010a). In addition, MiR-27a is overexpressed in several types of tumors. MiR-27a directly regulates ATM, promotes cell, and affects DSB-rejoining kinetics after irradiation (Di Francesco et al., 2013). MiR-18a expression is upregulated in both breast and colorectal cancer cell lines and tissue samples, and MiR-18a overexpression leads to the downregulation of ATM expression. The efficiency of DNA damage repair and homologous recombination increases when MiR-18a expression is deficient. Inhibition of MiRNA-18a also reduces cellular radiosensitivity (Sharma and Misteli, 2013; Tessitore et al., 2014; Zhou et al., 2015). Overexpression of MiR-100 and MiR-101, which are also targeted by ATM, correlates with aberrant DNA repair mechanisms and radiosensitivity in vitro and in vivo (Luan et al., 2015; Yan et al., 2010; Zhou et al., 2016b).

Unlike ATM, ATR responds to single-stranded DNA, double-stranded DNA adjacent to single-stranded DNA, adducts, cross-links, and DNA polymerase. MiR-185 negatively regulates ATR expression and enhances radiation-induced apoptosis and inhibition of proliferation. MiR-185, therefore, may be a useful tool for enhancing the radiosensitivity of tumor cells (Li et al., 2015b).

3.3 Effectors of the DNA Damage Response

3.3.1 *Effectors of Apoptosis*

If a DNA lesion cannot be repaired quickly, apoptosis is initiated to eliminate the damaged cell before it can undergo malignant transformation. P53 plays a crucial role in DSB-induced apoptosis. In response to DSBs, transcriptional activation of proapoptotic factors such as FAS, PUMA, and BAX is induced by ATM/CHK2 and ATR/CHK1, phosphorylating and stabilizing p53. CHK1 and CHK2 also stimulate expression of E2F1 and p73 to support the action of p53. This in turn transcribes BAX, PUMA, and NOXA. Several MiRNAs are involved in apoptotic processes. For example, MiR-125b is a potential MiRNA-binding site in the 3′ UTR of *TP53*. It is an important negative regulator of p53-induced apoptosis during development and during the cellular stress response (Amir et al., 2013; Yu et al., 2016). MiR-504 targets p53 in a similar manner (Hu et al., 2010b). In addition, MiRNA-375 downregulates p53 expression through an interaction with the 3′ UTR, desensitizing gastric cancer cells to ionizing radiation and etoposide-based treatment (Liu et al., 2013). MiRNA-605 was also recently identified as a key cofactor in the *TP53* regulatory network.

3.3.2 *Effectors of the Cell Cycle*

The major pathway controlling the DNA damage-induced G1/S-phase checkpoint and triggering G1-phase arrest is the ATM/ATR-CHK2/CHK1-p53/MDM2-p21 pathway. DSBs and single-strand breaks activate ATM/CHK2 and ATR/CHK1, thereby stabilizing p53 expression and inducing p21 expression, which inhibits the activity of the CCNE/CDK2 and CCND/CDK4/6 complexes. This signaling cascade leads to cell-cycle arrest at G1 phase. Cell-cycle arrest may also be induced via activation of CHK1/CHK2 and inactivation of the CDC25A phosphatase. It consequently inhibits expression of the cyclin E/CDK2 and cyclin A/CDK2 complexes and delays the cell cycle at the G1/S transition (Kastan and Bartek, 2004; Solier et al., 2012).

Two pathways trigger intra-S-phase arrest. In the first pathway, CHK1/2/CDC25A/cyclin E (A)/CDK2, CDC25A degradation blocks expression of cyclin E/CDK2 and cyclin A/CDK2 and delays replication by blocking the process of CDC45 onto chromatin. Consequently, this pathway facilitates DNA repair (Bartek et al., 2004). In the second pathway, ATM–NBS1–SMC1, ATM mediates phosphorylation of NBS1 and recruits to form a complex that regulates S-phase checkpoints (Antoccia et al., 2008; Yazdi et al., 2002).

DNA damage-induced G2/M-phase checkpoints are controlled by both p53-dependent and -independent pathways. Both types of pathway target the cyclin B/CDK1 complex. In the p53-dependent pathway, p53 induces p21 expression and cyclin B/CDK1 activity. In the p53-independent pathway, phosphorylation of CHK1/CHK2 inhibits expression of the CDC25 protein family (CDC25A-C) and PLK1. Furthermore, phosphorylation of CHK1/CHK2 upregulates the activity of the WEE1 kinase, which decreases phosphorylation of CDK1. Acting synergistically, these pathways cause cell-cycle arrest at the G2/M phase (Fukasawa, 2007). MiR-449a, MiR-449b, MiR-21, and MiR-16 also target CDC25A, and MiR-195, MiR-155, and family target WEE1. In the MiR-15 family, the MiRNA let-7 targets CHK1. Additionally, MiR-106b targets p21, and MiR-100 downregulates PLK1 (Nakamachi et al., 2009; Pothof et al., 2009; Pouliot et al., 2012; Qi et al., 2009; Shi et al., 2010; Wang and Taniguchi, 2013).

3.3.3 Effectors of DNA Repair

Nonhomologous end joining and homologous recombination repair are the two major DSB repair pathways. Nonhomologous end joining acts in the G0/G1 phase, and homologous recombination repair acts in the late S/G2 phase. Seven core proteins are required for nonhomologous end joining: Ku70, Ku80, DNA-PKcs, Artemis, XRCC4, XLF/Cernunnos, and DNA ligase IV. Ku70 and Ku80 bind to DSB ends and recruit DNA-PKcs, consequently recruiting XRCC4, Artemis, XLF, and DNA ligase IV to coordinate end processes (Lieber, 2008). DSB repair can also be mediated by homologous pathways. These pathways are initiated by the CtIP/BRCA1-complex-mediated resection and mediated by RAD52 and RAD51 paralogs (A, B, and C). Resynthesis of the broken DNA segment is then mediated by the RAD51/BRCA2 complex and RAD54 (San Filippo et al., 2008). Induction of MiR-210 expression inhibits RAD52 expression, whereas MiR-373 affects not only RAD52 but also NER and

RAD23B. These hypoxia-inducible MiRNAs influence the effects of hypoxia on DNA damage repair in cancer cells (Crosby et al., 2009; Huan et al., 2014).

MiR-182 also targets several genes involved in DNA repair, including BRCA1. BRCA1 forms a large multisubunit protein complex to recruit to DNA lesions, facilitating DNA repair and cell-cycle regulation (Chambon et al., 2003; Shapiro et al., 2016). MiR-182 functions on BRCA1 protein levels and protects breast tumor cells from ionizing radiation-induced cell death (Jin et al., 2009; Moskwa et al., 2011; Yao and Ventura, 2011). MiR-146a and MiR-146b-5p silence BRCA1 expression and play an important role in breast tumorigenesis (Garcia et al., 2011).

4. MiRNAs IN CANCER

Tumors form when cells abnormally proliferate or apoptosis triggers in cells fail. Studies demonstrate that MiRNAs are involved in the regulation of cell proliferation and apoptosis (Sanchez-Beato et al., 2003). Nairz et al. (2006) found that abnormal expression of MiR-278 can lead to inhibition of apoptosis and infinite proliferation of eye cells in *Drosophila* (Nairz et al., 2006). Their results suggested that the specific role of MiRNAs in tumor development and growth may be the induction of the regulation of cell proliferation and apoptosis of the target genes. Studies have shown that MiRNAs can be used as a new class of oncogene or tumor suppressor gene.

MiRNAs act as protooncogenes or tumor suppressor genes in many tumors (Tie and Fan, 2011) (Table 1; Fig. 3). Because MiRNAs negatively regulate target gene expression, if the target mRNA is a tumor suppressor gene, high MiRNA expression levels can promote tumorigenesis. Lai et al. (2012) reported that in nonsmall cell lung carcinoma cells, high expression of MiR-372 promoted the occurrence of tumors by targeting the *LATS2* tumor suppressor gene (Lai et al., 2012). In contrast, if the target mRNA is a protooncogene, MiRNAs can act as tumor suppressors (Table 1; Fig. 4). For instance, MiR-31 inhibits expression of the tumor suppressor genes *LATS2* and *PPP2R2A* and thereby suppresses the growth of lung cancer cells (Liu et al., 2010).

4.1 MiRNAs as Oncogenes

MiR-21 is encoded on the plus strand of chromosome 17 at q23.2 (Lagos-Quintana et al., 2001; Suh et al., 2004). To date, MiRNA-21 is the only

Table 1 Functions of MicroRNAs in Cancer

MiRNA Function	Target Genes	Tumor Type	References
Oncogene			
MiR-21	*PTEN*	Hepatocellular carcinoma cell lines	Meng et al. (2007)
	PDCD4	Colorectal cancer	Asangani et al. (2008)
	PDCD4	Breast cancer	Frankel et al. (2008)
	PDCD4	Human glioblastoma cell T98G	Chen et al. (2008)
	TPM1		Zhu et al. (2007)
	BCL2	Pancreatic cancer cell lines	Dong et al. (2011)
MiR-107	*NF1*	Gastric cancer	Wang et al. (2016a)
MiR-152	*PTEN*	Nasopharyngeal carcinoma	Huang et al. (2016)
MiR-155	*RHOA*	Breast cancer	Kong et al. (2008)
	TP53INP1	Pancreatic cancer	Gironella et al. (2007)
	SHIP1	Diffuse large B cell lymphoma	Pedersen et al. (2009)
MiR-17-92 cluster		Small-cell lung cancer	Hayashita et al. (2005)
	E2F1		O'Donnell et al. (2005)
MiR-10b	*HOXD10*	Breast cancer	Gee et al. (2008) and Ma et al. (2007)
MiR-19a	*PTEN*	Myeloma	Zhang et al. (2017)
MiR-194	*SUFU*	Gastric cancer	Peng et al. (2017)
MiR-215	*RB1*	Glioma	Wei et al. (2017)
MiR-221 and MiR-222	$p27^{Kip1}$	Human thyroid papillary carcinoma	Visone et al. (2007)
	$p27^{Kip1}$	Prostate carcinoma	Galardi et al. (2007) and Mercatelli et al. (2008)
	$p27^{Kip1}$	Other cancers	le Sage et al. (2007)

Table 1 Functions of MicroRNAs in Cancer—cont'd

MiRNA Function	Target Genes	Tumor Type	References
	CDKN1B/p27 and *CDKN1C*/p57	Hepatocellular carcinoma	Fornari et al. (2008)
	DDIT4	Hepatocellular carcinoma	Pineau et al. (2010)
MiR-487a		Hepatocellular carcinoma	Chang et al. (2016)
MiR-639	*CDKN1A*	Thyroid cancer	Lei et al. (2016)
Tumor suppressor			
Let-7	*RAS*	Lung tumors	Johnson et al. (2005)
	HMGA2	Various tumors, including lung cancers	Lee and Dutta (2007)
	Lin28		Bussing et al. (2008)
MiR-15a and MiR-16-1	*BCL2*		Cimmino et al. (2005)
	WNT3A		Bonci et al. (2008)
MiR-22	*YWHAZ*	Hepatocellular carcinoma	Chen et al. (2016b)
MiR-26a	*KRT19*	Cholangiocarcinoma	Wang and Lv (2016)
MiR-29b	*LAMC1*, *LASP1*, and *PPIC*	Melanoma	Andrews et al. (2016)
	DNMT3A and *DNMT3B*	Ovarian cancer	Teng et al. (2016)
MiR-93	*WNK1*	Triple-negative breast cancer	Shyamasundar et al. (2016)
MiR-128	*RELN* and *DCX*	Neuroblastoma	Evangelisti et al. (2009)
	BMI1	Glioma	Godlewski et al. (2008)
MiR-143	*MMP7*	Colorectal cancer	Yu et al. (2017)
MiR-193a	*WT1*-E-cadherin	Nonsmall cell lung cancer	Chen et al. (2016a)

Continued

Table 1 Functions of MicroRNAs in Cancer—cont'd

MiRNA Function	Target Genes	Tumor Type	References
MiR-199a-5p	*CCR7*	Bladder cancer	Zhou et al. (2016a)
MiR-199b-5p	*HES1*	Medulloblastoma	Garzia et al. (2009)
MiR-124-3p	*CBL*	Breast cancer	Wang et al. (2016b)
MiR-125a and MiR-125b	*ERBB2* and *ERBB3*	Breast cancer	Scott et al. (2007)
MiR-125b	*LIN28B*	Liver cancer	Liang et al. (2010)
MiR-34 family	*BCL2*, *NOTCH1*, and *NOTCH2*	Pancreatic cancer cell lines MiaPaCa2 and BxPC3	Ji et al. (2009)
	BCL2	Gastric cancer	Ji et al. (2008)
	CD44	Prostate cancer	Liu et al. (2011)
MiR-143 and MiR-145	*ERK5*	Human colon cancer cell lines DLD-1 and SW480	Akao et al. (2006)
	KRAS	Colorectal cancer	Chen et al. (2009b)
		Most B cell malignancies	Akao et al. (2007)
MiR-200 family	*ZEB1* and *ZEB2*	Various solid tumors	Korpal et al. (2008)
	ZEB1 and *ZEB2*	Breast cancer	Tryndyak et al. (2010)
	ZEB1 and *ZEB2*	Prostate cancer	Kong et al. (2009)
	BMI1	Breast cancer	Shimono et al. (2009)
	NOTCH1	PC3 prostate cancer cells overexpressing PDGF-D	Kong et al. (2010a)
MiR-205	*ZEB1* and *ZEB2*	Madin–Darby canine kidney epithelial cells	Gregory et al. (2008)
	ZEB1 and *ZEB2*	Breast cancer and esophageal squamous-cell cancer	Matsushima et al. (2011)
	E2F1 and *E2F5*	Melanoma	Dar et al. (2011)

Table 1 Functions of MicroRNAs in Cancer—cont'd

MiRNA Function	Target Genes	Tumor Type	References
MiR-326	Notch pathway	Glioma	Kefas et al. (2009)
MiR-589-5p	*MAP3K8*	Hepatocellular carcinoma	Zhang et al. (2016a)**
MiR-635	*YY1*	Nonsmall cell lung cancer	Zhang et al. (2016b)
MiR-4443	*NCOA1* and *TRAF4*	Colon cancer	Meerson and Yehuda (2016)
MiR-9600	*STAT3*	Nonsmall cell lung cancer	Sun et al. (2016)
Both oncogene and tumor suppressor			
MiR-135a and MiR-135b	*APC*	Colorectal adenomas and carcinomas	Nagel et al. (2008)
MiR-135a	*JAK2*	Classical Hodgkin lymphoma	Navarro et al. (2009)

MiRNA reported to be highly overexpressed in solid tumors, such as those of the brain, head and neck, esophagus, breast, lung, stomach, pancreas, colon, and prostate (Kumarswamy et al., 2011). The main mechanism of MiRNA-21 regulation is the targeting of tumor suppressor genes. For instance, PTEN negatively regulates the Akt/PKB signaling pathway to prevent cells from growing too rapidly (Milella et al., 2015). By targeting PTEN, MiR-21 increases the proliferation, migration, and invasion capacities of hepatocellular carcinoma cells (Meng et al., 2007; Wu et al., 2016). MiR-21 also targets the transformation suppressor PDCD4, which plays an important role in apoptosis (Sun et al., 2014). MiR-21 posttranscriptionally downregulates PDCD4 in colorectal cancer (Asangani et al., 2008), breast cancer (Frankel et al., 2008), and T98G human glioblastoma (Chen et al., 2008) cells.

Tropomyosins, a class of actin-binding proteins, are suppressed in many transformed cells; MiR-21 can target the tropomyosin TPM1 and regulate its expression at the translational level (Zhu et al., 2007). *BCL2* is an apoptosis-regulating gene belonging to the Bcl-2 family. When induced

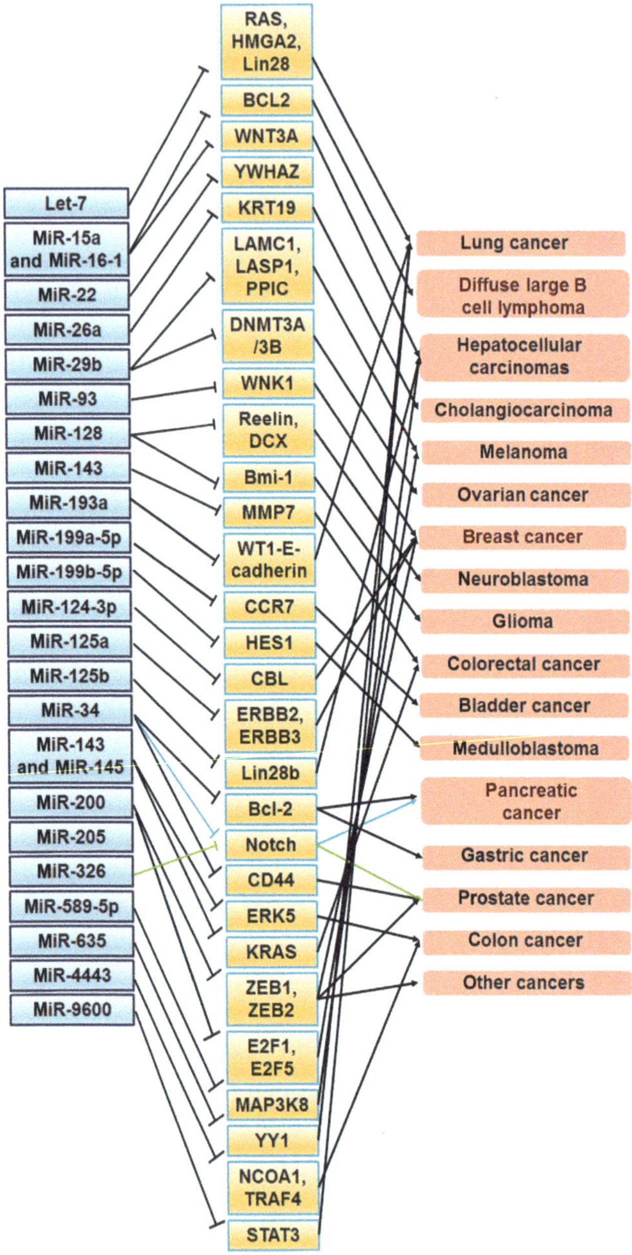

Fig. 4 MiRNAs act as tumor suppressors in various cancers through multiple signaling pathways.

by apoptotic signaling, Bax2 and Bak oligomerizate and move from the cytoplasm to the outer mitochondrial membrane, where they interact with voltage-dependent anion channels. These activated signaling pathways cause apoptotic factors such as cytochrome *c* to be released to the cytoplasm, inducing apoptosis. Bcl-2 can inhibit apoptosis by forming a heterodimer with Bax and Bak, thus inhibiting the opening of the mitochondrial membrane channel. This function implicates roles for Bcl-2 in various malignancies, including melanoma, breast cancer, prostate cancer, pancreatic cancer, and lung carcinoma. (Dong et al., 2011; Pan and Wu, 2012). MiR-21 also interacts with other cell regulators, including Sprouty1, Sprouty2, IL-12p35, JAG1, and HNRPK, to produce tumors (Li et al., 2012). However, the accommodation of different cell regulators by MiRNA-21 may be specific to the cell type (Mladinov et al., 2013).

MiR-155 is evolutionally conserved because of its crucial interactions with several cytokine-coding genes (Faraoni et al., 2009). The oncogene *MYC* regulates cell proliferation and apoptosis by binding to E-box sequences and recruiting histone acetyltransferases (Pan et al., 2014). By targeting genes that antagonize the MYC pathway, MiR-155 functions as an oncogene and induces the development of many kinds of tumors. Kong et al. (2008) demonstrated that MiR-155 expression may play a crucial role in the TGF-induced epithelial-to-mesenchymal transition and in cell migration and invasion in breast cancer cases. They detected a significantly elevated level of MiR-155 when NMuMG mouse mammary cells were treated with TGF-β (Kong et al., 2008). Knockdown of MiR-155 expression prevented cells from undergoing a TGF-β-induced epithelial-to-mesenchymal transition, tight junction dissolution, migration, and invasion (Liu et al., 2016). Using the RNA22 MiRNA target deletion and miRBase databases, Kong and colleagues also found that the 3′ UTR of the *RHOA* gene may contain conserved MiR-155-binding sites. To confirm this, they induced ectopically expression of MiR-155 in NMuMG cells. As expected, the RhoA protein expression level decreased, but the mRNA level did not, and tight junction formation was disrupted. A luciferase assay containing a mutant *RHOA* 3′ UTR reporter further demonstrated that *RHOA* is negatively regulated by MiR-155. Therefore, targeting MiR-155 may be a strategy for the treatment of breast cancer (Kong et al., 2010b).

MiR-155 also has important functions in pancreatic cancer cells. TP53INP1, which interacts with p53, can suppress tumor formation by inducing caspase 3-mediated apoptosis (Shahbazi et al., 2013). Gironella et al. (2007) found that MiR-155 can suppress TP53INP1 expression and induce tumor formation. Their experiments used MiaPaCa2 pancreatic carcinoma cells, which express TP53INP1 when activated by ponasterone A. MiaPaCa2 cells formed tumors in nude mice. When TP53INP1 expression was restored in these cells, the capacity of the cells to form tumors in nude mice was lost. Similarly, in pancreatic ductal adenocarcinoma cells, immunohistochemical analysis demonstrated that the expression of TP53INP1 was dramatically reduced when MiR-155 expression increased. Understanding the association of MiR-155 with TP53INP1 therefore provides a new view for research on pancreatic cancer.

Much evidence indicates that MiR-155 generates from sequences present in BIC RNA, which accumulates in Hodgkin lymphoma, diffuse large B-cell lymphoma, and primary mediastinal B cell lymphoma cells (Eis et al., 2005; Fernandez-Mercado et al., 2015; Kluiver et al., 2005; Tili et al., 2009). Pedersen et al. (2009) found that in diffuse large B cell lymphoma cells, MiR-155 targets the 3′ UTR of the inositol phosphatase *SHIP1* gene and that overexpression of MiR-155 is stimulated by tumor necrosis factor α (Pedersen et al., 2009). However, further investigations are needed to understand the regulation of MiR-155.

The MiR-17-92 cluster, which is composed of seven MiRNAs and is located in intron 3 of the *C13orf25* gene at 13q31.3, is markedly overexpressed in lung cancer cells, especially those in the most aggressive form, small cell lung cancer (Hayashita et al., 2005). Its potential targets include the transcription factor *E2F1*, the tumor suppressor gene *PTEN*, and the transcriptional repressor *RB2*. He et al.'s (2005) study demonstrated that tumors overexpressing both MYC and MiR-17-19 had stronger multiplication capacity and lower cell death rates in B-cell lymphoma cells than did tumors overexpressing MYC alone (He et al., 2005). When MYC was overexpressed, MiR-17-19 functioned as oncogenes and induced cancer. O'Donnell et al., 2005 demonstrated that the MiR-17-92 cluster may be a tumor-related component (O'Donnell et al., 2005). Using MiRNA array analysis, they put forward a hypothesis to support the model developed by He et al. O'Donnell and colleagues found that MYC induced the expression of MiR-17-92, thereby inhibiting the translation of E2F1, which mediates both cell proliferation and apoptosis. The MiR-17-92 cluster may block E2F1-induced apoptosis to promote MYC and induce cell proliferation (Gurtan and Sharp, 2013; Hong et al., 2010).

MiR-10b promotes the development of many kinds of malignancies, especially breast cancer (Ma, 2010; Singh et al., 2014). HOXD10 inhibits the transcription of several genes associated with cell migration, including *RHOC*, *ITGA3*, *MMP14*, and *PLAUR* (Tian et al., 2010; Xiao et al., 2014). Ma et al. (2007) found that HOXD10 is a direct and functional target of MiR-10b. Twist-induced MiR-10b expression translationally inhibited HOXD10 and caused tumor cell migration and invasion in several breast cancer models (Ahmad et al., 2014; Liu et al., 2012; Ma et al., 2007). However, Gee et al. (2008) provided a slightly different interpretation of Ma and colleagues' results. They argued that in Ma et al.'s study, the number of patient tumor samples was too small to yield a conclusive result associating MiR-10b with metastasis. Their study included 219 patients with early-stage breast cancer. Unlike Ma et al., Gee and colleagues found that MiR-10b expression levels did not correlate with the development of metastasis, nor were they associated with recurrence-free survival, distant-metastasis-free survival, or breast cancer-specific survival.

MiR-221 and MiR-222 have long been recognized as oncogenes. By targeting *p27Kip1* (*LAMTOR1*), which negatively regulates cell-cycle progression, MiR-221 and MiR-222 induce cell proliferation and tumorigenesis in human thyroid papillary carcinoma (Visone et al., 2007; Wang et al., 2012a), prostate carcinoma (Galardi et al., 2007; Mercatelli et al., 2008; Yang et al., 2014), and other cancers (Garofalo et al., 2012; le Sage et al., 2007; Lupini et al., 2013). In human hepatocellular carcinoma cells, MiR-221 downregulates both CDKN1B/p27 and CDKN1C/p57 expression, whereas transfection of the cells with an anti-miR-221 produces significant upregulation of these genes (Fornari et al., 2008; He et al., 2014). DDIT4, a modulator of the mammalian target of rapamycin pathway, is another a target of MiR-221 in liver carcinoma cells (Cheng et al., 2015; Pineau et al., 2010). MiR-221 targets the 3′ UTR of the mRNA of DDIT4 and thereby stimulates tumor growth (Chu et al., 2014).

4.2 MiRNAs as Tumor Suppressors

The MiRNA let-7 was first identified in the nematode *Caenorhabditis elegans*, in which it regulates cell proliferation and differentiation (Rausch et al., 2015). Let-7 is conserved in mammals, indicating that it plays an important role in cell fate. Several studies have shown that decreased let-7 expression is associated with increased tumorigenesis. The RAS protein, a membrane-associated GTPase that regulates cell growth, is targeted by let-7 (Johnson et al., 2005). The 3′ UTRs of RAS contain several let-7 complementary sites. Let-7 expression levels are higher and RAS expression levels are lower in lung tumors than in normal lung tissues. At the same time, overexpression of let-7 in vitro effectively suppresses the growth of lung tumors and the expression of RAS. Let-7 also regulates *HMGA2*, which is oncogenic for various malignancies, including lung cancer (Lee and Dutta, 2007). By targeting *HMGA2*, let-7 reduces its expression and suppresses the growth of tumors. The RNA-binding protein Lin28 and let-7 can mutually inhibit each others' expression*** (Nguyen and Zhu, 2015; Wang et al., 2012b, 2015). Lin28 binds to the primary and hairpin precursors of let-7 and inhibits primary let-7 processing in the early development of normal tissues. Let-7 can also bind to Lin28 and reduce its expression (Stefani et al., 2015; Van Wynsberghe et al., 2011). In sum, the expression of let-7 is associated with stemness maintenance, and the loss of let-7 causes dedifferentiation of normal differentiated cells, turning them into cancer stem cells (Barh et al., 2010; Bussing et al., 2008).

MiR-15a and MiR-16-1 are located at 13q14, a region that is deleted in many B cell chronic lymphocytic leukemias, pituitary adenomas, and prostate tumors (Aqeilan et al., 2010; Bottoni et al., 2005; Humplikova et al., 2013; Pekarsky and Croce, 2015). Both of these MiRNAs act as tumor suppressors. Downregulation of MiR-15a and MiR-16-1 has been found in many solid tumors, indicating that these MiRNAs play a crucial role in cell regulation (Garofalo et al., 2014). To inhibit cell proliferation, promote apoptosis, and suppress tumorigenesis, MiR-15a and MiR-16-1 target several oncogenes, such as *BCL2*, *CCND1*, and *WNT3A* (Liang et al., 2015; Yang et al., 2012). Recovery of MiR-15a and MiR-16-1 expression after their knockdown in cancer cells produces apoptosis and regression, suggesting that the antitumor function of these MiRNAs is associated with the inhibition of the transformation from normal cells to cancer stem cells (Kang et al., 2015).

MiR-128 is richly expressed in normal brain tissue but is expressed at lower levels in neuroblastoma cells. Evangelisti et al. (2009) demonstrated that in neural cells, MiR-128 downregulates reelin (*RELN*) and *DCX*, which are involved in cell migration. Ectopic expression of MiR-128 represses neuroblastoma cell motility, invasiveness, and growth (Evangelisti et al., 2009). MiR-128 also targets the 3′ UTR of the mRNA of the oncogene *BMI1*. The expression of MiR-128 in glioma cells causes downregulation of *BMI1*, methylation of histone H3K27me3, phosphorylation of Akt, and overexpression of *p21CIP1* (*CDKN1A*) (Godlewski et al., 2008), all of which lead to the suppression of cell proliferation and invasion.

BMI1 is necessary for the efficient self-renewing division of stem cells, so dysregulation of MiR-128 toward *BMI1* may encourage the development of stemness characteristics in cancer cells.

MiR-199b-5p plays an important role in cancer stem cells by downregulating the expression of the transcription factor HES1, which is involved in the Notch signaling pathway (Garzia et al., 2009). Notch signaling takes part in determining cell differentiation and regulates a subset of medulloblastoma cells that have cancer stem cell properties and promote tumor growth. MiR-199b-5p overexpression downregulates HES1 expression, suppresses the expression of several cancer stem cell genes, decreases the subpopulation of stem-cell-like (CD133+) medulloblastoma cells, and depresses the rate of medulloblastoma cell proliferation. All of these observations indicate that MiR-199b-5p inhibits the formation of cancer stem cells.

MiR-125a and MiR-125b are suppressed in breast cancers that overexpress the receptor tyrosine kinases ERBB2 and ERBB3 (Bischoff et al., 2015; Wang et al., 2013). By targeting the 3′ UTR of *ERBB2* and *ERBB3*, overexpressed MiR-125a and MiR-125b decrease their transcription and protein levels and suppress the phosphorylation of ERK1,ERK2, and AKT in the human breast cancer cell line SKBR3 (Scott et al., 2007). These effects reduce the migration and invasion capacities of breast cancer cells. MiR-125b is dysregulated in many cancers, such as oral and bladder cancers, but the regulation mechanism in these tumors demands further exploration (Wu et al., 2015). In liver cancer cells, MiR-125b directly targets the 3′ UTR of Lin28B and thereby inhibits cancer cell proliferation and invasion (Zhao and Wang, 2015).

The MiR-34 family plays crucial roles in cell-cycle regulation and tumor suppression (Zhang et al., 2014). MiR-34 regulates numerous proteins and efficiently accommodates cellular activities. Researchers have shown that MiR-34 can directly target Notch, HMGA2, c-Met, and Bcl-2, which are involved in cancer stem cell self-renewal and cell survival (Ji et al., 2009). Ji et al. (2009) demonstrated that in the p53-mutant human pancreatic cancer cell lines MiaPaCa2 and BxPC3, restoring MiR-34 expression significantly inhibited cell growth and invasion and decreased the population of CD44+ and CD133+ cancer stem cells (Ji et al., 2009). An earlier study by Ji et al. (2008) had similar results for gastric cancers and brain tumors (Ji et al., 2008). Liu et al. (2011) provided an intuitive illustration of the functions of MiR-34a in cancer stem cells (Li et al., 2015a; Liu et al., 2011). They determined that overexpression of MiR-34a in purified CD44+ prostate cancer cells inhibits clonogenic expansion, tumor development, and metastasis by targeting CD44 (Hart et al., 2016; Hermeking, 2010; Mackiewicz et al., 2011).

MiR-143 and MiR-145 regulate many kinds of solid tumors. In colon adenocarcinoma, their expression decreased steadily (Gomes et al., 2016; Michael et al., 2003). Chen et al. (2009b) compared normal cells and colorectal cancer cells in vivo and found an inverse correlation between expression levels of the KRAS protein and MiR-143. To confirm that MiR-143 inhibited KRAS expression, they treated LoVo metastatic colon cancer cells with a MiR-143 mimic. They found that KRAS expression was completely abrogated in the presence of the mimic, whereas introduction of a MiR-143 inhibitor increased KRAS protein levels. Furthermore, MiR-143 suppressed the growth of LoVo cells, whereas its inhibitor had the opposite effect (Chen et al., 2009b). For most of the B cell malignancies, including chronic lymphocytic leukemia, B cell lymphoma, Epstein–Barr virus-transformed B cell lines, and Burkitt lymphoma cell lines, Yukihiro et al. reported decreased expression of MiR-143 and MiR-145. In addition, Akao et al. (2007) demonstrated that transfection with either precursor or mature MiR-143 and MiR-145 inhibited the growth of Raji Burkitt lymphoma cells in a dose-dependent manner (Akao et al., 2007). This evidence demonstrates that MiR-143 and MiR-145 have important roles in the self-renewal of cancer stem cells.

The MiR-200 family has five members: MiR-141, MiR-200a, MiR-200b, MiR-200c, and MiR-429. ZEB1 and ZEB2, mesenchymal markers of fibronectin, are upregulated in cells undergoing epithelial-to-mesenchymal transition, which may be the initial step of metastasis. Several reports have shown that the MiR-200 family targets ZEB1 and ZEB2 to inhibit the proliferation and invasion of solid tumors such as breast cancer and prostate cancer (Ding, 2014; Hawa et al., 2016; Humphries and Yang, 2015). Ectopic expression of the MiR-200 family in mesenchymal cells can also induce the mesenchymal-to-epithelial transition, reflecting the vital role of MiR-200 in regulating tumorigenesis. The MiR-200 family also modulates the expression of BMI1, a regulator of stem cell self-renewal, to link with cancer stem-cell signatures at the same time (Croset et al., 2015; Shimono et al., 2009). Shimono et al. (2009) compared the expression of the MiR-200 family in breast cancer stem cells with that found in normal stem cells. Their results demonstrated that the MiR-200 family is downregulated in breast cancer stem cells (Shimono et al., 2009). By targeting BMI1, ectopic expression of MiR-200c suppressed cancer cell growth, inhibited the clonogenicity and tumorigenicity of cancer stem cells, and induced cell differentiation (Feng et al., 2015). MiR-200 can also downregulate Notch1 expression to suppress the colony-formation and sphere-formation abilities of PC-3 metastatic prostate cancer cells overexpressing PDGF-D (Kong et al., 2010a).

MiR-205 acts as a tumor suppressor by targeting several factors. Gregory et al. (2008) found that Madin–Darby canine kidney epithelial cells treated with TGF-β or ectopically expressing the protein tyrosine phosphatase Pez underwent epithelial-to-mesenchymal transition and exhibited decreased MiR-205 expression levels (Gregory et al., 2008; Liu et al., 2016; Majid et al., 2011). MiR-205 binds to the 3′ UTRs of *ZEB1* and *ZEB2* to regulate their expression. Data showing that ectopic expression of MiR-205 in mesenchymal cells in the initial stages of mesenchymal-to-epithelial transition further demonstrated that MiR-205 can inhibit ZEB1 and ZEB2. Matsushima et al. (2011) identified the same mechanism in breast and esophageal squamous cancer cells (Matsushima et al., 2011). MiR-205 also targets E2F, another important regulator of DNA synthesis, the cell cycle, and apoptosis. E2F dysregulation can make normal cells invasive and proliferative like cancer stem cells. In melanoma cells, ectopically expressed MiR-205 targets the 3′ UTRs of E2F1 and E2F5 to reduce their protein expression levels and suppress the cells' proliferative capacity (Dar et al., 2011).

Finally, MiR-326 mutually regulates the Notch signaling pathway. It decreases glioma cell proliferation and the development, proliferation, and invasiveness of glioblastoma stem-cell-like lines (Kefas et al., 2009) (Table 1).

4.3 MiRNAs as Both Oncogenes and Tumor Suppressor Genes

MiR-135a and MiR-135b are markedly upregulated in colorectal adenomas and carcinomas (DeSano and Xu, 2009). They interact with APC, a tumor suppressor gene that regulates several cell activities, including division. APC controls β-catenin to regulate the Wnt signaling pathway. Inactivation of *APC* initiates colorectal tumorigenesis. Almost all *APC* mutations are nonsense mutations, which generate truncated proteins that lack β-catenin-binding sites. When not bound by APC, β-catenin takes part in the Wnt signaling pathway and activates transcription of target genes. By targeting the 3′ UTR of *APC*, MiR-135a and MiR-135b suppress β-catenin expression and induce the downstream Wnt pathway. However, MiR-135a and MiR-135b have been shown to be upregulated in colorectal adenomas and carcinomas regardless of the mutational status of *APC* (Nagel et al., 2008).

The regulation of APC by MiR-135a and MiR-135b is specific to humans, indicating that new therapies for colorectal cancer could target these MiRNAs.

However, in different tumor tissues, MiR-135 has the opposite functions. Classical Hodgkin lymphoma is characterized by the presence of relatively few tumor cells in a nonneoplastic microenvironment and an orderly spread from one lymph node group to another. Alfons et al. observed alterations in 25 MiRNAs in 89 patients with classical Hodgkin lymphoma who received the same treatments. They found that expression levels of MiR-135a were linked with a higher rate of disease-free survival and a lower probability of relapse. They also found that MiR-135a directly regulates JAK2, a component of the JAK/STAT pathway. Phosphorylated JAK2 activates STAT5 nuclear translocation; in the nucleus, they function as transcription factors and induce expression of *BCL-XL*, an antiapoptotic gene. MiR-135a targets the 3′ UTR of *JAK2* and reduces its protein expression levels. In turn, the mRNA and protein expression levels of Bcl-xL decrease, and apoptosis rates increase. These findings confirm that MiR-135a plays a crucial role in cell survival and suggest that MiR-135a is a novel therapeutic target for patients with classical Hodgkin lymphoma (Dang et al., 2014; Navarro et al., 2009).

5. THE ROLE OF MiRNAs IN DRUG RESISTANCE

Chemotherapy is one of the cornerstones of cancer treatment. However, chemotherapy often fails to eliminate all tumor cells because cancer cells can develop intrinsic or acquired drug resistance (Broxterman et al., 2009; Fojo, 2007). Recent studies have suggested that MiRNAs play important roles in the development of drug resistance through two mechanisms: abnormal regulation of gene expression and the presence of MiRNA polymorphisms or mutations (Calin and Croce, 2006). Different drug-resistant tumor cells may display different mechanisms of resistance.

The most common types of abnormal expression of MiRNAs in tumor cells are (1) location of the MiRNA targeting site in the vicinity of a cancer-related gene, (2) epigenetic regulation of MiRNA genes, and (3) abnormal expression of genes and proteins that process MiRNAs (Calin and Croce, 2006). Other mechanisms, including genetic variation and transcriptional regulation, may also cause abnormal MiRNA expression. Upregulation or downregulation of MiRNAs affects expression of target genes, which in turn influences the expression of a variety of downstream proteins; these changes can affect the drug sensitivity of cancer cells. Most MiRNAs regulate tumor cell survival pathways and/or apoptosis pathways (Zheng et al., 2010). In addition, some MiRNAs regulate drug delivery mechanisms and drug metabolism regulation (Fojo, 2007).

5.1 MiRNAs Regulate Drug Resistance Through Survival and/or Apoptosis Pathways

Several types of antitumor drugs regulate apoptosis pathways. The balance between survival signaling and apoptosis signaling is a key feature of normal cells. In cancer cells, especially drug-resistant ones, the balance of survival, cell-cycle, and apoptosis pathways is upset. Several MiRNAs are involved in regulating these pathways (Table 1).

5.2 MiRNAs Regulate Drug Resistance Through the Drug-Targeting and DNA-Repair Pathways

In addition to regulating survival- and apoptosis-related signaling pathways, MiRNAs can mediate drug resistance by interfering with drug targets and DNA-repair mechanisms. One example is modulation of TUBB3, which is abnormally expressed in some cancers (Seve and Dumontet, 2008). Cochrane and colleagues used microarray analysis to predict several target proteins of MiR-200c, including TUBB3, in ovarian and endometrial cancer cells (Zhao et al., 2016). Other experiments demonstrated that ectopic expression of MiR-200c increases the sensitivity of microtubule-targeted drugs, paclitaxel, vincristine, and epothilone B by up to 85% via downregulation of TUBB3. However, they found no evidence that MiR-200c enhances sensitivity to DNA-damaging agents such as cisplatin, doxorubicin, and mitomycin C or to apoptosis inducers such as TRAIL and FasL, suggesting that the mechanism by which MiR-200c restores chemotherapy resistance is reversal of the effects of TUBB3-targeting drugs. However, MiR-200c can also inhibit the expression of E-cadherin indirectly via inhibiting *ZEB1* and *ZEB2* expression, and restoration of MiR-200c expression and recovery of E-cadherin function can reduce the migration and invasion capacities of ovarian and endometrial cancer cells.

MiRNAs are also involved in the repair of DNA damage produced by radiation, including ultraviolet radiation (Pothof et al., 2009). By using microarray technology to compare the expression of MiRNAs before and after irradiation, Weidhaas et al. (2007) found that the MiRNAs in the let-7 family—except let-7g—were downregulated in both irradiated lung cancer cells and normal lung cells (Weidhaas et al., 2007). The key oncogene regulated by the let-7 family of MiRNAs is *RAS*, which is downregulated in lung cancer. Interestingly, overexpression of let-7a and let-7b can promote cancer cells' sensitivity to radiation, but overexpression of let-7g protects cells from the DNA-damaging effects of radiation, suggesting that different members of the let-7 family affect the survival of cells differently after irradiation. Furthermore, germ-cell death induced by radiation in a *C. elegans* model suggests that the let-7 family can alter the response to radiation therapy by modulating RAS expression and DNA damage response pathways. Similarly, Josson et al. (2008)) found that the expression of MiR-512 is downregulated in irradiated prostate cancer cells. Overexpression of MiR-512 can increase the sensitivity of prostate cancer cells to radiation therapy, and inhibition of MiR-512 expression induces resistance to radiation. In addition, inhibition of MiR-512 expression is involved in reducing the expression of radiation-responsive repair proteins, such as CSA and MnSOD, suggesting that alteration of DNA repair proteins is one of the mechanisms by which MiR-512 mediates radiation sensitivity (Table 2).

5.3 MiRNAs Regulate Drug Resistance Through Regulating Drug Delivery and Drug Metabolism

Increased expression of drug efflux pumps leads to the development of resistance to anticancer drugs in tumor cells. MDR1 and ABCG2 are two important families of ATP-binding domain transport proteins, which transport drugs from the intracellular to the extracellular space. Recent evidence suggests that these two drug efflux pumps can be directly regulated by MiRNAs in different types of tumors.

Zhu et al. (2008) compared MiRNA expression in resistant and parental strains of human ovarian and cervical cancer cells. They found that both MiR-27a and MiR-451 activate tumor cells by modulating the expression of the drug resistance protein MDR1/P-gp (Zhu et al., 2008). Mimics of MiR-27a and MiR-451 activate the MDR1/P-gp pathway. MiRNA inhibitors suppress the development of drug resistance by silencing the expression of P-gp, thereby enhancing intracellular accumulation of cytotoxic drugs. However, MiR-451 expression is distinctive in breast cancer cells; it regulates the expression of P-gp in the opposite direction, which may increase the sensitivity of breast cancer cells to treatment with doxorubicin (Adriamycin) (Kovalchuk et al., 2008). The different effects of MiR-451 in different tissue types show the complexity of MiRNA-mediated drug resistance.

Table 2 Effects of MicroRNAs on Survival and Apoptosis Pathways Affecting Anticancer Drug Sensitivity

MiRNA	Pathway	Target	Therapy Agents	Cancer Type	References
Let-7a	Induces resistance to apoptosis triggered by interleukin γ, doxorubicin, and paclitaxel	Caspase-3	Interleukin γ, doxorubicin, paclitaxel	Liver cancer; squamous cell carcinoma	Tsang and Kwok (2008)
Let-7a/b/g	Let-7a and Let-7b increase sensitivity to radiation. Let-7g prompts resistance	RAS	Radiation	Lung cancer	Pothof et al. (2009)
MiR-1	Increases sensitivity to doxorubicin	Mcl-1	Doxorubicin	Nonsmall cell lung cancer	Nasser et al. (2008)
MiR-15b/16	Increases sensitivity to several antitumor drugs	Bcl-2	TRAIL	Gastric cancer	Koschny et al. (2007)
MiR-27a	Induces resistance to paclitaxel	HIPK2	Paclitaxel	Ovarian cancer	Kovalchuk et al. (2008)
MiR-27a/451	Induces resistance to doxorubicin and vincristine	MDR1	Doxorubicin, vincristine	Ovarian cancer; cervical cancer	Li et al. (2016)
MiR-27b	Loss causes carcinogenic metabolism	CYPIB1	Doxorubicin	Breast cancer; cervical cancer	Murray et al. (1997)
MiR-29a/181a/221	Special expression in patients with resistance to fludarabine	Unknown	Fludarabine	Chronic lymphocytic leukemia	Moussay et al. (2010)
MiR-34a	Downregulation represses apoptosis	SIRT1	Camptothecin	Prostate cancer	Fujita et al. (2008)
MiR-98	Overexpression under hypoxic conditions promotes drug resistance	HMGA2	Doxorubicin, cisplatin	Head and neck squamous cell carcinoma	Boo et al. (2005)

Continued

Table 2 Effects of MicroRNAs on Survival and Apoptosis Pathways Affecting Anticancer Drug Sensitivity—cont'd

MiRNA	Pathway	Target	Therapy Agents	Cancer Type	References
MiR-122	Strengthens p53 expression and stability	Cyclin G1	Doxorubicin	Liver cancer	Fornari et al. (2009)
MiR-125b	Causes resistance to androgen-blocking agents; upregulation causes drug resistance of breast cancer	BAK1	Paclitaxel	Prostate cancer; breast cancer	DeVere White et al. (2009) and Zhou et al. (2010)
MiR-140	Promotes resistance to methotrexate and 5-fluorouracil	HDAC4	Methotrexate, 5-fluorouracil	Colorectal cancer; osteosarcoma	Song et al. (2009)
MiR-143	Induces anti-Fas antibody-based treatment sensitivity	ERK5	CH-11	T-cell leukemia	Akao et al. (2009)
MiR-148a	Overexpression in prostate cancer promotes drug resistance; downregulation enhances drug metabolism	MSK1, PXR	Paclitaxel	Prostate cancer; colorectal cancer	Fujita et al. (2008) and Honkakoski and Negishi (2000)
MiR-155	Involved in drug resistance	FOXO3a	Paclitaxel, doxorubicin	Breast cancer	Kong et al. (2016)
MiR-192/215	Promotes drug resistance	TS	5-Fluorouracil	Colorectal cancer	Yang et al. (2008)
MiR-199a-3p	Loss inhibits doxorubicin-induced apoptosis	mTOR, c-Met	Doxorubicin	Liver cancer	Fornari et al. (2010) and Kim et al. (2008)
MiR-200c	Downregulation produces resistance to microtubule-targeting drugs	TUBB3	Taxane	Ovarian cancer, breast cancer	Seve and Dumontet (2008)
MiR-204	Downregulation promotes drug resistance	Mcl-1	5-Fluorouracil	Cholangiocarcinoma	Chen et al. (2009a)

MiR-205	Loss of expression leads to resistance to tyrosine kinase inhibitors	HER3	Gefitinib, lapatinib	Breast cancer	Iorio et al. (2009)
MiR-212	Loss of expression leads to resistance to apoptin 2-induced apoptosis	PED	TRAIL	Nonsmall cell lung cancer	Incoronato et al. (2010)
MiR-214	Causes drug resistance	PTEN	Cisplatin	Ovarian cancer	Yang et al. (2008)
MiR-221/222	Upregulation leads to resistance to apoptin 2-induced apoptosis; ectopic expression in breast cancer induces tamoxifen resistance	P27 ERα	Taxanes, tamoxifen, TRAIL	Nonsmall cell lung cancer; breast cancer	Garofalo et al. (2008), Rao et al. (2011), and Seve and Dumontet (2008)
MiR-320	Downregulation promotes drug resistance	Bcl-2	5-Fluorouracil	Cholangiocarcinoma	Chen et al. (2009a)
MiR-328	Enhances drug sensitivity	ABCG2	Mitoxantrone	Breast cancer	Chen et al. (2013)
MiR-451	Ectopic expression recovers drug sensitivity	Unknown	Doxorubicin	Breast cancer	Zhu et al. (2008)
MiR-512	Loss of expression leads to resistance to radiation therapy	CSA	Radiation	Prostate cancer	Weidhaas et al. (2007)

ABCG2, also known as the breast cancer resistance protein, is another efflux-pump protein that promotes drug resistance (Li et al., 2016). A recent study demonstrated that MiR-328 directly inhibits ABCG2 expression in breast cancer cells and increases their sensitivity to mitoxantrone (Pan et al., 2009). The expression of ABCG2 is also inhibited by MiR-519c in the S1 colon cancer cell line; in turn, inhibition of ABCG2 protein expression reverses the resistance of colon cancer cells (To et al., 2008). The superfamily of cytochrome P450 enzymes, which catalyze drug metabolism, is also a major target for the MiRNAs in this family.

5.4 MiRNAs Regulate Drug Resistance in Pharmacogenomics

Pharmacogenomic research on MiRNAs is a new and promising avenue for the development of new targeted drug therapies. MiRNA pharmacogenomics aims to guide the choice of chemotherapeutic drugs by identifying MiRNAs according to their target genes, polymorphisms, and single-nucleotide polymorphisms (miRSNPs) (Mishra et al., 2008). MiRNA polymorphisms affect genes involved in the absorption, metabolism, and management of drugs and are suspected to lead to drug resistance (Tsuchiya et al., 2006).

MiRSNPs also play a role in drug resistance and drug sensitization (Bertino et al., 2007; Mishra et al., 2008). The miRSNP/MiRNA polymorphism was found to be associated with methotrexate resistance. The presence of miRSNP 829C–T in the 3′ UTR of the MiR-24 attachment site of dihydrofolate reductase results in loss of MiR-24 function, which consequently leads to overexpression of dihydrofolate reductase and resistance to methotrexate (Mishra et al., 2007). This finding may have prognostic significance for the clinical outcomes of patients given methotrexate. Interestingly, a SNP allele in the vicinity of a MiRNA-linker site of dihydrofolate reductase is found in 14.2% of the Japanese population (Mishra et al., 2007). Identifying the miRSNPs in an individual patient's genome will enable oncologists to better understand that patient's response to specific chemotherapeutic drugs and to predict toxicity. Many of these functional miRSNPs or MiRNA polymorphisms are present in the human genome, leading to loss of MiRNA function and, consequently, changes in protein expression. These changes may lead to drug resistance if drug targets are affected (Bertino et al., 2007).

6. CONCLUSION

Because MiRNAs regulate the cellular DNA damage response, they play important roles in various cancers. Although significant progress has been made in the treatment of many types of cancer, the development of drug resistance remains a major obstacle to curing malignancies and often leads to poor patient outcomes. A better understanding of the mechanisms by which drug resistance develops is necessary to improve upon current chemotherapy regimens. More and more studies are revealing that abnormal expression of MiRNAs in cancer cells is highly correlated with drug resistance. Understanding the mechanisms by which MiRNAs regulate DNA damage and chemoresistance is an urgently needed step toward improving outcomes for patients with cancer.

REFERENCES

Aguda, B.D., Kim, Y., Piper-Hunter, M.G., Friedman, A., Marsh, C.B., 2008. MicroRNA regulation of a cancer network: consequences of the feedback loops involving miR-17-92, E2F, and Myc. Proc. Natl. Acad. Sci. U.S.A. 105, 19678–19683.

Ahmad, A., Sethi, S., Chen, W., Ali-Fehmi, R., Mittal, S., Sarkar, F.H., 2014. Up-regulation of microRNA-10b is associated with the development of breast cancer brain metastasis. Am. J. Transl. Res. 6, 384–390.

Akao, Y., Nakagawa, Y., Naoe, T., 2006. MicroRNAs 143 and 145 are possible common onco-microRNAs in human cancers. Oncol. Rep. 16, 845–850.

Akao, Y., Nakagawa, Y., Kitade, Y., Kinoshita, T., Naoe, T., 2007. Downregulation of microRNAs-143 and -145 in B-cell malignancies. Cancer Sci. 98, 1914–1920.

Akao, Y., Nakagawa, Y., Iio, A., Naoe, T., 2009. Role of microRNA-143 in Fas-mediated apoptosis in human T-cell leukemia Jurkat cells. Leuk. Res. 33, 1530–1538.

Amir, S., Ma, A.H., Shi, X.B., Xue, L., Kung, H.J., Devere White, R.W., 2013. Oncomir miR-125b suppresses p14(ARF) to modulate p53-dependent and p53-independent apoptosis in prostate cancer. PLoS One 8. e61064.

Andrews, M.C., Cursons, J., Hurley, D.G., Anaka, M., Cebon, J.S., Behren, A., Crampin, E.J., 2016. Systems analysis identifies miR-29b regulation of invasiveness in melanoma. Mol. Cancer 15, 72.

Antoccia, A., Sakamoto, S., Matsuura, S., Tauchi, H., Komatsu, K., 2008. NBS1 prevents chromatid-type aberrations through ATM-dependent interactions with SMC1. Radiat. Res. 170, 345–352.

Aqeilan, R.I., Calin, G.A., Croce, C.M., 2010. MiR-15a and miR-16-1 in cancer: discovery, function and future perspectives. Cell Death Differ. 17, 215–220.

Asangani, I.A., Rasheed, S.A., Nikolova, D.A., Leupold, J.H., Colburn, N.H., Post, S., Allgayer, H., 2008. MicroRNA-21 (miR-21) post-transcriptionally downregulates tumor suppressor Pdcd4 and stimulates invasion, intravasation and metastasis in colorectal cancer. Oncogene 27, 2128–2136.

Bader, A.G., 2012. MiR-34—a microRNA replacement therapy is headed to the clinic. Front. Genet. 3, 120.

Barh, D., Malhotra, R., Ravi, B., Sindhurani, P., 2010. MicroRNA let-7: an emerging next-generation cancer therapeutic. Curr. Oncol. 17, 70–80.
Bartek, J., Lukas, C., Lukas, J., 2004. Checking on DNA damage in S phase. Nat. Rev. Mol. Cell Biol. 5, 792–804.
Behm-Ansmant, I., Rehwinkel, J., Doerks, T., Stark, A., Bork, P., Izaurralde, E., 2006. mRNA degradation by miRNAs and GW182 requires both CCR4:NOT deadenylase and DCP1:DCP2 decapping complexes. Genes Dev. 20, 1885–1898.
Berezikov, E., Chung, W.J., Willis, J., Cuppen, E., Lai, E.C., 2007. Mammalian mirtron genes. Mol. Cell 28, 328–336.
Bertino, J.R., Banerjee, D., Mishra, P.J., 2007. Pharmacogenomics of microRNA: a miRSNP towards individualized therapy. Pharmacogenomics 8, 1625–1627.
Bischoff, A., Bayerlova, M., Strotbek, M., Schmid, S., Beissbarth, T., Olayioye, M.A., 2015. A global microRNA screen identifies regulators of the ErbB receptor signaling network. Cell Commun. Signal. 13, 5.
Bonci, D., Coppola, V., Musumeci, M., Addario, A., Giuffrida, R., Memeo, L., D'Urso, L., Pagliuca, A., Biffoni, M., Labbaye, C., Bartucci, M., Muto, G., Peschle, C., De Maria, R., 2008. The miR-15a-miR-16-1 cluster controls prostate cancer by targeting multiple oncogenic activities. Nat. Med. 14, 1271–1277.
Boo, L.M., Lin, H.H., Chung, V., Zhou, B., Louie, S.G., O'Reilly, M.A., Yen, Y., Ann, D.K., 2005. High mobility group A2 potentiates genotoxic stress in part through the modulation of basal and DNA damage-dependent phosphatidylinositol 3-kinase-related protein kinase activation. Cancer Res. 65, 6622–6630.
Boominathan, L., 2010. The tumor suppressors p53, p63, and p73 are regulators of micro-RNA processing complex. PLoS One 5. e10615.
Bottoni, A., Piccin, D., Tagliati, F., Luchin, A., Zatelli, M.C., Degli Uberti, E.C., 2005. MiR-15a and miR-16-1 down-regulation in pituitary adenomas. J. Cell. Physiol. 204, 280–285.
Bouwman, P., Jonkers, J., 2012. The effects of deregulated DNA damage signalling on cancer chemotherapy response and resistance. Nat. Rev. Cancer 12, 587–598.
Braun, C.J., Zhang, X., Savelyeva, I., Wolff, S., Moll, U.M., Schepeler, T., Orntoft, T.F., Andersen, C.L., Dobbelstein, M., 2008. P53-Responsive micrornas 192 and 215 are capable of inducing cell cycle arrest. Cancer Res. 68, 10094–10104.
Broxterman, H.J., Gotink, K.J., Verheul, H.M., 2009. Understanding the causes of multidrug resistance in cancer: a comparison of doxorubicin and sunitinib. Drug Resist. Updat. 12, 114–126.
Bussing, I., Slack, F.J., Grosshans, H., 2008. Let-7 microRNAs in development, stem cells and cancer. Trends Mol. Med. 14, 400–409.
Calin, G.A., Croce, C.M., 2006. MicroRNA signatures in human cancers. Nat. Rev. Cancer 6, 857–866.
Cannell, I.G., Kong, Y.W., Johnston, S.J., Chen, M.L., Collins, H.M., Dobbyn, H.C., Elia, A., Kress, T.R., Dickens, M., Clemens, M.J., Heery, D.M., Gaestel, M., Eilers, M., Willis, A.E., Bushell, M., 2010. p38 MAPK/MK2-mediated induction of miR-34c following DNA damage prevents Myc-dependent DNA replication. Proc. Natl. Acad. Sci. U.S.A. 107, 5375–5380.
Cha, Y.H., Kim, N.H., Park, C., Lee, I., Kim, H.S., Yook, J.I., 2012. MiRNA-34 intrinsically links p53 tumor suppressor and Wnt signaling. Cell Cycle 11, 1273–1281.
Chambon, M., Nirde, P., Gleizes, M., Roger, P., Vignon, F., 2003. Localization of BRCA1 protein in human breast cancer cells. Breast Cancer Res. Treat. 79, 107–119.
Chang, T.C., Yu, D., Lee, Y.S., Wentzel, E.A., Arking, D.E., West, K.M., Dang, C.V., Thomas-Tikhonenko, A., Mendell, J.T., 2008. Widespread microRNA repression by Myc contributes to tumorigenesis. Nat. Genet. 40, 43–50.
Chang, R.M., Xiao, S., Lei, X., Yang, H., Fang, F., Yang, L.Y., 2016. MiRNA-487a promotes proliferation and metastasis in hepatocellular carcinoma. Clin. Cancer Res. http://dx.doi.org/10.1158/1078-0432.CCR-16-085.

Chen, Y., Liu, W., Chao, T., Zhang, Y., Yan, X., Gong, Y., Qiang, B., Yuan, J., Sun, M., Peng, X., 2008. MicroRNA-21 down-regulates the expression of tumor suppressor PDCD4 in human glioblastoma cell T98G. Cancer Lett. 272, 197–205.

Chen, L., Yan, H.X., Yang, W., Hu, L., Yu, L.X., Liu, Q., Li, L., Huang, D.D., Ding, J., Shen, F., Zhou, W.P., Wu, M.C., Wang, H.Y., 2009a. The role of microRNA expression pattern in human intrahepatic cholangiocarcinoma. J. Hepatol. 50, 358–369.

Chen, X., Guo, X., Zhang, H., Xiang, Y., Chen, J., Yin, Y., Cai, X., Wang, K., Wang, G., Ba, Y., Zhu, L., Wang, J., Yang, R., Zhang, Y., Ren, Z., Zen, K., Zhang, J., Zhang, C.Y., 2009b. Role of miR-143 targeting KRAS in colorectal tumorigenesis. Oncogene 28, 1385–1392.

Chen, Z., Ma, T., Huang, C., Zhang, L., Lv, X., Xu, T., Hu, T., Li, J., 2013. MiR-27a modulates the MDR1/P-glycoprotein expression by inhibiting FZD7/beta-catenin pathway in hepatocellular carcinoma cells. Cell. Signal. 25, 2693–2701.

Chen, X., Wang, Y., Zang, W., Du, Y., Li, M., Zhao, G., 2015. MiR-194 targets RBX1 gene to modulate proliferation and migration of gastric cancer cells. Tumour Biol. 36, 2393–2401.

Chen, J., Gao, S., Wang, C., Wang, Z., Zhang, H., Huang, K., Zhou, B., Li, H., Yu, Z., Wu, J., Chen, C., 2016a. Pathologically decreased expression of miR-193a contributes to metastasis by targeting WT1-E-cadherin axis in non-small cell lung cancers. J. Exp. Clin. Cancer Res. 35, 173.

Chen, M., Hu, W., Xiong, C.L., Qu, Z., Yin, C.Q., Wang, Y.H., Luo, C.L., Guan, Q., Yuan, C.H., Wang, F.B., 2016b. MiR-22 targets YWHAZ to inhibit metastasis of hepatocellular carcinoma and its down-regulation predicts a poor survival. Oncotarget 7 (49), 80751–80764.

Cheng, C.J., Bahal, R., Babar, I.A., Pincus, Z., Barrera, F., Liu, C., Svoronos, A., Braddock, D.T., Glazer, P.M., Engelman, D.M., Saltzman, W.M., Slack, F.J., 2015. MicroRNA silencing for cancer therapy targeted to the tumour microenvironment. Nature 518, 107–110.

Chu, R., Mo, G., Duan, Z., Huang, M., Chang, J., Li, X., Liu, P., 2014. MiRNAs affect the development of hepatocellular carcinoma via dysregulation of their biogenesis and expression. Cell Commun. Signal. 12, 45.

Cimmino, A., Calin, G.A., Fabbri, M., Iorio, M.V., Ferracin, M., Shimizu, M., Wojcik, S.E., Aqeilan, R.I., Zupo, S., Dono, M., Rassenti, L., Alder, H., Volinia, S., Liu, C.G., Kipps, T.J., Negrini, M., Croce, C.M., 2005. MiR-15 and miR-16 induce apoptosis by targeting BCL2. Proc. Natl. Acad. Sci. U.S.A. 102, 13944–13949.

Croce, C.M., 2008. Oncogenes and cancer. N. Engl. J. Med. 358, 502–511.

Crosby, M.E., Kulshreshtha, R., Ivan, M., Glazer, P.M., 2009. MicroRNA regulation of DNA repair gene expression in hypoxic stress. Cancer Res. 69, 1221–1229.

Croset, M., Kan, C., Clezardin, P., 2015. Tumour-derived miRNAs and bone metastasis. Bonekey Rep. 4, 688.

Czech, B., Hannon, G.J., 2011. Small RNA sorting: matchmaking for Argonautes. Nat. Rev. Genet. 12, 19–31.

Dang, Z., Xu, W.H., Lu, P., Wu, N., Liu, J., Ruan, B., Zhou, L., Song, W.J., Dou, K.F., 2014. MicroRNA-135a inhibits cell proliferation by targeting Bmi1 in pancreatic ductal adenocarcinoma. Int. J. Biol. Sci. 10, 733–745.

Dar, A.A., Majid, S., de Semir, D., Nosrati, M., Bezrookove, V., Kashani-Sabet, M., 2011. MiRNA-205 suppresses melanoma cell proliferation and induces senescence via regulation of E2F1 protein. J. Biol. Chem. 286, 16606–16614.

Deng, Y., Huang, Z., Xu, Y., Jin, J., Zhuo, W., Zhang, C., Zhang, X., Shen, M., Yan, X., Wang, L., Wang, X., Kang, Y., Si, J., Zhou, T., 2014. MiR-215 modulates gastric cancer cell proliferation by targeting RB1. Cancer Lett. 342, 27–35.

Dent, P., Yacoub, A., Fisher, P.B., Hagan, M.P., Grant, S., 2003. MAPK pathways in radiation responses. Oncogene 22, 5885–5896.

DeSano, J.T., Xu, L., 2009. MicroRNA regulation of cancer stem cells and therapeutic implications. AAPS J. 11, 682–692.
DeVere White, R.W., Vinall, R.L., Tepper, C.G., Shi, X.B., 2009. MicroRNAs and their potential for translation in prostate cancer. Urol. Oncol. 27, 307–311.
Di Francesco, A., De Pitta, C., Moret, F., Barbieri, V., Celotti, L., Mognato, M., 2013. The DNA-damage response to gamma-radiation is affected by miR-27a in A549 cells. Int. J. Mol. Sci. 14, 17881–17896.
Ding, X.M., 2014. MicroRNAs: regulators of cancer metastasis and epithelial-mesenchymal transition (EMT). Chin. J. Cancer 33, 140–147.
Dong, J., Zhao, Y.P., Zhou, L., Zhang, T.P., Chen, G., 2011. Bcl-2 upregulation induced by miR-21 via a direct interaction is associated with apoptosis and chemoresistance in MIA PaCa-2 pancreatic cancer cells. Arch. Med. Res. 42, 8–14.
Eis, P.S., Tam, W., Sun, L., Chadburn, A., Li, Z., Gomez, M.F., Lund, E., Dahlberg, J.E., 2005. Accumulation of miR-155 and BIC RNA in human B cell lymphomas. Proc. Natl. Acad. Sci. U.S.A. 102, 3627–3632.
Evangelisti, C., Florian, M.C., Massimi, I., Dominici, C., Giannini, G., Galardi, S., Bue, M.C., Massalini, S., McDowell, H.P., Messi, E., Gulino, A., Farace, M.G., Ciafre, S.A., 2009. MiR-128 up-regulation inhibits Reelin and DCX expression and reduces neuroblastoma cell motility and invasiveness. FASEB J. 23, 4276–4287.
Faraoni, I., Antonetti, F.R., Cardone, J., Bonmassar, E., 2009. MiR-155 gene: a typical multifunctional microRNA. Biochim. Biophys. Acta 1792, 497–505.
Felekkis, K., Touvana, E., Stefanou, C., Deltas, C., 2010. microRNAs: a newly described class of encoded molecules that play a role in health and disease. Hippokratia 14, 236–240.
Feng, Z.M., Qiu, J., Chen, X.W., Liao, R.X., Liao, X.Y., Zhang, L.P., Chen, X., Li, Y., Chen, Z.T., Sun, J.G., 2015. Essential role of miR-200c in regulating self-renewal of breast cancer stem cells and their counterparts of mammary epithelium. BMC Cancer 15, 645.
Fernandez-Mercado, M., Manterola, L., Lawrie, C.H., 2015. MicroRNAs in lymphoma: regulatory role and biomarker potential. Curr. Genomics 16, 349–358.
Filipowicz, W., Bhattacharyya, S.N., Sonenberg, N., 2008. Mechanisms of post-transcriptional regulation by microRNAs: are the answers in sight? Nat. Rev. Genet. 9, 102–114.
Fojo, T., 2007. Multiple paths to a drug resistance phenotype: mutations, translocations, deletions and amplification of coding genes or promoter regions, epigenetic changes and microRNAs. Drug Resist. Updat. 10, 59–67.
Fornari, F., Gramantieri, L., Ferracin, M., Veronese, A., Sabbioni, S., Calin, G.A., Grazi, G.L., Giovannini, C., Croce, C.M., Bolondi, L., Negrini, M., 2008. MiR-221 controls CDKN1C/p57 and CDKN1B/p27 expression in human hepatocellular carcinoma. Oncogene 27, 5651–5661.
Fornari, F., Gramantieri, L., Giovannini, C., Veronese, A., Ferracin, M., Sabbioni, S., Calin, G.A., Grazi, G.L., Croce, C.M., Tavolari, S., Chieco, P., Negrini, M., Bolondi, L., 2009. MiR-122/cyclin G1 interaction modulates p53 activity and affects doxorubicin sensitivity of human hepatocarcinoma cells. Cancer Res. 69, 5761–5767.
Fornari, F., Milazzo, M., Chieco, P., Negrini, M., Calin, G.A., Grazi, G.L., Pollutri, D., Croce, C.M., Bolondi, L., Gramantieri, L., 2010. MiR-199a-3p regulates mTOR and c-Met to influence the doxorubicin sensitivity of human hepatocarcinoma cells. Cancer Res. 70, 5184–5193.
Frankel, L.B., Christoffersen, N.R., Jacobsen, A., Lindow, M., Krogh, A., Lund, A.H., 2008. Programmed cell death 4 (PDCD4) is an important functional target of the microRNA miR-21 in breast cancer cells. J. Biol. Chem. 283, 1026–1033.

Fujita, Y., Kojima, K., Hamada, N., Ohhashi, R., Akao, Y., Nozawa, Y., Deguchi, T., Ito, M., 2008. Effects of miR-34a on cell growth and chemoresistance in prostate cancer PC3 cells. Biochem. Biophys. Res. Commun. 377, 114–119.

Fukasawa, K., 2007. Oncogenes and tumour suppressors take on centrosomes. Nat. Rev. Cancer 7, 911–924.

Galardi, S., Mercatelli, N., Giorda, E., Massalini, S., Frajese, G.V., Ciafre, S.A., Farace, M.G., 2007. MiR-221 and miR-222 expression affects the proliferation potential of human prostate carcinoma cell lines by targeting p27Kip1. J. Biol. Chem. 282, 23716–23724.

Garcia, A.I., Buisson, M., Bertrand, P., Rimokh, R., Rouleau, E., Lopez, B.S., Lidereau, R., Mikaelian, I., Mazoyer, S., 2011. Down-regulation of BRCA1 expression by miR-146a and miR-146b-5p in triple negative sporadic breast cancers. EMBO Mol. Med. 3, 279–290.

Garofalo, M., Quintavalle, C., Di Leva, G., Zanca, C., Romano, G., Taccioli, C., Liu, C.G., Croce, C.M., Condorelli, G., 2008. MicroRNA signatures of TRAIL resistance in human non-small cell lung cancer. Oncogene 27, 3845–3855.

Garofalo, M., Quintavalle, C., Romano, G., Croce, C.M., Condorelli, G., 2012. MiR221/222 in cancer: their role in tumor progression and response to therapy. Curr. Mol. Med. 12, 27–33.

Garofalo, M., Leva, G.D., Croce, C.M., 2014. MicroRNAs as anti-cancer therapy. Curr. Pharm. Des. 20, 5328–5335.

Garzia, L., Andolfo, I., Cusanelli, E., Marino, N., Petrosino, G., De Martino, D., Esposito, V., Galeone, A., Navas, L., Esposito, S., Gargiulo, S., Fattet, S., Donofrio, V., Cinalli, G., Brunetti, A., Vecchio, L.D., Northcott, P.A., Delattre, O., Taylor, M.D., Iolascon, A., Zollo, M., 2009. MicroRNA-199b-5p impairs cancer stem cells through negative regulation of HES1 in medulloblastoma. PLoS One 4. e4998.

Gee, H.E., Camps, C., Buffa, F.M., Colella, S., Sheldon, H., Gleadle, J.M., Ragoussis, J., Harris, A.L., 2008. MicroRNA-10b and breast cancer metastasis. Nature 455. E8–9, author reply E9.

Georges, S.A., Biery, M.C., Kim, S.Y., Schelter, J.M., Guo, J., Chang, A.N., Jackson, A.L., Carleton, M.O., Linsley, P.S., Cleary, M.A., Chau, B.N., 2008. Coordinated regulation of cell cycle transcripts by p53-inducible microRNAs, miR-192 and miR-215. Cancer Res. 68, 10105–10112.

Gironella, M., Seux, M., Xie, M.J., Cano, C., Tomasini, R., Gommeaux, J., Garcia, S., Nowak, J., Yeung, M.L., Jeang, K.T., Chaix, A., Fazli, L., Motoo, Y., Wang, Q., Rocchi, P., Russo, A., Gleave, M., Dagorn, J.C., Iovanna, J.L., Carrier, A., Pebusque, M.J., Dusetti, N.J., 2007. Tumor protein 53-induced nuclear protein 1 expression is repressed by miR-155, and its restoration inhibits pancreatic tumor development. Proc. Natl. Acad. Sci. U.S.A. 104, 16170–16175.

Godlewski, J., Nowicki, M.O., Bronisz, A., Williams, S., Otsuki, A., Nuovo, G., Raychaudhury, A., Newton, H.B., Chiocca, E.A., Lawler, S., 2008. Targeting of the Bmi-1 oncogene/stem cell renewal factor by microRNA-128 inhibits glioma proliferation and self-renewal. Cancer Res. 68, 9125–9130.

Gomes, S.E., Simoes, A.E., Pereira, D.M., Castro, R.E., Rodrigues, C.M., Borralho, P.M., 2016. MiR-143 or miR-145 overexpression increases cetuximab-mediated antibody-dependent cellular cytotoxicity in human colon cancer cells. Oncotarget 7, 9368–9387.

Gregory, P.A., Bert, A.G., Paterson, E.L., Barry, S.C., Tsykin, A., Farshid, G., Vadas, M.A., Khew-Goodall, Y., Goodall, G.J., 2008. The miR-200 family and miR-205 regulate epithelial to mesenchymal transition by targeting ZEB1 and SIP1. Nat. Cell Biol. 10, 593–601.

Gu, S., Kay, M.A., 2010. How do miRNAs mediate translational repression? Silence 1, 11.

Gurtan, A.M., Sharp, P.A., 2013. The role of miRNAs in regulating gene expression networks. J. Mol. Biol. 425, 3582–3600.

Ha, T.Y., 2011. The role of microRNAs in regulatory T cells and in the immune response. Immune Netw. 11, 11–41.
Ha, M., Kim, V.N., 2014. Regulation of microRNA biogenesis. Nat. Rev. Mol. Cell Biol. 15, 509–524.
Han, C., Wan, G., Langley, R.R., Zhang, X., Lu, X., 2012. Crosstalk between the DNA damage response pathway and microRNAs. Cell. Mol. Life Sci. 69, 2895–2906.
Hart, M., Rheinheimer, S., Leidinger, P., Backes, C., Menegatti, J., Fehlmann, T., Grasser, F., Keller, A., Meese, E., 2016. Identification of miR-34a-target interactions by a combined network based and experimental approach. Oncotarget 7, 34288–34299.
Hawa, Z., Haque, I., Ghosh, A., Banerjee, S., Harris, L., Banerjee, S.K., 2016. The miRacle in pancreatic cancer by miRNAs: tiny angels or devils in disease progression. Int. J. Mol. Sci. 17 (6). pii: E809. http://dx.doi.org/10.3390/ijms17060809.
Hayashita, Y., Osada, H., Tatematsu, Y., Yamada, H., Yanagisawa, K., Tomida, S., Yatabe, Y., Kawahara, K., Sekido, Y., Takahashi, T., 2005. A polycistronic microRNA cluster, miR-17-92, is overexpressed in human lung cancers and enhances cell proliferation. Cancer Res. 65, 9628–9632.
He, L., Thomson, J.M., Hemann, M.T., Hernando-Monge, E., Mu, D., Goodson, S., Powers, S., Cordon-Cardo, C., Lowe, S.W., Hannon, G.J., Hammond, S.M., 2005. A microRNA polycistron as a potential human oncogene. Nature 435, 828–833.
He, X.X., Guo, A.Y., Xu, C.R., Chang, Y., Xiang, G.Y., Gong, J., Dan, Z.L., Tian, D.A., Liao, J.Z., Lin, J.S., 2014. Bioinformatics analysis identifies miR-221 as a core regulator in hepatocellular carcinoma and its silencing suppresses tumor properties. Oncol. Rep. 32, 1200–1210.
Hermeking, H., 2010. The miR-34 family in cancer and apoptosis. Cell Death Differ. 17, 193–199.
Hermeking, H., 2012. MicroRNAs in the p53 network: micromanagement of tumour suppression. Nat. Rev. Cancer 12, 613–626.
Hong, L., Lai, M., Chen, M., Xie, C., Liao, R., Kang, Y.J., Xiao, C., Hu, W.Y., Han, J., Sun, P., 2010. The miR-17-92 cluster of microRNAs confers tumorigenicity by inhibiting oncogene-induced senescence. Cancer Res. 70, 8547–8557.
Honkakoski, P., Negishi, M., 2000. Regulation of cytochrome P450 (CYP) genes by nuclear receptors. Biochem. J. 347, 321–337.
Hu, H., Gatti, R.A., 2011. MicroRNAs: new players in the DNA damage response. J. Mol. Cell Biol. 3, 151–158.
Hu, H., Du, L., Nagabayashi, G., Seeger, R.C., Gatti, R.A., 2010a. ATM is down-regulated by N-Myc-regulated microRNA-421. Proc. Natl. Acad. Sci. U.S.A. 107, 1506–1511.
Hu, W., Chan, C.S., Wu, R., Zhang, C., Sun, Y., Song, J.S., Tang, L.H., Levine, A.J., Feng, Z., 2010b. Negative regulation of tumor suppressor p53 by microRNA miR-504. Mol. Cell 38, 689–699.
Huan, L.C., Wu, J.C., Chiou, B.H., Chen, C.H., Ma, N., Chang, C.Y., Tsen, Y.K., Chen, S.C., 2014. MicroRNA regulation of DNA repair gene expression in 4-aminobiphenyl-treated HepG2 cells. Toxicology 322, 69–77.
Huang, S., Li, X., Zhu, H., 2016. MicroRNA-152 targets phosphatase and tensin homolog to inhibit apoptosis and promote cell migration of nasopharyngeal carcinoma cells. Med. Sci. Monit. 22, 4330–4337.
Huen, M.S., Chen, J., 2010. Assembly of checkpoint and repair machineries at DNA damage sites. Trends Biochem. Sci. 35, 101–108.
Humphries, B., Yang, C., 2015. The microRNA-200 family: small molecules with novel roles in cancer development, progression and therapy. Oncotarget 6, 6472–6498.
Humplikova, L., Kollinerova, S., Papajik, T., Pikalova, Z., Holzerova, M., Prochazka, V., Divoka, M., Modriansky, M., Indrak, K., Jarosova, M., 2013. Expression of miR-15a and miR-16-1 in patients with chronic lymphocytic leukemia. Biomed. Pap. Med. Fac. Univ. Palacky Olomouc Czech Repub. 157, 284–293.

Incoronato, M., Garofalo, M., Urso, L., Romano, G., Quintavalle, C., Zanca, C., Iaboni, M., Nuovo, G., Croce, C.M., Condorelli, G., 2010. MiR-212 increases tumor necrosis factor-related apoptosis-inducing ligand sensitivity in non-small cell lung cancer by targeting the antiapoptotic protein PED. Cancer Res. 70, 3638–3646.

Iorio, M.V., Casalini, P., Piovan, C., Di Leva, G., Merlo, A., Triulzi, T., Menard, S., Croce, C.M., Tagliabue, E., 2009. microRNA-205 regulates HER3 in human breast cancer. Cancer Res. 69, 2195–2200.

Jansson, M.D., Lund, A.H., 2012. MicroRNA and cancer. Mol. Oncol. 6, 590–610.

Ji, Q., Hao, X., Meng, Y., Zhang, M., Desano, J., Fan, D., Xu, L., 2008. Restoration of tumor suppressor miR-34 inhibits human p53-mutant gastric cancer tumorspheres. BMC Cancer 8, 266.

Ji, Q., Hao, X., Zhang, M., Tang, W., Yang, M., Li, L., Xiang, D., Desano, J.T., Bommer, G.T., Fan, D., Fearon, E.R., Lawrence, T.S., Xu, L., 2009. MicroRNA miR-34 inhibits human pancreatic cancer tumor-initiating cells. PLoS One 4. e6816.

Jin, Z.B., Hirokawa, G., Gui, L., Takahashi, R., Osakada, F., Hiura, Y., Takahashi, M., Yasuhara, O., Iwai, N., 2009. Targeted deletion of miR-182, an abundant retinal microRNA. Mol. Vis. 15, 523–533.

Jin, Y., Chen, Z., Liu, X., Zhou, X., 2013. Evaluating the microRNA targeting sites by luciferase reporter gene assay. Methods Mol. Biol. 936, 117–127.

Johnson, S.M., Grosshans, H., Shingara, J., Byrom, M., Jarvis, R., Cheng, A., Labourier, E., Reinert, K.L., Brown, D., Slack, F.J., 2005. RAS is regulated by the let-7 microRNA family. Cell 120, 635–647.

Josson, S., Sung, S.Y., Lao, K., Chung, L.W., Johnstone, P.A., 2008. Radiation modulation of microRNA in prostate cancer cell lines. Prostate 68, 1599–1606.

Kang, W., Tong, J.H., Lung, R.W., Dong, Y., Zhao, J., Liang, Q., Zhang, L., Pan, Y., Yang, W., Pang, J.C., Cheng, A.S., Yu, J., Tome, K.F., 2015. Targeting of YAP1 by microRNA-15a and microRNA-16-1 exerts tumor suppressor function in gastric adenocarcinoma. Mol. Cancer 14, 52.

Kastan, M.B., Bartek, J., 2004. Cell-cycle checkpoints and cancer. Nature 432, 316–323.

Kefas, B., Comeau, L., Floyd, D.H., Seleverstov, O., Godlewski, J., Schmittgen, T., Jiang, J., diPierro, C.G., Li, Y., Chiocca, E.A., Lee, J., Fine, H., Abounader, R., Lawler, S., Purow, B., 2009. The neuronal microRNA miR-326 acts in a feedback loop with notch and has therapeutic potential against brain tumors. J. Neurosci. 29, 15161–15168.

Khalili, M., Vasei, M., Khalili, D., Alimoghaddam, K., Sadeghizadeh, M., Mowla, S.J., 2015. Downregulation of the genes involved in reprogramming (SOX2, c-MYC, miR-302, miR-145, and P21) in gastric adenocarcinoma. J. Gastrointest. Cancer 46, 251–258.

Kim, S., Lee, U.J., Kim, M.N., Lee, E.J., Kim, J.Y., Lee, M.Y., Choung, S., Kim, Y.J., Choi, Y.C., 2008. MicroRNA miR-199a* regulates the MET proto-oncogene and the downstream extracellular signal-regulated kinase 2 (ERK2). J. Biol. Chem. 283, 18158–18166.

Kluiver, J., Poppema, S., de Jong, D., Blokzijl, T., Harms, G., Jacobs, S., Kroesen, B.J., van den Berg, A., 2005. BIC and miR-155 are highly expressed in Hodgkin, primary mediastinal and diffuse large B cell lymphomas. J. Pathol. 207, 243–249.

Kong, W., Yang, H., He, L., Zhao, J.J., Coppola, D., Dalton, W.S., Cheng, J.Q., 2008. MicroRNA-155 is regulated by the transforming growth factor beta/Smad pathway and contributes to epithelial cell plasticity by targeting RhoA. Mol. Cell. Biol. 28, 6773–6784.

Kong, D., Li, Y., Wang, Z., Banerjee, S., Ahmad, A., Kim, H.R., Sarkar, F.H., 2009. MiR-200 regulates PDGF-D-mediated epithelial-mesenchymal transition, adhesion, and invasion of prostate cancer cells. Stem Cells 27, 1712–1721.

Kong, D., Banerjee, S., Ahmad, A., Li, Y., Wang, Z., Sethi, S., Sarkar, F.H., 2010a. Epithelial to mesenchymal transition is mechanistically linked with stem cell signatures in prostate cancer cells. PLoS One 5. e12445.

Kong, W., He, L., Coppola, M., Guo, J., Esposito, N.N., Coppola, D., Cheng, J.Q., 2010b. MicroRNA-155 regulates cell survival, growth, and chemosensitivity by targeting FOXO3a in breast cancer. J. Biol. Chem. 285, 17869–17879.

Kong, W., He, L., Coppola, M., Guo, J., Esposito, N.N., Coppola, D., Cheng, J.Q., 2016. MicroRNA-155 regulates cell survival, growth, and chemosensitivity by targeting FOXO3a in breast cancer. J. Biol. Chem. 291, 22855.

Korpal, M., Lee, E.S., Hu, G., Kang, Y., 2008. The miR-200 family inhibits epithelial-mesenchymal transition and cancer cell migration by direct targeting of E-cadherin transcriptional repressors ZEB1 and ZEB2. J. Biol. Chem. 283, 14910–14914.

Koschny, R., Walczak, H., Ganten, T.M., 2007. The promise of TRAIL—potential and risks of a novel anticancer therapy. J. Mol. Med. (Berl) 85, 923–935.

Kovalchuk, O., Filkowski, J., Meservy, J., Ilnytskyy, Y., Tryndyak, V.P., Chekhun, V.F., Pogribny, I.P., 2008. Involvement of microRNA-451 in resistance of the MCF-7 breast cancer cells to chemotherapeutic drug doxorubicin. Mol. Cancer Ther. 7, 2152–2159.

Kozomara, A., Griffiths-Jones, S., 2011. MiRBase: integrating microRNA annotation and deep-sequencing data. Nucleic Acids Res. 39, D152–D157.

Krause, C.J., Popp, O., Thirunarayanan, N., Dittmar, G., Lipp, M., Muller, G., 2016. MicroRNA-34a promotes genomic instability by a broad suppression of genome maintenance mechanisms downstream of the oncogene KSHV-vGPCR. Oncotarget 7, 10414–10432.

Krell, J., Stebbing, J., Carissimi, C., Dabrowska, A.F., de Giorgio, A., Frampton, A.E., Harding, V., Fulci, V., Macino, G., Colombo, T., Castellano, L., 2016. TP53 regulates miRNA association with AGO2 to remodel the miRNA-mRNA interaction network. Genome Res. 26, 331–341.

Kumarswamy, R., Volkmann, I., Thum, T., 2011. Regulation and function of miRNA-21 in health and disease. RNA Biol. 8, 706–713.

Lagos-Quintana, M., Rauhut, R., Lendeckel, W., Tuschl, T., 2001. Identification of novel genes coding for small expressed RNAs. Science 294, 853–858.

Lai, J.H., She, T.F., Juang, Y.M., Tsay, Y.G., Huang, A.H., Yu, S.L., Chen, J.J., Lai, C.C., 2012. Comparative proteomic profiling of human lung adenocarcinoma cells (CL 1-0) expressing miR-372. Electrophoresis 33, 675–688.

Lal, A., Navarro, F., Maher, C.A., Maliszewski, L.E., Yan, N., O'Day, E., Chowdhury, D., Dykxhoorn, D.M., Tsai, P., Hofmann, O., Becker, K.G., Gorospe, M., Hide, W., Lieberman, J., 2009a. MiR-24 inhibits cell proliferation by targeting E2F2, MYC, and other cell-cycle genes via binding to "seedless" 3'UTR microRNA recognition elements. Mol. Cell 35, 610–625.

Lal, A., Pan, Y., Navarro, F., Dykxhoorn, D.M., Moreau, L., Meire, E., Bentwich, Z., Lieberman, J., Chowdhury, D., 2009b. MiR-24-mediated downregulation of H2AX suppresses DNA repair in terminally differentiated blood cells. Nat. Struct. Mol. Biol. 16, 492–498.

Lee, Y.S., Dutta, A., 2007. The tumor suppressor microRNA let-7 represses the HMGA2 oncogene. Genes Dev. 21, 1025–1030.

Lee, M.K., Teoh, W.W., Phang, B.H., Tong, W.M., Wang, Z.Q., Sabapathy, K., 2012. Cell-type, dose, and mutation-type specificity dictate mutant p53 functions in vivo. Cancer Cell 22, 751–764.

Lei, S.T., Shen, F., Chen, J.W., Feng, J.H., Cai, W.S., Shen, L., Hu, Z.W., Xu, B., 2016. MiR-639 promoted cell proliferation and cell cycle in human thyroid cancer by suppressing CDKN1A expression. Biomed. Pharmacother. 84, 1834–1840.

le Sage, C., Nagel, R., Egan, D.A., Schrier, M., Mesman, E., Mangiola, A., Anile, C., Maira, G., Mercatelli, N., Ciafre, S.A., Farace, M.G., Agami, R., 2007. Regulation of the p27(Kip1) tumor suppressor by miR-221 and miR-222 promotes cancer cell proliferation. EMBO J. 26, 3699–3708.

Lewis, B.P., Burge, C.B., Bartel, D.P., 2005. Conserved seed pairing, often flanked by adenosines, indicates that thousands of human genes are microRNA targets. Cell 120, 15–20.

Li, S., Liang, Z., Xu, L., Zou, F., 2012. MicroRNA-21: a ubiquitously expressed pro-survival factor in cancer and other diseases. Mol. Cell. Biochem. 360, 147–158.

Li, Y., Casey, S.C., Choi, P.S., Felsher, D.W., 2014. MiR-17-92 explains MYC oncogene addiction. Mol. Cell. Oncol. 1. e970092.

Li, J., Lam, M., Reproducibility Project: Cancer Biology, 2015a. Registered report: the microRNA miR-34a inhibits prostate cancer stem cells and metastasis by directly repressing CD44. Elife 4. e06434.

Li, S., Ma, Y., Hou, X., Liu, Y., Li, K., Xu, S., Wang, J., 2015b. MiR-185 acts as a tumor suppressor by targeting AKT1 in non-small cell lung cancer cells. Int. J. Clin. Exp. Pathol. 8, 11854–11862.

Li, W., Zhang, H., Assaraf, Y.G., Zhao, K., Xu, X., Xie, J., Yang, D.H., Chen, Z.S., 2016. Overcoming ABC transporter-mediated multidrug resistance: molecular mechanisms and novel therapeutic drug strategies. Drug Resist. Updat. 27, 14–29.

Liang, L., Wong, C.M., Ying, Q., Fan, D.N., Huang, S., Ding, J., Yao, J., Yan, M., Li, J., Yao, M., Ng, I.O., He, X., 2010. MicroRNA-125b suppressesed human liver cancer cell proliferation and metastasis by directly targeting oncogene LIN28B2. Hepatology 52, 1731–1740.

Liang, H., Fu, Z., Jiang, X., Wang, N., Wang, F., Wang, X., Zhang, S., Wang, Y., Yan, X., Guan, W.X., Zhang, C.Y., Zen, K., Zhang, Y., Chen, X., Zhou, G., 2015. MiR-16 promotes the apoptosis of human cancer cells by targeting FEAT. BMC Cancer 15, 448.

Lieber, M.R., 2008. The mechanism of human nonhomologous DNA end joining. J. Biol. Chem. 283, 1–5.

Liu, X., Sempere, L.F., Ouyang, H., Memoli, V.A., Andrew, A.S., Luo, Y., Demidenko, E., Korc, M., Shi, W., Preis, M., Dragnev, K.H., Li, H., Direnzo, J., Bak, M., Freemantle, S.J., Kauppinen, S., Dmitrovsky, E., 2010. MicroRNA-31 functions as an oncogenic microRNA in mouse and human lung cancer cells by repressing specific tumor suppressors. J. Clin. Invest. 120, 1298–1309.

Liu, C., Kelnar, K., Liu, B., Chen, X., Calhoun-Davis, T., Li, H., Patrawala, L., Yan, H., Jeter, C., Honorio, S., Wiggins, J.F., Bader, A.G., Fagin, R., Brown, D., Tang, D.G., 2011. The microRNA miR-34a inhibits prostate cancer stem cells and metastasis by directly repressing CD44. Nat. Med. 17, 211–215.

Liu, Y., Zhao, J., Zhang, P.Y., Zhang, Y., Sun, S.Y., Yu, S.Y., Xi, Q.S., 2012. MicroRNA-10b targets E-cadherin and modulates breast cancer metastasis. Med. Sci. Monit. 18, BR299–BR308.

Liu, Y., Xing, R., Zhang, X., Dong, W., Zhang, J., Yan, Z., Li, W., Cui, J., Lu, Y., 2013. MiR-375 targets the p53 gene to regulate cellular response to ionizing radiation and etoposide in gastric cancer cells. DNA Repair (Amst) 12, 741–750.

Liu, S., Long, J., Yuan, B., Zheng, M., Xiao, M., Xu, J., Lin, X., Feng, X.H., 2016. SUMO modification reverses inhibitory effects of Smad nuclear interacting protein-1 in TGF-beta responses. J. Biol. Chem. 291, 24418–24430.

Lu, X., Nannenga, B., Donehower, L.A., 2005. PPM1D dephosphorylates Chk1 and p53 and abrogates cell cycle checkpoints. Genes Dev. 19, 1162–1174.

Luan, Y., Zhang, S., Zuo, L., Zhou, L., 2015. Overexpression of miR-100 inhibits cell proliferation, migration, and chemosensitivity in human glioblastoma through FGFR3. Onco Targets Ther. 8, 3391–3400.

Lupini, L., Bassi, C., Ferracin, M., Bartonicek, N., D'Abundo, L., Zagatti, B., Callegari, E., Musa, G., Moshiri, F., Gramantieri, L., Corrales, F.J., Enright, A.J., Sabbioni, S., Negrini, M., 2013. MiR-221 affects multiple cancer pathways by modulating the level of hundreds messenger RNAs. Front. Genet. 4, 64.

Ma, L., 2010. Role of miR-10b in breast cancer metastasis. Breast Cancer Res. 12, 210.

Ma, L., Teruya-Feldstein, J., Weinberg, R.A., 2007. Tumour invasion and metastasis initiated by microRNA-10b in breast cancer. Nature 449, 682–688.
Macfarlane, L.A., Murphy, P.R., 2010. MicroRNA: biogenesis, function and role in cancer. Curr. Genomics 11, 537–561.
Mackiewicz, M., Huppi, K., Pitt, J.J., Dorsey, T.H., Ambs, S., Caplen, N.J., 2011. Identification of the receptor tyrosine kinase AXL in breast cancer as a target for the human miR-34a microRNA. Breast Cancer Res. Treat. 130, 663–679.
Madka, V., Mohammed, A., Li, Q., Zhang, Y., Biddick, L., Patlolla, J.M., Lightfoot, S., Towner, R.A., Wu, X.R., Steele, V.E., Kopelovich, L., Rao, C.V., 2016. Targeting mTOR and p53 signaling inhibits muscle invasive bladder cancer in vivo. Cancer Prev. Res. (Phila.) 9, 53–62.
Majid, S., Saini, S., Dar, A.A., Hirata, H., Shahryari, V., Tanaka, Y., Yamamura, S., Ueno, K., Zaman, M.S., Singh, K., Chang, I., Deng, G., Dahiya, R., 2011. MicroRNA-205 inhibits Src-mediated oncogenic pathways in renal cancer. Cancer Res. 71, 2611–2621.
Majidinia, M., Yousefi, B., 2016. DNA damage response regulation by microRNAs as a therapeutic target in cancer. DNA Repair (Amst) 47, 1–11.
Marsico, A., Huska, M.R., Lasserre, J., Hu, H., Vucicevic, D., Musahl, A., Orom, U., Vingron, M., 2013. PROmiRNA: a new miRNA promoter recognition method uncovers the complex regulation of intronic miRNAs. Genome Biol. 14, R84.
Martinez-Outschoorn, U.E., Whitaker-Menezes, D., Pavlides, S., Chiavarina, B., Bonuccelli, G., Casey, T., Tsirigos, A., Migneco, G., Witkiewicz, A., Balliet, R., Mercier, I., Wang, C., Flomenberg, N., Howell, A., Lin, Z., Caro, J., Pestell, R.G., Sotgia, F., Lisanti, M.P., 2010. The autophagic tumor stroma model of cancer or "battery-operated tumor growth": a simple solution to the autophagy paradox. Cell Cycle 9, 4297–4306.
Matsushima, K., Isomoto, H., Yamaguchi, N., Inoue, N., Machida, H., Nakayama, T., Hayashi, T., Kunizaki, M., Hidaka, S., Nagayasu, T., Nakashima, M., Ujifuku, K., Mitsutake, N., Ohtsuru, A., Yamashita, S., Korpal, M., Kang, Y., Gregory, P.A., Goodall, G.J., Kohno, S., Nakao, K., 2011. MiRNA-205 modulates cellular invasion and migration via regulating zinc finger E-box binding homeobox 2 expression in esophageal squamous cell carcinoma cells. J. Transl. Med. 9, 30.
Meerson, A., Yehuda, H., 2016. Leptin and insulin up-regulate miR-4443 to suppress NCOA1 and TRAF4, and decrease the invasiveness of human colon cancer cells. BMC Cancer 16, 882.
Meng, F., Henson, R., Wehbe-Janek, H., Ghoshal, K., Jacob, S.T., Patel, T., 2007. MicroRNA-21 regulates expression of the PTEN tumor suppressor gene in human hepatocellular cancer. Gastroenterology 133, 647–658.
Mercatelli, N., Coppola, V., Bonci, D., Miele, F., Costantini, A., Guadagnoli, M., Bonanno, E., Muto, G., Frajese, G.V., De Maria, R., Spagnoli, L.G., Farace, M.G., Ciafre, S.A., 2008. The inhibition of the highly expressed miR-221 and miR-222 impairs the growth of prostate carcinoma xenografts in mice. PLoS One 3. e4029.
Michael, M.Z., O'Connor, S.M., van Holst Pellekaan, N.G., Young, G.P., James, R.J., 2003. Reduced accumulation of specific microRNAs in colorectal neoplasia. Mol. Cancer Res. 1, 882–891.
Milella, M., Falcone, I., Conciatori, F., Cesta Incani, U., Del Curatolo, A., Inzerilli, N., Nuzzo, C.M., Vaccaro, V., Vari, S., Cognetti, F., Ciuffreda, L., 2015. PTEN: multiple functions in human malignant tumors. Front. Oncol. 5, 24.
Mishra, P.J., Humeniuk, R., Longo-Sorbello, G.S., Banerjee, D., Bertino, J.R., 2007. A miR-24 microRNA binding-site polymorphism in dihydrofolate reductase gene leads to methotrexate resistance. Proc. Natl. Acad. Sci. U.S.A. 104, 13513–13518.
Mishra, P.J., Banerjee, D., Bertino, J.R., 2008. MiRSNPs or MiR-polymorphisms, new players in microRNA mediated regulation of the cell: introducing microRNA pharmacogenomics. Cell Cycle 7, 853–858.

Mladinov, D., Liu, Y., Mattson, D.L., Liang, M., 2013. MicroRNAs contribute to the maintenance of cell-type-specific physiological characteristics: miR-192 targets Na+/K+-ATPase beta1. Nucleic Acids Res. 41, 1273–1283.

Mogilyansky, E., Rigoutsos, I., 2013. The miR-17/92 cluster: a comprehensive update on its genomics, genetics, functions and increasingly important and numerous roles in health and disease. Cell Death Differ. 20, 1603–1614.

Molitoris, J.K., McColl, K.S., Distelhorst, C.W., 2011. Glucocorticoid-mediated repression of the oncogenic microRNA cluster miR-17~92 contributes to the induction of Bim and initiation of apoptosis. Mol. Endocrinol. 25, 409–420.

Moskwa, P., Buffa, F.M., Pan, Y., Panchakshari, R., Gottipati, P., Muschel, R.J., Beech, J., Kulshrestha, R., Abdelmohsen, K., Weinstock, D.M., Gorospe, M., Harris, A.L., Helleday, T., Chowdhury, D., 2011. MiR-182-mediated downregulation of BRCA1 impacts DNA repair and sensitivity to PARP inhibitors. Mol. Cell 41, 210–220.

Moussay, E., Palissot, V., Vallar, L., Poirel, H.A., Wenner, T., El Khoury, V., Aouali, N., Van Moer, K., Leners, B., Bernardin, F., Muller, A., Cornillet-Lefebvre, P., Delmer, A., Duhem, C., Ries, F., van Dyck, E., Berchem, G., 2010. Determination of genes and microRNAs involved in the resistance to fludarabine in vivo in chronic lymphocytic leukemia. Mol. Cancer 9, 115.

Murray, G.I., Taylor, M.C., McFadyen, M.C., McKay, J.A., Greenlee, W.F., Burke, M.D., Melvin, W.T., 1997. Tumor-specific expression of cytochrome P450 CYP1B1. Cancer Res. 57, 3026–3031.

Nagel, R., le Sage, C., Diosdado, B., van der Waal, M., Oude Vrielink, J.A., Bolijn, A., Meijer, G.A., Agami, R., 2008. Regulation of the adenomatous polyposis coli gene by the miR-135 family in colorectal cancer. Cancer Res. 68, 5795–5802.

Nairz, K., Rottig, C., Rintelen, F., Zdobnov, E., Moser, M., Hafen, E., 2006. Overgrowth caused by misexpression of a microRNA with dispensable wild-type function. Dev. Biol. 291, 314–324.

Nakamachi, Y., Kawano, S., Takenokuchi, M., Nishimura, K., Sakai, Y., Chin, T., Saura, R., Kurosaka, M., Kumagai, S., 2009. MicroRNA-124a is a key regulator of proliferation and monocyte chemoattractant protein 1 secretion in fibroblast-like-synoviocytes from patients with rheumatoid arthritis. Arthritis Rheum. 60, 1294–1304.

Nasser, M.W., Datta, J., Nuovo, G., Kutay, H., Motiwala, T., Majumder, S., Wang, B., Suster, S., Jacob, S.T., Ghoshal, K., 2008. Down-regulation of micro-RNA-1 (miR-1) in lung cancer. Suppression of tumorigenic property of lung cancer cells and their sensitization to doxorubicin-induced apoptosis by miR-1. J. Biol. Chem. 283, 33394–33405.

Navarro, A., Diaz, T., Martinez, A., Gaya, A., Pons, A., Gel, B., Codony, C., Ferrer, G., Martinez, C., Montserrat, E., Monzo, M., 2009. Regulation of JAK2 by miR-135a: prognostic impact in classic Hodgkin lymphoma. Blood 114, 2945–2951.

Nguyen, L.H., Zhu, H., 2015. Lin28 and let-7 in cell metabolism and cancer. Transl. Pediatr. 4, 4–11.

O'Donnell, K.A., Wentzel, E.A., Zeller, K.I., Dang, C.V., Mendell, J.T., 2005. c-Myc-regulated microRNAs modulate E2F1 expression. Nature 435, 839–843.

Okamura, K., Hagen, J.W., Duan, H., Tyler, D.M., Lai, E.C., 2007. The mirtron pathway generates microRNA-class regulatory RNAs in Drosophila. Cell 130, 89–100.

Ota, H., Sakurai, M., Gupta, R., Valente, L., Wulff, B.E., Ariyoshi, K., Iizasa, H., Davuluri, R.V., Nishikura, K., 2013. ADAR1 forms a complex with Dicer to promote microRNA processing and RNA-induced gene silencing. Cell 153, 575–589.

Pan, T.H., Wu, X.Y., 2012. Regarding article 'Bcl-2 upregulation induced by miR-21 via a direct interaction is associated with apoptosis and chemoresistance in MIA PaCa-2 pancreatic cancer cells'. Arch. Med. Res. 43, 252, author reply 253.

Pan, Y.Z., Morris, M.E., Yu, A.M., 2009. MicroRNA-328 negatively regulates the expression of breast cancer resistance protein (BCRP/ABCG2) in human cancer cells. Mol. Pharmacol. 75, 1374–1379.

Pan, X.N., Chen, J.J., Wang, L.X., Xiao, R.Z., Liu, L.L., Fang, Z.G., Liu, Q., Long, Z.J., Lin, D.J., 2014. Inhibition of c-Myc overcomes cytotoxic drug resistance in acute myeloid leukemia cells by promoting differentiation. PLoS One 9. e105381.

Park, E.Y., Chang, E., Lee, E.J., Lee, H.W., Kang, H.G., Chun, K.H., Woo, Y.M., Kong, H.K., Ko, J.Y., Suzuki, H., Song, E., Park, J.H., 2014. Targeting of miR34a-NOTCH1 axis reduced breast cancer stemness and chemoresistance. Cancer Res. 74, 7573–7582.

Pavlides, S., Tsirigos, A., Migneco, G., Whitaker-Menezes, D., Chiavarina, B., Flomenberg, N., Frank, P.G., Casimiro, M.C., Wang, C., Pestell, R.G., Martinez-Outschoorn, U.E., Howell, A., Sotgia, F., Lisanti, M.P., 2010. The autophagic tumor stroma model of cancer: role of oxidative stress and ketone production in fueling tumor cell metabolism. Cell Cycle 9, 3485–3505.

Pedersen, I.M., Otero, D., Kao, E., Miletic, A.V., Hother, C., Ralfkiaer, E., Rickert, R.C., Gronbaek, K., David, M., 2009. Onco-miR-155 targets SHIP1 to promote TNFalpha-dependent growth of B cell lymphomas. EMBO Mol. Med. 1, 288–295.

Pekarsky, Y., Croce, C.M., 2015. Role of miR-15/16 in CLL. Cell Death Differ. 22, 6–11.

Peng, Y., Zhang, X., Ma, Q., Yan, R., Qin, Y., Zhao, Y., Cheng, Y., Yang, M., Wang, Q., Feng, X., Huang, Y., Huang, W., Zhao, Z., Wang, L., Wei, Y., He, Z., Fan, X., Li, S., Jin, Z., Meltzer, S.J., 2017. MiRNA-194 activates the Wnt/beta-catenin signaling pathway in gastric cancer by targeting the negative Wnt regulator, SUFU. Cancer Lett. 385, 117–127.

Pineau, P., Volinia, S., McJunkin, K., Marchio, A., Battiston, C., Terris, B., Mazzaferro, V., Lowe, S.W., Croce, C.M., Dejean, A., 2010. MiR-221 overexpression contributes to liver tumorigenesis. Proc. Natl. Acad. Sci. U.S.A. 107, 264–269.

Pothof, J., Verkaik, N.S., van, I.W., Wiemer, E.A., Ta, V.T., van der Horst, G.T., Jaspers, N.G., van Gent, D.C., Hoeijmakers, J.H., Persengiev, S.P., 2009. MicroRNA-mediated gene silencing modulates the UV-induced DNA-damage response. EMBO J. 28, 2090–2099.

Pouliot, L.M., Chen, Y.C., Bai, J., Guha, R., Martin, S.E., Gottesman, M.M., Hall, M.D., 2012. Cisplatin sensitivity mediated by WEE1 and CHK1 is mediated by miR-155 and the miR-15 family. Cancer Res. 72, 5945–5955.

Psathas, J.N., Doonan, P.J., Raman, P., Freedman, B.D., Minn, A.J., Thomas-Tikhonenko, A., 2013. The Myc-miR-17-92 axis amplifies B-cell receptor signaling via inhibition of ITIM proteins: a novel lymphomagenic feed-forward loop. Blood 122, 4220–4229.

Qi, J., Yu, J.Y., Shcherbata, H.R., Mathieu, J., Wang, A.J., Seal, S., Zhou, W., Stadler, B.M., Bourgin, D., Wang, L., Nelson, A., Ware, C., Raymond, C., Lim, L.P., Magnus, J., Ivanovska, I., Diaz, R., Ball, A., Cleary, M.A., Ruohola-Baker, H., 2009. microRNAs regulate human embryonic stem cell division. Cell Cycle 8, 3729–3741.

Rao, X., Di Leva, G., Li, M., Fang, F., Devlin, C., Hartman-Frey, C., Burow, M.E., Ivan, M., Croce, C.M., Nephew, K.P., 2011. MicroRNA-221/222 confers breast cancer fulvestrant resistance by regulating multiple signaling pathways. Oncogene 30, 1082–1097.

Rausch, M., Ecsedi, M., Bartake, H., Mullner, A., Grosshans, H., 2015. A genetic interactome of the let-7 microRNA in C. elegans. Dev. Biol. 401, 276–286.

Ren, G., Yu, B., 2012. Post-transcriptional control of miRNA abundance in Arabidopsis. Plant Signal. Behav. 7, 1443–1446.

Rodriguez, A., Griffiths-Jones, S., Ashurst, J.L., Bradley, A., 2004. Identification of mammalian microRNA host genes and transcription units. Genome Res. 14, 1902–1910.

Ruby, J.G., Jan, C.H., Bartel, D.P., 2007. Intronic microRNA precursors that bypass Drosha processing. Nature 448, 83–86.

Sachdeva, M., Zhu, S., Wu, F., Wu, H., Walia, V., Kumar, S., Elble, R., Watabe, K., Mo, Y.Y., 2009. P53 represses c-Myc through induction of the tumor suppressor miR-145. Proc. Natl. Acad. Sci. U.S.A. 106, 3207–3212.

San Filippo, J., Sung, P., Klein, H., 2008. Mechanism of eukaryotic homologous recombination. Annu. Rev. Biochem. 77, 229–257.

Sanchez-Beato, M., Sanchez-Aguilera, A., Piris, M.A., 2003. Cell cycle deregulation in B-cell lymphomas. Blood 101, 1220–1235.

Scott, G.K., Goga, A., Bhaumik, D., Berger, C.E., Sullivan, C.S., Benz, C.C., 2007. Coordinate suppression of ERBB2 and ERBB3 by enforced expression of micro-RNA miR-125a or miR-125b. J. Biol. Chem. 282, 1479–1486.

Seve, P., Dumontet, C., 2008. Is class III beta-tubulin a predictive factor in patients receiving tubulin-binding agents? Lancet Oncol. 9, 168–175.

Shahbazi, J., Lock, R., Liu, T., 2013. Tumor Protein 53-Induced Nuclear. Protein 1 Enhances p53 Function and Represses Tumorigenesis. Front. Genet. 4, 80.

Shapiro, A.M., Miller-Pinsler, L., Wells, P.G., 2016. Breast cancer 1 (BRCA1)-deficient embryos develop normally but are more susceptible to ethanol-initiated DNA damage and embryopathies. Redox Biol. 7, 30–38.

Sharma, V., Misteli, T., 2013. Non-coding RNAs in DNA damage and repair. FEBS Lett. 587, 1832–1839.

Shi, W., Alajez, N.M., Bastianutto, C., Hui, A.B., Mocanu, J.D., Ito, E., Busson, P., Lo, K.W., Ng, R., Waldron, J., O'Sullivan, B., Liu, F.F., 2010. Significance of Plk1 regulation by miR-100 in human nasopharyngeal cancer. Int. J. Cancer 126, 2036–2048.

Shimono, Y., Zabala, M., Cho, R.W., Lobo, N., Dalerba, P., Qian, D., Diehn, M., Liu, H., Panula, S.P., Chiao, E., Dirbas, F.M., Somlo, G., Pera, R.A., Lao, K., Clarke, M.F., 2009. Downregulation of miRNA-200c links breast cancer stem cells with normal stem cells. Cell 138, 592–603.

Shyamasundar, S., Lim, J.P., Bay, B.H., 2016. MiR-93 inhibits the invasive potential of triple-negative breast cancer cells in vitro via protein kinase WNK1. Int. J. Oncol. 49 (6), 2629–2636.

Simone, N.L., Soule, B.P., Ly, D., Saleh, A.D., Savage, J.E., Degraff, W., Cook, J., Harris, C.C., Gius, D., Mitchell, J.B., 2009. Ionizing radiation-induced oxidative stress alters miRNA expression. PLoS One 4. e6377.

Singh, R., Pochampally, R., Watabe, K., Lu, Z., Mo, Y.Y., 2014. Exosome-mediated transfer of miR-10b promotes cell invasion in breast cancer. Mol. Cancer 13, 256.

Solier, S., Zhang, Y.W., Ballestrero, A., Pommier, Y., Zoppoli, G., 2012. DNA damage response pathways and cell cycle checkpoints in colorectal cancer: current concepts and future perspectives for targeted treatment. Curr. Cancer Drug Targets 12, 356–371.

Song, B., Wang, Y., Kudo, K., Gavin, E.J., Xi, Y., Ju, J., 2008. MiR-192 regulates dihydrofolate reductase and cellular proliferation through the p53-microRNA circuit. Clin. Cancer Res. 14, 8080–8086.

Song, B., Wang, Y., Xi, Y., Kudo, K., Bruheim, S., Botchkina, G.I., Gavin, E., Wan, Y., Formentini, A., Kornmann, M., Fodstad, O., Ju, J., 2009. Mechanism of chemoresistance mediated by miR-140 in human osteosarcoma and colon cancer cells. Oncogene 28, 4065–4074.

Stefani, G., Chen, X., Zhao, H., Slack, F.J., 2015. A novel mechanism of LIN-28 regulation of let-7 microRNA expression revealed by in vivo HITS-CLIP in C. elegans. RNA 21, 985–996.

Su, X., Chakravarti, D., Cho, M.S., Liu, L., Gi, Y.J., Lin, Y.L., Leung, M.L., El-Naggar, A., Creighton, C.J., Suraokar, M.B., Wistuba, I., Flores, E.R., 2010. TAp63 suppresses metastasis through coordinate regulation of Dicer and miRNAs. Nature 467, 986–990.

Suh, M.R., Lee, Y., Kim, J.Y., Kim, S.K., Moon, S.H., Lee, J.Y., Cha, K.Y., Chung, H.M., Yoon, H.S., Moon, S.Y., Kim, V.N., Kim, K.S., 2004. Human embryonic stem cells express a unique set of microRNAs. Dev. Biol. 270, 488–498.

Sun, Z., Li, S., Kaufmann, A.M., Albers, A.E., 2014. MiR-21 increases the programmed cell death 4 gene-regulated cell proliferation in head and neck squamous carcinoma cell lines. Oncol. Rep. 32, 2283–2289.

Sun, C.C., Li, S.J., Zhang, F., Zhang, Y.D., Zuo, Z.Y., Xi, Y.Y., Wang, L., Li, D.J., 2016. The novel miR-9600 suppresses tumor progression and promotes paclitaxel sensitivity in non-small-cell lung cancer through altering STAT3 expression. Mol. Ther. Nucleic Acids 5. e387.

Sundaram, P., Hultine, S., Smith, L.M., Dews, M., Fox, J.L., Biyashev, D., Schelter, J.M., Huang, Q., Cleary, M.A., Volpert, O.V., Thomas-Tikhonenko, A., 2011. P53-responsive miR-194 inhibits thrombospondin-1 and promotes angiogenesis in colon cancers. Cancer Res. 71, 7490–7501.

Surova, O., Zhivotovsky, B., 2013. Various modes of cell death induced by DNA damage. Oncogene 32, 3789–3797.

Suzuki, H.I., Yamagata, K., Sugimoto, K., Iwamoto, T., Kato, S., Miyazono, K., 2009. Modulation of microRNA processing by p53. Nature 460, 529–533.

Teng, Y., Zuo, X., Hou, M., Zhang, Y., Li, C., Luo, W., Li, X., 2016. A double-negative feedback interaction between MicroRNA-29b and DNMT3A/3B contributes to ovarian cancer progression. Cell. Physiol. Biochem. 39, 2341–2352.

Tessitore, A., Cicciarelli, G., Del Vecchio, F., Gaggiano, A., Verzella, D., Fischietti, M., Vecchiotti, D., Capece, D., Zazzeroni, F., Alesse, E., 2014. MicroRNAs in the DNA damage/repair network and cancer. Int. J. Genomics 2014, 820248.

Thomson, D.W., Bracken, C.P., Goodall, G.J., 2011. Experimental strategies for microRNA target identification. Nucleic Acids Res. 39, 6845–6853.

Tian, Y., Luo, A., Cai, Y., Su, Q., Ding, F., Chen, H., Liu, Z., 2010. MicroRNA-10b promotes migration and invasion through KLF4 in human esophageal cancer cell lines. J. Biol. Chem. 285, 7986–7994.

Tie, J., Fan, D., 2011. Big roles of microRNAs in tumorigenesis and tumor development. Histol. Histopathol. 26, 1353–1361.

Tili, E., Croce, C.M., Michaille, J.J., 2009. MiR-155: on the crosstalk between inflammation and cancer. Int. Rev. Immunol. 28, 264–284.

To, K.K., Zhan, Z., Litman, T., Bates, S.E., 2008. Regulation of ABCG2 expression at the 3′ untranslated region of its mRNA through modulation of transcript stability and protein translation by a putative microRNA in the S1 colon cancer cell line. Mol. Cell. Biol. 28, 5147–5161.

Tryndyak, V.P., Beland, F.A., Pogribny, I.P., 2010. E-cadherin transcriptional down-regulation by epigenetic and microRNA-200 family alterations is related to mesenchymal and drug-resistant phenotypes in human breast cancer cells. Int. J. Cancer 126, 2575–2583.

Tsang, W.P., Kwok, T.T., 2008. Let-7a microRNA suppresses therapeutics-induced cancer cell death by targeting caspase-3. Apoptosis 13, 1215–1222.

Tsuchiya, Y., Nakajima, M., Takagi, S., Taniya, T., Yokoi, T., 2006. MicroRNA regulates the expression of human cytochrome P450 1B1. Cancer Res. 66, 9090–9098.

Valeri, N., Gasparini, P., Fabbri, M., Braconi, C., Veronese, A., Lovat, F., Adair, B., Vannini, I., Fanini, F., Bottoni, A., Costinean, S., Sandhu, S.K., Nuovo, G.J., Alder, H., Gafa, R., Calore, F., Ferracin, M., Lanza, G., Volinia, S., Negrini, M., McIlhatton, M.A., Amadori, D., Fishel, R., Croce, C.M., 2010. Modulation of mismatch repair and genomic stability by miR-155. Proc. Natl. Acad. Sci. U.S.A. 107, 6982–6987.

van Attikum, H., Gasser, S.M., 2009. Crosstalk between histone modifications during the DNA damage response. Trends Cell Biol. 19, 207–217.

van Wynsberghe, P.M., Kai, Z.S., Massirer, K.B., Burton, V.H., Yeo, G.W., Pasquinelli, A.E., 2011. LIN-28 co-transcriptionally binds primary let-7 to regulate miRNA maturation in Caenorhabditis elegans. Nat. Struct. Mol. Biol. 18, 302–308.

Visone, R., Russo, L., Pallante, P., De Martino, I., Ferraro, A., Leone, V., Borbone, E., Petrocca, F., Alder, H., Croce, C.M., Fusco, A., 2007. MicroRNAs (miR)-221 and miR-222, both overexpressed in human thyroid papillary carcinomas, regulate p27Kip1 protein levels and cell cycle. Endocr. Relat. Cancer 14, 791–798.

Wahid, F., Shehzad, A., Khan, T., Kim, Y.Y., 2010. MicroRNAs: synthesis, mechanism, function, and recent clinical trials. Biochim. Biophys. Acta 1803, 1231–1243.

Wan, G., Mathur, R., Hu, X., Zhang, X., Lu, X., 2011. MiRNA response to DNA damage. Trends Biochem. Sci. 36, 478–484.

Wang, P., Lv, L., 2016. MiR-26a induced the suppression of tumor growth of cholangiocarcinoma via KRT19 approach. Oncotarget 7 (49), 81367–81376.

Wang, J., Martin, J.F., 2014. Macro advances in microRNAs and myocardial regeneration. Curr. Opin. Cardiol. 29, 207–213.

Wang, Y., Taniguchi, T., 2013. MicroRNAs and DNA damage response: implications for cancer therapy. Cell Cycle 12, 32–42.

Wang, Y., Huang, J.W., Li, M., Cavenee, W.K., Mitchell, P.S., Zhou, X., Tewari, M., Furnari, F.B., Taniguchi, T., 2011. MicroRNA-138 modulates DNA damage response by repressing histone H2AX expression. Mol. Cancer Res. 9, 1100–1111.

Wang, D., Lu, J.G., Wang, Q., Du, X.L., Dong, R., Wang, P., Zhao, L., Jiang, X., Yuan, L.J., 2012a. Increased immunohistochemical expression of YKL-40 in the spleen of patients with portal hypertension. Braz. J. Med. Biol. Res. 45, 264–272.

Wang, X., Cao, L., Wang, Y., Wang, X., Liu, N., You, Y., 2012b. Regulation of let-7 and its target oncogenes (Review). Oncol. Lett. 3, 955–960.

Wang, S., Huang, J., Lyu, H., Lee, C.K., Tan, J., Wang, J., Liu, B., 2013. Functional cooperation of miR-125a, miR-125b, and miR-205 in entinostat-induced downregulation of erbB2/erbB3 and apoptosis in breast cancer cells. Cell Death Dis. 4. e556.

Wang, T., Wang, G., Hao, D., Liu, X., Wang, D., Ning, N., Li, X., 2015. Aberrant regulation of the LIN28A/LIN28B and let-7 loop in human malignant tumors and its effects on the hallmarks of cancer. Mol. Cancer 14, 125.

Wang, S., Ma, G., Zhu, H., Lv, C., Chu, H., Tong, N., Wu, D., Qiang, F., Gong, W., Zhao, Q., Tao, G., Zhou, J., Zhang, Z., Wang, M., 2016a. MiR-107 regulates tumor progression by targeting NF1 in gastric cancer. Sci. Rep. 6, 36531.

Wang, Y., Chen, L., Wu, Z., Wang, M., Jin, F., Wang, N., Hu, X., Liu, Z., Zhang, C.Y., Zen, K., Chen, J., Liang, H., Zhang, Y., Chen, X., 2016b. MiR-124-3p functions as a tumor suppressor in breast cancer by targeting CBL. BMC Cancer 16, 826.

Wei, Y., Sun, J., Li, X., 2017. MicroRNA-215 enhances invasion and migration by targeting retinoblastoma tumor suppressor gene 1 in high-grade glioma. Biotechnol. Lett. 39 (2), 197–205.

Weidhaas, J.B., Babar, I., Nallur, S.M., Trang, P., Roush, S., Boehm, M., Gillespie, E., Slack, F.J., 2007. MicroRNAs as potential agents to alter resistance to cytotoxic anticancer therapy. Cancer Res. 67, 11111–11116.

Wong, K.Y., Yu, L., Chim, C.S., 2011. DNA methylation of tumor suppressor miRNA genes: a lesson from the miR-34 family. Epigenomics 3, 83–92.

Wouters, M.D., van Gent, D.C., Hoeijmakers, J.H., Pothof, J., 2011. MicroRNAs, the DNA damage response and cancer. Mutat. Res. 717, 54–66.

Wu, L., Fan, J., Belasco, J.G., 2006. MicroRNAs direct rapid deadenylation of mRNA. Proc. Natl. Acad. Sci. U.S.A. 103, 4034–4039.

Wu, S., Liu, F., Xie, L., Peng, Y., Lv, X., Zhu, Y., Zhang, Z., He, X., 2015. MiR-125b suppresses proliferation and invasion by targeting MCL1 in gastric cancer. Biomed. Res. Int. 2015, 365273.

Wu, Y.R., Qi, H.J., Deng, D.F., Luo, Y.Y., Yang, S.L., 2016. MicroRNA-21 promotes cell proliferation, migration, and resistance to apoptosis through PTEN/PI3K/AKT signaling pathway in esophageal cancer. Tumour Biol. 37 (9), 12061–12070.

Xiao, H., Li, H., Yu, G., Xiao, W., Hu, J., Tang, K., Zeng, J., He, W., Zeng, G., Ye, Z., Xu, H., 2014. MicroRNA-10b promotes migration and invasion through KLF4 and HOXD10 in human bladder cancer. Oncol. Rep. 31, 1832–1838.

Xiong, H., Pang, J., Yang, H., Dai, M., Liu, Y., Ou, Y., Huang, Q., Chen, S., Zhang, Z., Xu, Y., Lai, L., Zheng, Y., 2015. Activation of miR-34a/SIRT1/p53 signaling contributes to cochlear hair cell apoptosis: implications for age-related hearing loss. Neurobiol. Aging 36, 1692–1701.

Yan, H.L., Xue, G., Mei, Q., Wang, Y.Z., Ding, F.X., Liu, M.F., Lu, M.H., Tang, Y., Yu, H.Y., Sun, S.H., 2009. Repression of the miR-17-92 cluster by p53 has an important function in hypoxia-induced apoptosis. EMBO J. 28, 2719–2732.

Yan, D., Ng, W.L., Zhang, X., Wang, P., Zhang, Z., Mo, Y.Y., Mao, H., Hao, C., Olson, J.J., Curran, W.J., Wang, Y., 2010. Targeting DNA-PKcs and ATM with miR-101 sensitizes tumors to radiation. PLoS One 5. e11397.

Yang, H., Kong, W., He, L., Zhao, J.J., O'Donnell, J.D., Wang, J., Wenham, R.M., Coppola, D., Kruk, P.A., Nicosia, S.V., Cheng, J.Q., 2008. MicroRNA expression profiling in human ovarian cancer: miR-214 induces cell survival and cisplatin resistance by targeting PTEN. Cancer Res. 68, 425–433.

Yang, H., Zhang, H., Zhu, L., Wang, J., Zhang, C., Li, D., 2012. Pathway analysis of cancer-associated microRNA targets. Int. J. Oncol. 41, 2213–2226.

Yang, X., Yang, Y., Gan, R., Zhao, L., Li, W., Zhou, H., Wang, X., Lu, J., Meng, Q.H., 2014. Down-regulation of mir-221 and mir-222 restrain prostate cancer cell proliferation and migration that is partly mediated by activation of SIRT1. PLoS One 9. e98833.

Yang, Z., Zheng, B., Zhang, Y., He, M., Zhang, X.H., Ma, D., Zhang, R.N., Wu, X.L., Wen, J.K., 2015. MiR-155-dependent regulation of mammalian sterile 20-like kinase 2 (MST2) coordinates inflammation, oxidative stress and proliferation in vascular smooth muscle cells. Biochim. Biophys. Acta 1852, 1477–1489.

Yao, E., Ventura, A., 2011. A new role for miR-182 in DNA repair. Mol. Cell 41, 135–137.

Yazdi, P.T., Wang, Y., Zhao, S., Patel, N., Lee, E.Y., Qin, J., 2002. SMC1 is a downstream effector in the ATM/NBS1 branch of the human S-phase checkpoint. Genes Dev. 16, 571–582.

Yu, G., Zhan, X., Zhang, Z., Li, Y., 2016. Overexpression of miR-125b promotes apoptosis of macrophages. Xi Bao Yu Fen Zi Mian Yi Xue Za Zhi 32, 958–962.

Yu, B., Liu, X., Chang, H., 2017. MicroRNA-143 inhibits colorectal cancer cell proliferation through targeting MMP7. Minerva Med. 108 (1), 13–19.

Zhang, D.G., Zheng, J.N., Pei, D.S., 2014. P53/microRNA-34-induced metabolic regulation: new opportunities in anticancer therapy. Mol. Cancer 13, 115.

Zhang, X., Ni, Z., Duan, Z., Xin, Z., Wang, H., Tan, J., Wang, G., Li, F., 2015. Overexpression of E2F mRNAs associated with gastric cancer progression identified by the transcription factor and miRNA co-regulatory network analysis. PLoS One 10. e0116979.

Zhang, X., Jiang, P., Shuai, L., Chen, K., Li, Z., Zhang, Y., Jiang, Y., Li, X., 2016a. MiR-589-5p inhibits MAP3K8 and suppresses CD90+ cancer stem cells in hepatocellular carcinoma. J. Exp. Clin. Cancer Res. 35, 176.

Zhang, Y., Sun, Z., Fu, T., Liu, C., Liu, Y., Lin, Y., 2016b. The microRNA-635 suppresses tumorigenesis in non-small cell lung cancer. Biomed. Pharmacother. 84, 1274–1281.

Zhang, X., Chen, Y., Zhao, P., Zang, L., Zhang, Z., Wang, X., 2017. MicroRNA-19a functions as an oncogene by regulating PTEN/AKT/pAKT pathway in myeloma. Leuk. Lymphoma 58, 932–940.

Zhao, L., Wang, W., 2015. MiR-125b suppresses the proliferation of hepatocellular carcinoma cells by targeting Sirtuin7. Int. J. Clin. Exp. Med. 8, 18469–18475.

Zhao, J.J., Lin, J., Yang, H., Kong, W., He, L., Ma, X., Coppola, D., Cheng, J.Q., 2016. MicroRNA-221/222 negatively regulates estrogen receptor alpha and is associated with tamoxifen resistance in breast cancer. J. Biol. Chem. 291, 22859.

Zheng, T., Wang, J., Chen, X., Liu, L., 2010. Role of microRNA in anticancer drug resistance. Int. J. Cancer 126, 2–10.

Zhou, M., Liu, Z., Zhao, Y., Ding, Y., Liu, H., Xi, Y., Xiong, W., Li, G., Lu, J., Fodstad, O., Riker, A.I., Tan, M., 2010. MicroRNA-125b confers the resistance of breast cancer cells to paclitaxel through suppression of pro-apoptotic Bcl-2 antagonist killer 1 (Bak1) expression. J. Biol. Chem. 285, 21496–21507.

Zhou, Z., Liu, H., Wang, C., Lu, Q., Huang, Q., Zheng, C., Lei, Y., 2015. Long non-coding RNAs as novel expression signatures modulate DNA damage and repair in cadmium toxicology. Sci. Rep. 5, 15293.

Zhou, M., Wang, S., Hu, L., Liu, F., Zhang, Q., Zhang, D., 2016a. MiR-199a-5p suppresses human bladder cancer cell metastasis by targeting CCR7. BMC Urol. 16, 64.

Zhou, S.M., Zhang, F., Chen, X.B., Jun, C.M., Jing, X., Wei, D.X., Xia, Y., Zhou, Y.B., Xiao, X.Q., Jia, R.Q., Li, J.T., Sheng, W., Zeng, Y., 2016b. MiR-100 suppresses the proliferation and tumor growth of esophageal squamous cancer cells via targeting CXCR7. Oncol. Rep. 35, 3453–3459.

Zhu, S., Si, M.L., Wu, H., Mo, Y.Y., 2007. MicroRNA-21 targets the tumor suppressor gene tropomyosin 1 (TPM1). J. Biol. Chem. 282, 14328–14336.

Zhu, H., Wu, H., Liu, X., Evans, B.R., Medina, D.J., Liu, C.G., Yang, J.M., 2008. Role of MicroRNA miR-27a and miR-451 in the regulation of MDR1/P-glycoprotein expression in human cancer cells. Biochem. Pharmacol. 76, 582–588.

CHAPTER TWO

MicroRNAs as Key Effectors in the p53 Network

Frauke Goeman, Sabrina Strano, Giovanni Blandino[1]
Oncogenomic and Epigenetic Unit, Regina Elena National Cancer Institute, Rome, Italy
[1]Corresponding author: e-mail address: giovanni.blandino@ifo.gov.it

Contents

Abstract

The guardian of the genome p53 is embedded in a fine-spun network of MicroRNAs. p53 is able to activate or repress directly the transcription of MicroRNAs that are participating in the tumor-suppressive mission of p53. On the other hand, the expression of p53 is under tight control of MicroRNAs that are either targeting directly p53 or factors that are modifying its protein level or activity. Although the most important function of p53 is suggested to be transcriptional regulation, there are several nontranscriptional functions described. One of those regards the modulation of MicroRNA biogenesis. Wild-type p53 is increasing the maturation of selected MicroRNAs from the primary transcript to the precursor MiRNA by interacting with the Microprocessor complex. Furthermore, p53 is modulating the mRNA accessibility for certain MicroRNAs by association with the RISC complex and transcriptional regulation of RNA-binding proteins. In this way p53 is able to remodel the MiRNA–mRNA interaction network. As wild-type p53 is employing MicroRNAs to suppress cancer development, gain-of-function mutant p53 proteins use MicroRNAs to confer oncogenic properties like chemoresistance and the ability to drive metastasis. Like its wild-type counterpart mutant p53 is able to regulate MicroRNAs transcriptionally and posttranscriptionally. Mutant p53 affects the MiRNA

International Review of Cell and Molecular Biology, Volume 333
ISSN 1937-6448
http://dx.doi.org/10.1016/bs.ircmb.2017.04.003

processing at two cleavage steps through interfering with the Microprocessor complex and by downregulating Dicer and KSRP, a modulator of MiRNA biogenesis. Thus, MicroRNAs are essential components in the p53 pathway, contributing substantially to combat or enhance tumor development depending on the wild-type or mutant p53 context.

1. INTRODUCTION

p53 is considered to be among the most important proteins in tumor prevention. David P. Lane described it in 1992 as the "guardian of the genome," which became a p53 definition since then (Lane, 1992). It was first discovered in 1979 by two independent groups of David P. Lane and Arnold Levine and described as a tumor antigen (Lane and Crawford, 1979; Linzer and Levine, 1979). In fact, the first 10 years p53 was considered to be an oncogene, when the existence of tumor suppressors was still unknown and the p53 protein was found to be bound to the SV40 large T antigen, overexpressed in cancers and capable of cooperating with oncogenic proteins like HRAS in transforming cells (Levine and Oren, 2009). Only several years later it became evident that the observed transforming ability of p53 was due to gain-of-function (GOF) mutations in the various tested p53 constructs. Nowadays there is no doubt about the pivotal role of p53 as tumor suppressor. It is the most frequently altered gene in human cancer, harboring deletions or point mutations in approximately 50%–70% of all tumors (Hollstein et al., 1991; Kandoth et al., 2013). Moreover, it was estimated that many of the tumors with wild-type p53 still display a loss of the p53 pathway through other mechanisms (Liu et al., 2017). In normal and nonstressed cells p53 protein levels are barely detectable due to constant protein degradation. Upon cellular stress like DNA damage, hypoxia, telomere erosion, and oncogene activation p53 becomes posttranslationally modified which leads to a stabilization and activation of the protein. The most prominent and key role of p53 is its ability to regulate transcription (Vousden and Prives, 2009). However, several nontranscriptional and cytoplasmic p53 activities are reported. Triggering apoptosis without transcriptional activation of target genes is the best-studied cytoplasmic p53 function. In response to a variety of stress stimuli p53 can translocate to the mitochondria where it promotes mitochondrial outer membrane permeabilization, provoking the release of proapoptotic factors (Galluzzi et al., 2011). Inhibition of autophagy is another cytoplasmic activity of p53 which was also described to be a baseline function of p53 (Green and Kroemer, 2009). But also induction of necrosis and the regulation of endocytosis represent important extranuclear p53 activities (Comel et al., 2014).

In the nucleus p53 exerts its central role by acting as a transcription factor. The target genes regulated by stress-activated p53 mediate cell cycle arrest, apoptosis, senescence, DNA repair, control of metabolic pathways and suppression of stem cell properties, angiogenesis, invasion, and metastasis (Bieging et al., 2014; Kruiswijk et al., 2015; Levine and Oren, 2009; Spike and Wahl, 2011). The spectrum of target genes comprises protein-coding genes but also long noncoding (lncRNA) and MicroRNAs. MicroRNAs are small 18–24-nucleotide noncoding RNAs with important impact on virtually all biological processes including those that play a role in cancer development like cell proliferation, apoptosis, stress response, maintenance of stem cell potency, and metabolism. Many efforts have been made to analyze the remarkable quantitative and qualitative changes of MicroRNA expression in various diseases, especially cancer (Gulyaeva and Kushlinskiy, 2016). MicroRNAs are able to posttranscriptionally regulate entire sets of genes by promoting RNA degradation or inhibiting mRNA translation. In addition, each gene is regulated by several MicroRNAs. It was estimated that more than 60% of all human protein-coding genes are MicroRNA targets, underlining their important role in fine-tuning gene expression (Friedman et al., 2009).

2. MicroRNAs, SMALL HELPERS WITH BIG EFFECTS IN THE p53 PATHWAY

2.1 p53 Transcriptionally Upregulates MicroRNAs

The first MicroRNAs that were shown to be directly regulated by p53 belong to the MiR-34 gene family (Fig. 1) (Bommer et al., 2007; Chang et al., 2007; Corney et al., 2007; He et al., 2007; Raver-Shapira et al., 2007; Tarasov et al., 2007). Ionizing radiations, DNA damage after doxorubicin treatment, or oncogenic stress was able to induce MiR-34a, MiR-34b, and MiR-34c in a p53-dependent manner. p53 was demonstrated to bind to p53 consensus sequences in the MiR-34a and MiR-34b/c promoter regions. Ectopic expression of MiR-34a or MiR-34b/c induced cell cycle arrest in primary and cancer cell lines. The cell cycle arrest in the primary cells was accompanied by an increase of cellular senescence (He et al., 2007). Especially MiR-34a was shown to be a critical regulator of p53 function, mediating cell cycle arrest and apoptosis. Among its target

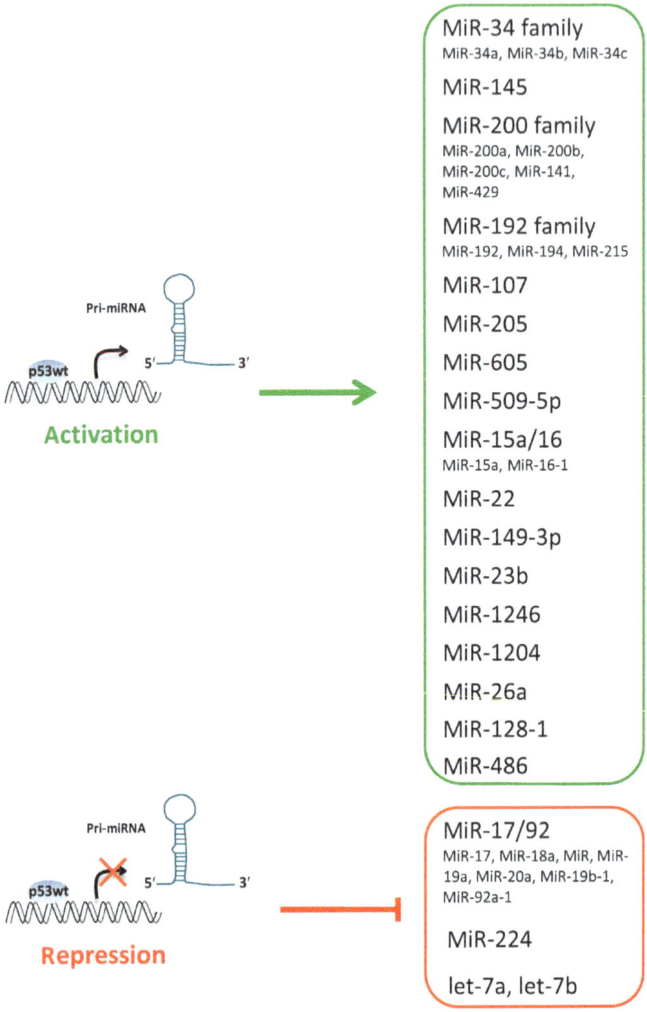

Fig. 1 Overview of MicroRNAs directly transcriptionally up- or downregulated by wild-type p53.

genes regulating the cell cycle are cyclin-dependent kinase 4 (CDK4), CDK6, cyclin E2, CCND1, E2F3, MYC, and NMYC (Hermeking, 2012). Targets promoting apoptosis are BCL2 (Bommer et al., 2007) and survivin (BIRC5) (Chen et al., 2015). Furthermore, it represses epithelial–mesenchymal transition (EMT) and metastasis by downregulating the WNT signaling-related gene TCF7, MET (Hwang et al., 2011; Migliore et al., 2008), SNAIL, and IL6 (Rokavec et al., 2014; Siemens et al., 2011). MiR-34 was also shown to be involved in cellular senescence, in part by downregulating SIRT1 (He et al., 2007; Ito et al., 2010; Tazawa et al., 2007). In addition, MiR-34a is involved in the suppression of stemness by directly targeting NOTCH1, NANOG, SOX2, and the cancer stem cell marker CD44 (Choi et al., 2011; Kang et al., 2015; Liu et al., 2011; Park et al., 2014). A rather recent finding of MiR-34a function regards its role in regulating cell metabolism. By the direct downregulation of lactate dehydrogenase A in breast cancer cells it is capable of restricting aerobic glycolysis and subsequently cell proliferation (Xiao et al., 2016). It was demonstrated to contribute substantially to the p53 downstream effects and that its expression is epigenetically silenced in various cancers. Moreover, MiR-34a was shown to modulate the tumor immune response by directly targeting PDL1, a protein that is crucial for immune evasion and one of the targets in cancer immunotherapy (Cortez et al., 2016; Wang et al., 2015). MiR-34a is forming an important positive feedback loop with p53 by targeting several of its negative regulators like MDM4, SIRT1, MTA2, HDAC1, YY1, and NR4A2, reinforcing therefore the p53 activity (Beard et al., 2016; Chen et al., 2011; Kaller et al., 2011; Mandke et al., 2012; Yamakuchi et al., 2008). Of note, MiR-34a, MiR-34b, and MiR-34c were recently shown to be activated in a rapid way by 5′ phosphorylation in response to DNA damage. Salzman et al. demonstrated that there is an inactive mature pool of MiR-34 that lacks a 5′-phosphate. Upon DNA damage MiR-34 becomes 5′-end phosphorylated involving ATM and the RNA kinase hClp1 which leads to MiR-34 loading into the AGO2 complex, rendering the MicroRNA active. This rapid activation occurs before transcriptional activation by p53 (Salzman et al., 2016).

MiR-145 is another transcriptionally activated target of p53. It is a MicroRNA that plays a significant role in inhibiting tumor initiation and progression and is commonly downregulated in various types of cancer by DNA hypermethylation (Cui et al., 2014; Suh et al., 2011). Like MiR-34a it is targeting the G1-S-phase cell cycle regulators MYC and CDK6 (Sachdeva et al., 2009; Shao et al., 2013). It is impairing pluripotency in human embryonic stem cells and restricting angiogenesis in various cancer types (Xu et al., 2009, 2012; Zou et al., 2012). It becomes evident that MiR-145 plays an important role in metastasis, being able to inhibit brain metastasis from lung cancer and bone metastasis from prostate cancer (Donzelli et al., 2015; Peng et al., 2011). The MiR-200 family comprises another directly upregulated target of p53. It consists of five members, MiR-200a, MiR-200b, MiR-200c, MiR-141, and MiR-429. Their most prominent role is the inhibition of EMT by targeting the EMT transcription factors ZEB1 and ZEB2 that are downregulating E-cadherin. Loss of E-cadherin represents a hallmark of EMT (Gregory et al., 2008; Park et al., 2008). Interestingly, there is a negative feedback loop between ZEB1 and the MiR-200 family. ZEB1 is able to repress the MiR-141/200c cluster (Hill et al., 2013). In pancreatic and breast cancer the MiR-200 family was shown to exert a dual role in regulating metastasis. In both cancer types the expression of MiR-200 family members positively correlated with metastatic spread. Therefore, it was hypothesized that the MiR-200 family hinders the detachment, migration, and intravasation of tumor cells into the bloodstream, suppressing EMT, while in later steps it might favor colonization of distant organs, in line with promoting mesenchymal–epithelial transition (MET) and therefore supporting an EMT-MET switch which might be necessary for the colonization of tumor cells (Hill et al., 2013). According to that, the MiR-200 family was shown to negatively regulate the mRNA of SEC23A, a protein that mediates secretion of metastasis-suppressive proteins (Korpal et al., 2011). Many other targets are identified in the meanwhile that are involved in cell cycle regulation, angiogenesis, and stemness (Humphries and Yang, 2015). The MiR-192 family displays another group of MicroRNAs that is directly upregulated by p53 (Braun et al., 2008). It is comprised of two MicroRNAs, MiR-192 and MiR-215, that share a common seed sequence. They are able to induce cell cycle arrest upon ectopic expression and are downregulated in cancers like renal cell and hepatocellular carcinoma (HCC), colon, lung, and ovarian cancer, and multiple myeloma (Lian et al., 2016; Liu et al., 2017; Wu et al., 2016). Several cell cycle regulators of the G1 but also G2 phase were identified to be direct targets of MiR-192 and MiR-215 (Georges et al., 2008). In human HCC cells MiR-192/215 was shown to suppress metastasis. In line with that, low expression of MiR-192/215 correlated with poor outcome in HCC patients (Lian et al., 2016). Other targets of MiR-192/215 with a crucial role of driving metastasis are ZEB2 or interleukin-17 (Geng et al., 2014; Sun et al., 2014). A recent publication highlights the role of MiR-192 in angiogenesis of ovarian and renal cancers which seems to be specific for MiR-192 but not MiR-215. Here, the authors showed that MiR-192 directly targets the key transcription factors EGR1 and HOXB9 which leads to the global downregulation of the angiogenic pathway (Wu et al., 2016). MiR-194 is coexpressed with both MiR-192 and MiR-215 that are coded on two different clusters (Braun et al., 2008). Therefore, it is also upregulated by p53 and is able to inhibit growth and colony formation in multiple myeloma cells. Furthermore, it participates in the important targeting of MDM2, mediated by all three MicroRNAs of these homologue clusters, MiR-192, MiR194, and MiR-215. In this way they are able to enhance the protein level of p53 and create a positive feedback loop (Khella et al., 2013; Pichiorri et al., 2010). Another target of p53 is MiR-107 which was first described by Yamakuchi et al. to be directly upregulated by p53 (Yamakuchi et al., 2010). The role of MiR-107 is contradictory and seems to be tissue dependent. In colon carcinoma cells MiR-107 targets important cell cycle regulators of the G1 phase, like CDK6 (Bohlig et al., 2011). It is reducing the tumor volume of injected colon cancer cells in nude mice and suppressing hypoxia-induced angiogenesis by directly targeting the mRNA of HIF-1β. In line with that in colon cancer specimen the expression of MiR-107 is anticorrelated with VEGF, a HIF-1β target gene (Yamakuchi et al., 2010). Also in glioma and lung cancer MiR-107 was described as tumor suppressor (Chen et al., 2013; Takahashi et al., 2009; Zhang et al., 2014). Instead, in breast and gastric cancer MiR-107 seems to have an oncogenic role. It is correlated with metastasis and poor outcome targeting the endonuclease DICER which is essential for MicroRNA maturation (Li et al., 2011; Martello et al., 2010).

MiR-205 represents another MicroRNA that is directly regulated by p53. It harbors a p53-responsive element in front of its genomic region, showing p53 binding and transcriptional activation upon p53 induction (Piovan et al., 2012). In general, MiR-205 was described as tumor suppressor as well as oncomir, depending on the tissue context. In melanoma (Dar et al., 2011), renal (Majid et al., 2011), breast, and prostate cancer it is acting as tumor suppressor. It is downregulated in breast cancer and upon ectopic expression able to inhibit cell growth, invasion, and metastasis (Gregory et al., 2008; Piovan et al., 2012; Wu et al., 2009). Prostate carcinoma showed also lower MiR-205 expression levels compared to its normal counterpart and its downregulation correlates with poor outcome (Gandellini et al., 2009; Tucci et al., 2012). In human cervical and endometrial cancer and aggressive squamous cell carcinoma MiR-205 seems to have a role as oncomir. Interestingly, although MiR-205 was shown to be downregulated in breast cancer, in normal mammary gland stem cells and breast cancer stem cells it was found to be overexpressed, leading to an increase of replicative competence in the first and targeted therapy resistance in the latter (De Cola et al., 2015; Greene et al., 2010). Recently, two more MicroRNAs of the p53 network were identified, MiR-605 and MiR-509-5p. They are directly upregulated by p53 and form a positive feedback loop by targeting MDM2

(Ren et al., 2014; Xiao et al., 2011b). Upon stress MiR-605 becomes transcriptionally activated by p53. Xiao et al. claim that MiR-605 is rather increasing the transactivation activity of p53 by inhibiting MDM2 than affecting the p53 protein level. In line with the participation in the p53 network, MiR-605 increases apoptosis only in the presence of p53 (Xiao et al., 2011a). In cervical cancer and HCC cell lines doxorubicin treatment or serum starvation elicited a p53-dependent upregulation of MiR-509-5p. Ectopic expression of MiR-509-5p led to a suppression of cell growth, migration, and invasion. MDM2 was identified to be targeted by MiR-509-5p. Consequently, ectopic expression of MiR-509-5p decreased p53 protein ubiquitination, increasing p53 protein accumulation. In line with that, in cervical cancer and HCC MiR-509-5p and p53 are downregulated, while MDM2 is upregulated compared to normal tissue (Ren et al., 2014).

The MiR-15a/16 cluster is also directly transcriptionally activated by p53 in conjunction with its host gene DLEU2. It is located in a genomic region that is frequently deleted in chronic lymphocytic leukemia (CLL) (Calin et al., 2002; Fabbri et al., 2011). MiR-15a/16 is posttranscriptionally repressing antiapoptotic factors like BCL2, inducing in this way apoptosis and reduced engraftment of leukemia cells in nude mice (Calin et al., 2008; Cimmino et al., 2005). In colon cancer the MiR-15a/16 cluster is downregulated. Ectopic expression of MiR15a/16 in colorectal cancer cells leads to a reduction of proliferation, migration, and invasion and to a suppression of lung metastasis upon injection into nude mice. For this antimetastatic effect the posttranscriptional downregulation of AP4, an EMT-inducing transcription factor, was shown to be essential (Shi et al., 2014). Tsuchiya et al. characterized MiR-22 as another MicroRNA directly upregulated by p53. p21, an inhibitor of cell proliferation and important mediator of p53-induced cell cycle arrest, was described to be posttranscriptionally downregulated by MiR-22, sensitizing thereby cells to apoptosis in a p53 wild-type context. Interestingly, p53 upregulated MiR-22 only upon severe stress like strong DNA damage, showing that MiR-22 can function as a molecular switch that determines p53-dependent cell fate switching from cell cycle arrest to apoptosis (Tsuchiya et al., 2011). Generally MiR-22 seems to act as tumor suppressor. It was shown to be downregulated in osteosarcoma, de novo acute myeloid leukemia, rhabdomyosarcoma, cervical, prostate, and lung cancer and to be able to suppress tumor growth and metastasis in animal models (Bersani et al., 2016; Jiang et al., 2016; Xin et al., 2016). Instead, in breast cancer and myelodysplastic syndrome it was reported to be frequently overexpressed and to correlate with poor survival

(Pandey et al., 2015; Song et al., 2013a, b). In melanoma p53 is rarely mutated and frequently overexpressed in its wild-type form, indicating that the tumor-suppressor pathway is not effective. After ER stress p53 was shown to transcriptionally upregulate MiR-149-3p in melanoma cells that subsequently targeted glycogen synthase kinase-3α which led to a stabilization of Mcl-1, a member of the Bcl-2 family. Therefore, in melanoma p53 seems to drive a prosurvival pathway via MiR-149-3p (Jin et al., 2011). MiR-23b acts as a tumor suppressor in cervical cancer and is another MicroRNA that is under transcriptional control of p53 (Au Yeung et al., 2011). In human papilloma virus-associated cervical cancer MiR-23b is often downregulated mediated by the viral oncoprotein E6 that inactivates p53 and therefore also the p53-mediated upregulation of MiR-23b. One important target of miR23b is the urokinase-type plasminogen activator that is involved in the stimulation of tumor cell invasion and metastasis. MiR-1246, originally identified as p63 target, was revealed to be induced by the whole p53 family. p53 was shown to bind to its promoter and to be able to upregulate MiR-1246 upon DNA damage or ectopic p53 expression (Zhang et al., 2011c). Strikingly, MiR-1246 posttranscriptionally represses DYRK1A, a Down syndrome-associated kinase that links therefore p53 to the Down syndrome. DYRK1A is inactivating the transcriptional factor NFATc1 by phosphorylation and therefore excluding it from the nucleus. Downregulation of DYRK1A by MiR-1246 leads to activation of NFATc1 and finally to apoptosis (Zhang et al., 2011c). MiR-1204 that is encoded in the nonprotein-coding region PVT1 was also reported to be directly upregulated by p53. Notably, MiR-1204 was shown to form a positive feedback loop, upregulating p53 protein levels which in turn promoted an increase of apoptosis in colon cancer cells (Barsotti et al., 2012). Another MicroRNA that was claimed to be upregulated by p53 is MiR-26a. In osteosarcoma cells p53 was described to bind to upstream regions of the MiR-26A1 gene harboring several putative p53-binding sites which is inducing a transcriptional upregulation. Ectopic expression of MiR-26a led to an augmentation of cell cycle arrest in G1 phase and apoptosis after doxorubicin treatment. Moreover, high MiR-26a expression levels correlate with better survival rates in breast and prostate cancer (Lezina et al., 2013). In cardiomyocytes p53 plays an important role in apoptosis related to cardiac pathology. Stimulation of rat cardiomyocytes with hydrogen peroxide or doxorubicin induces apoptosis along with an increase of MiR-23a. MiR-23a sensitizes cardiomyocytes to undergo apoptosis after hydrogen peroxide or doxorubicin treatment at least in part through p53 which was shown to be a downstream effector of MiR-23a. On the one hand MiR-23a was exhibited to bind to p53; on the other hand it was revealed to enhance p53 binding to the promoter of MiR-128-1, leading to an enhanced expression of MiR-128-1. Prohibitin was identified as direct MiR-128 target. Its downregulation sensitizes cells to apoptosis. Therefore, in cardiomyocytes MiR-23a, p53, MiR-128-1, and prohibitin are forming an axis in promoting apoptosis (Li et al., 2015b). A recent complex work combining global mRNA, MicroRNA, proteomic, and p53 genomic binding analysis identified MiR-486 as a novel direct p53 target (Hunten et al., 2015). The citron rho-interacting serine/threonine kinase (CIT) was established to be targeted and downregulated by MiR-486. CIT was found to be upregulated in various tumors including colon, breast, lung, pancreas, and gastric cancer and to be of prognostic relevance together with KLF12 and HMBG1 in colorectal cancer. In this multiple omics approach the authors identified five additional putative MicroRNAs that were regulated upon ectopic p53 expression and harbor a p53-binding site nearby. MiR-1293 and MiR-641 were upregulated, while MiR-3191, MiR-3928, and MiR-1908 were downregulated and are waiting to be confirmed to be direct p53-targeted MicroRNAs.

2.2 p53 Transcriptionally Downregulates a Subset of MicroRNAs

p53 is also able to transcriptionally downregulate MicroRNAs. The most prominent example is the oncogenic cluster MiR-17/92 which is among the best-studied MicroRNA clusters. It codes for six different members, namely, MiR-17, MiR-18a, MiR-19a, MiR-20a, MiR-19b-1, and MiR-92a-1. The p53 response element is located within the proximal promoter region and overlapping with the binding site of the TATA-binding protein (TBP). Upon hypoxia or DNA damage p53 is binding and therefore competing with TBP which leads to a downregulation of the MiR-17-92 cluster. The downregulation was shown to be important for hypoxia-induced apoptosis. Furthermore, in colorectal carcinoma the expression of the MiR-17/92 cluster is correlated with the p53 status, being higher in p53-mutated cancers (Yan et al., 2009). Moreover, p53 cooperates with the NFκB p65 subunit to downregulate MiR-224 together with its host gene upon TGF-β-1 treatment. SMAD4 was identified to be post-transcriptionally repressed by MiR-224 (Liang et al., 2013). In HCC MiR-224 is frequently overexpressed and its expression anticorrelated with SMAD4 protein levels and patient survival (Wang et al., 2013c). In colon cancer cells p53 is downregulating let-7a and let-7b upon genotoxic stress.

Following irradiation p53 was shown to bind upstream of the let-7a3 and let-7b genes to a p53 consensus sequence which is necessary for the subsequent downregulation (Saleh et al., 2011).

2.3 p53 Modulates MicroRNA Maturation and mRNA Accessibility

As mentioned earlier, although most p53-mediated activities are made by transcriptional regulation, there are also several nontranscriptional activities of p53. One of those was recently highlighted by Suzuki et al., showing that p53 takes further part in the MicroRNA network by influencing the maturation of MicroRNAs (Suzuki et al., 2009). The biogenesis of MicroRNAs comprises several steps (Fig. 2). The primary transcript of the MicroRNAs is called pri-miRNA and ranges from several hundred up to thousands of nucleotides. It is usually generated by RNA-polymerase II, although there are also some reports showing a role of RNA-polymerase III (Borchert et al., 2006; Ozsolak et al., 2008). The pri-miRNA contains a hairpin structure and is recognized by the Microprocessor complex consisting of Drosha, an RNAse III enzyme, and the double-stranded RNA-binding protein DGCR8 (Pasha) as main components. The pri-miRNAs are cleaved at the stem of the hairpin loop into the precursor MiRNAs (pre-miRNAs) that comprise around 70 nucleotides still harboring the hairpin structure. Subsequently, the pre-miRNAs are transported into the cytoplasm by Exportin-5 (XPO5) together with Ran-GTP. Here, the next endonucleolytic step takes place driven by Dicer that generates the short double-stranded MiRNA duplex of 22–24 nucleotides followed by the loading into the RNA-induced silencing complex (RISC). The RISC complex includes the Argonaute protein (AGO) that mediates the removal of one RNA strand resulting in the mature RISC complex that is ready to target the corresponding mRNAs leading either to translational repression or mRNA degradation (Fig. 2) (Gurtner et al., 2016; Ha and Kim, 2014). Interestingly, in human cancers there is a global decrease of mature MicroRNAs which displays a feature that was shown to enhance tumorigenesis (Kumar et al., 2007; Lu et al., 2005). Suzuki et al. showed that the tumor suppressor p53 is modulating the abundance of selected MicroRNAs not only transcriptionally but also post-transcriptionally (Suzuki et al., 2009). It is associating with the Drosha Microprocessor complex, probably by interacting directly with p68, a DEAD-box RNA helicase, that is an auxiliary component of the Microprocessor complex and thought to stimulate processing of MicroRNAs (Bates et al., 2005; Gregory et al., 2004). This association enhances the binding and

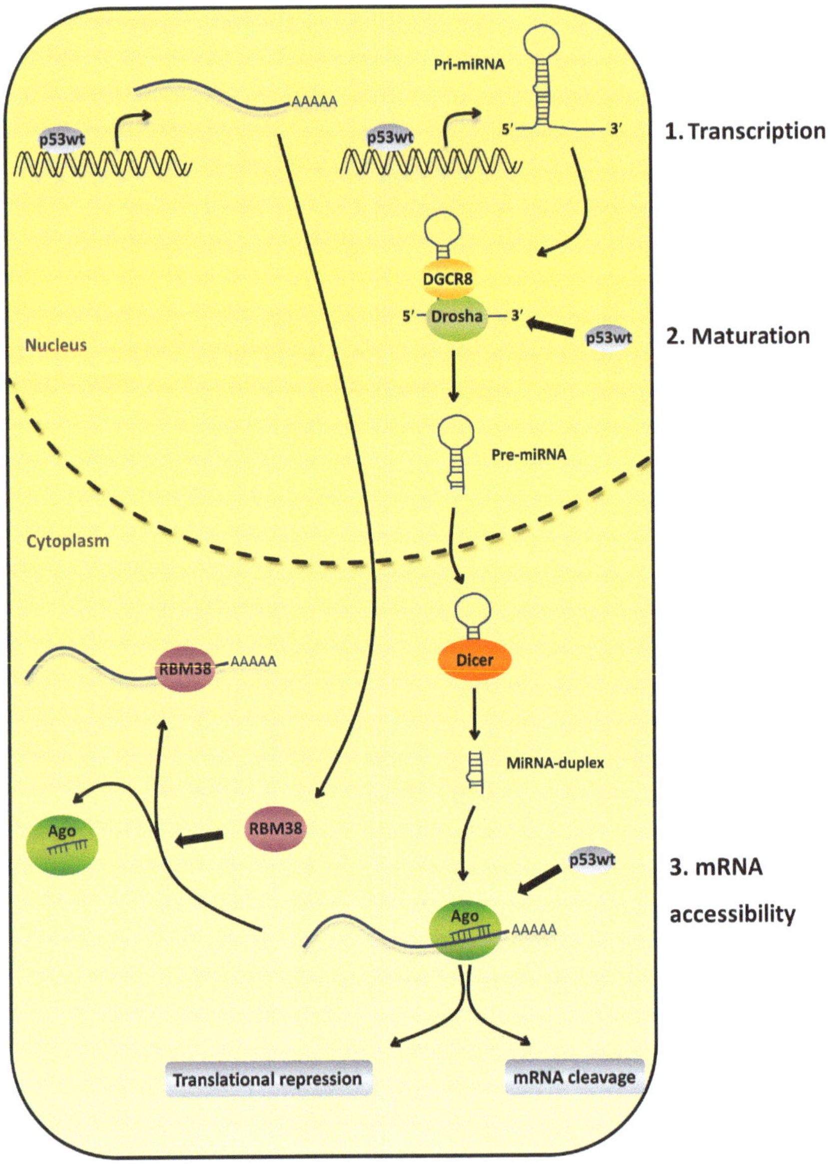

Fig. 2 Wild-type p53 is modulating the transcription, maturation, and mRNA accessibility for MicroRNAs. *1. Transcription*: p53 is able to transcriptionally activate and repress the expression of MicroRNAs. *2. Maturation*: The processing of selected MicroRNAs from pri- to pre-microRNAs is increased by an interaction of p53 with the Drosha complex. *3. mRNA accessibility for MicroRNAs*: p53 is able to increase the association of AGO2 with certain MicroRNAs and their corresponding targets. In addition, p53 controls MicroRNA access to their target mRNAs via the RNA-binding protein RBM38 that is transcriptionally activated by p53.

processing activity of the Drosha complex for certain pri-miRNAs, which finally gives rise to a higher level of mature MiR-16-1, MiR-143, MiR-135, MiR-206, MiR-103, and MiR-203 (Fig. 3) (Chang et al., 2013; Suzuki et al., 2009). MiR-1915 displays another MicroRNA that is post-transcriptionally augmented by p53. After DNA damage p53 increases the level of pre- and mature MiR-1915, leading to a downregulation of Bcl2 that is a direct target of MiR-1915 and to an increase of apoptosis (Nakazawa et al., 2014; Xu et al., 2013). Acetylation of p53 at lysine 120 after genotoxic stress was shown to be crucial for the association of p53 with the Microprocessor complex and the modulation of MicroRNA maturation.

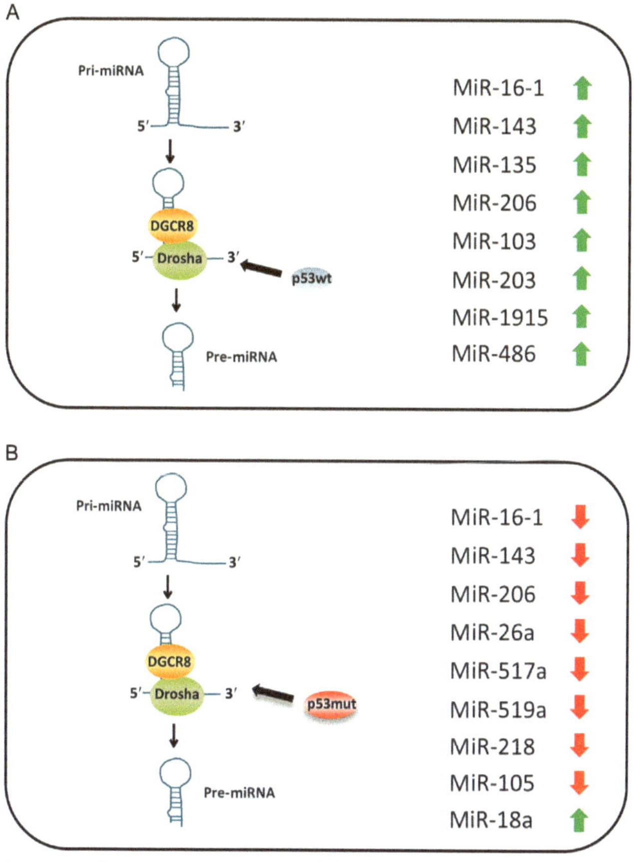

Fig. 3 Wild-type and mutant p53 are modulating the activity of the Microprocessor complex. Schematic representation of the impact of wild-type (A) and mutant p53 (B) on the processing of specific MicroRNAs from the primary transcript (pri-miRNA) to the precursor MicroRNA (pre-miRNA).

The acetylation of K120 drove cells toward apoptosis by augmenting mature MiR-203 which finally led to a downregulation of its target Bcl-w (Chang et al., 2013).

Moreover, p53 is also intervening at later steps of the MicroRNA regulatory network. Following DNA damage p53 was reported to bind to AGO2 and to influence the association of certain MicroRNAs with the RISC complex and their corresponding mRNA targets (Krell et al., 2016). After doxorubicin treatment p53 increased for instance the association of AGO2 with almost the entire let-7 family apart from let-7d without notably modulating their cellular abundance. Remarkably, the association of p53 with AGO2 but also the Drosha Microprocessor complex was abolished after RNAse treatment, showing that the presence of RNA is necessary for the stability of the complexes. The RNA-binding protein RBM38 is adding another mechanism of MiRNA regulation orchestrated by p53 (Leveille et al., 2011). It is directly upregulated by p53 and decreases MiRNA accessibility on various p53-induced transcripts like p21, PCNA, DDIT4, TNFRSF10B, LATS2, and IER5, reinforcing therefore p53 response. Conversely, knockdown of RBM38 resulted in a reduction of G1 arrest after ionizing radiation or Nutlin-3 treatment. In breast cancer the expression of RBM38 is lower in the subset that retained wild-type p53, underlining its importance in participating in the p53 tumor-suppressor pathway. Its downregulation might reduce the functionality of the p53 pathway in the wild-type context (Leveille et al., 2011; Shu et al., 2006). Interestingly, Mudhasani et al. showed that ablation of Dicer resulted in increased DNA damage which subsequently enhanced p53 protein levels. The decrease of Dicer and therefore also the overall amount of mature MicroRNAs activated p53 leading to cellular senescence in fibroblasts but also in vivo in the developing limb and adult skin of conditional knockout mice (Mudhasani et al., 2008). Therefore, p53 is not only modulating the biogenesis of MicroRNAs at different steps but also monitoring their maturation.

2.4 p53 Is a Direct Target of MicroRNAs

The expression of the "guardian of the genome" is tightly regulated, especially at the posttranscriptional level via ubiquitin ligases such as MDM2 and MDM4 (Levine and Oren, 2009). The MicroRNAs that in the recent years were identified to bind directly to the mRNA of p53 are extending our knowledge about the fine-tuning of p53 expression (Fig. 4). The MiRNAs that are targeting p53 are usually also downregulating its mutant counterpart.

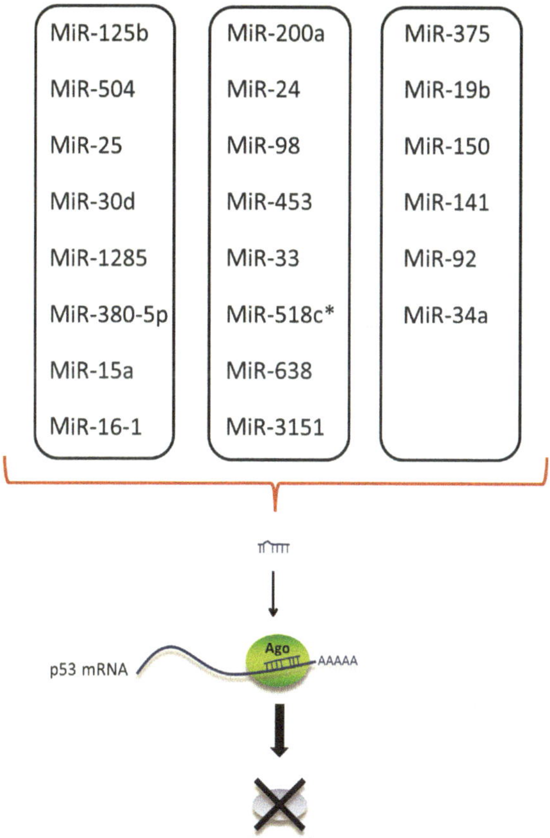

Fig. 4 Overview of MicroRNAs posttranscriptionally repressing p53.

MiR-125b was the first MicroRNA that was demonstrated to directly repress p53 posttranscriptionally (Le et al., 2009). Ectopic expression of MiR-125b in human lung fibroblasts led to a translational repression of p53, resulting in a reduction of p53 protein levels and a decrease of apoptosis. Knockdown of MiR-125b had the opposite effect, increasing p53 protein levels and subsequently apoptosis. In zebrafish, MiR-125b was shown to be downregulated after DNA damage allowing therefore p53 to accumulate. The RNA-binding protein HuR antagonizes the MiR-125b-mediated downregulation of p53 protein. Binding of HuR to the 3′UTR of the p53 mRNA inhibits the binding of the RISC complex and therefore the MiR-125b translational repression of p53 (Ahuja et al., 2016). In colon cancer MiR-125b was demonstrated to be an independent prognostic factor linking high expression with poor prognosis (Nishida et al., 2011). Also in myeloid leukemia, B-cell leukemia, nonsmall cell lung cancer, glioblastoma, clear-cell renal carcinoma, prostate cancer, pancreatic cancer, gastric cancer, and oligodendroglioma MiR-125b is supposed to act as oncomir, while in other cancer types it was shown to behave like a tumor suppressor (Chang et al., 2016; Yin et al., 2015). The second MicroRNA that was identified to target p53 was MiR-504. In breast, lung, and colon cancer cells MiR-504 downregulated the p53 protein via two binding sites in the 3′UTR of the p53 mRNA (Hu et al., 2010). After etoposide treatment MiR-504 was able to reduce p53 protein levels and counteract p53-induced apoptosis or cell cycle arrest. Moreover, ectopic expression of MiR-504 in HCT116 cells led to an increase in tumor volume when subcutaneously inoculated into nude mice (Hu et al., 2010). Very recently, MiR-504 itself was shown to be controlled by p53, establishing therefore another negative feedback loop. The p53-mediated downregulation of MiR-504 was presumed to be indirect as a direct p53 binding was not detected (Bublik et al., 2017). MiR-25 and MiR-30d were identified together by Kumar et al. to post-transcriptionally target p53 in kidney, lung, and colon cancer cells. They could downregulate both p53 mRNA and protein level and reduce apoptosis and cell cycle arrest in the presence as well as in the absence of etoposide treatment (Kumar et al., 2011). Vice versa, downregulation of MiR-25 and MiR-30d increases p53 expression together with p53 target genes like p21, Bax, and Puma and augments apoptosis. Ectopic expression of MiR-25 or MiR-30d led to a decrease of the senescent marker SA-β-galactosidase in a p53-dependent manner, indicating also a role of these MicroRNAs in the regulation of cellular senescence (Kumar et al., 2011). Like MiR-25 and MiR-30d, also MiR-1285 was identified in a screen for candidate MicroRNAs that target p53. The mRNA of p53 harbors two functional MiR-1285-binding sites in its 3′UTR that are responsible for the MiR-1285-mediated downregulation of p53 mRNA and protein. The posttranscriptional repression was accompanied by a downregulation of the p53 target gene p21 (Tian et al., 2010). Primary neuroblastoma usually retains p53 wild-type status, suggesting that the p53 tumor-suppressor pathway might be interrupted at another step. MiR-380-5p was found to be highly expressed in neuroblastoma and its expression level is correlated with poor outcome in MYCN-amplified neuroblastoma. It was shown to directly target the 3′UTR of p53 and to inhibit its translation. Ectopic expression of MiR-380-5p led to a reduction of cell death after genotoxic stress and in cooperation with HRAS to a transformation of primary mouse mammary epithelial cells. On the contrary, inhibition of MiR-380-5p led to a reduction of tumor growth in an orthotopic mouse model of neuroblastoma (Swarbrick et al., 2010). As mentioned before, p53 is controlling the expression of the MiR-15a/16-1 cluster. On the other hand, MiR-15a as well as MiR-16-1 is posttranscriptionally repressing p53 by targeting its 3′UTR which leads to a downregulation of p53 mRNA and protein, forming therefore a negative feedback loop vs p53. This feedback circuitry between MiR-15a/16-1 and p53 plays an important role in B-cell CLL with 13q deletions that show reduced MiR-15a/16-1 expression and high p53 and MiR-34a expression. Notably, this deletion is associated with an indolent form of CLL (Fabbri et al., 2011). MiR-200a was suggested to represent another MicroRNA that is upregulated by p53 and is exerting a negative feedback loop vs p53. It was reported to downregulate p53 protein levels by targeting its 3′UTR and to decrease G1 cell cycle arrest and apoptosis (Becker et al., 2012). Also MiR-24 was recently shown to target directly p53 and to be upregulated in breast, colon, and head and neck cancer which inversely correlated with overall survival. Downregulation of MiR-24 and as a consequence upregulation of p53 were revealed to be an important player in mediating anticancer effects of melatonin (Mori et al., 2016). MiR-98 and MiR-453 were identified to be additional MicroRNAs that are posttranscriptionally repressing p53 (Zhang et al., 2011a). Treatment of lung adenocarcinoma cells with cisplatin displayed a downregulation of MiR-98 which was suggested to aid in the increase of p53 protein levels and subsequently in cisplatin-induced growth inhibition. MiR-33, another MicroRNA that targets p53 posttranscriptionally, was shown to play an important role in the hematopoietic stem cell (HSC) self-renewal, increasing the HSC self-renewal capacity by p53 downregulation (Herrera-Merchan et al., 2010). Furthermore, ectopic expression of MiR-33 reduced apoptosis in several cancer cell lines after DNA damage. Retroviral transduction of MiR-33 together with RasV12 in mouse embryonic fibroblasts led to an oncogenic transformation, increasing foci formation and anchorage-independent growth and tumor formation in nude mice. MiR-518c* and MiR-638 were discovered to target both PTEN and p53. Therefore, they are able to inactivate two important tumor-suppressor genes simultaneously. Notably, they are not targeting only the classical 3′UTR. MiR-518c* inhibits p53 by binding exclusively to its 5′UTR, while MiR-638 is targeting both, the coding sequence and the 3′UTR of p53. These MicroRNAs are upregulated in prostate, colon, and kidney cancer and were shown to increase migration, invasion, proliferation, and anchorage-independent growth upon overexpression. Moreover, MiR-638 was reported to augment the tumor growth of colon cancer cells upon injection into nude mice (Tay et al., 2014). MiR-3151 is another MicroRNA that is downregulating p53, although the final proof whether this occurs directly is still missing. It was demonstrated to be upregulated in malignant melanoma and papillary thyroid cancer due to BRAF-activating mutations which is the most frequent aberration in these cancer types. Therefore, MiR-3151 is a downstream effector of mutated BRAF, linking it to the p53 tumor-suppressor pathway. In malignant melanoma BRAF inhibition by targeted therapy is inevitably followed by the emergence of resistance, evidencing the requirement of combination therapy. Simultaneous inhibition of mutated BRAF and MiR-3151 led to a dramatic growth reduction of malignant melanoma and papillary thyroid cancer cells, rendering MiR-3151 inhibition an attractive target for the development of new therapeutic approaches (Lankenau et al., 2015). MiR-375 was found to be upregulated in recurrent gastric cancer which was correlated with poor survival (Zhang et al., 2011b). Ectopic expression of MiR-375 posttranscriptionally represses p53 by targeting directly its 3′UTR antagonizing cell cycle arrest and apoptosis after DNA damage (Liu et al., 2013). MiR-19b was identified in a MiRNA screen to target p53. MiR-19b overexpression reduces the protein level of p53 by targeting directly its 3′UTR which gives rise to translational inhibition. The decrease of p53 was accompanied by a down-regulation of p53 targets like p21 and Bax. Interestingly, MiR-19a that shares the same seed region with MiR-19b and differs only at one nucleotide position was not able to posttranscriptionally repress p53. Ectopic MiR-19b expression reduced apoptosis and senescence, while it increased proliferation, migration, and invasion in cervical and breast cancer cell lines. Cervical cancer cells overexpressing MiR-19b revealed after subcutaneous injection into nude mice larger tumors and lung metastasis (Fan et al., 2014). MiR-150 seems to be a critical p53 regulator in lung cancer. Its expression is upregulated compared to normal tissue and correlates negatively with p53 protein levels. It targets directly the 3′UTR of p53 and modulates via p53 the apoptosis and growth rate of lung cancer cells (Wang et al., 2013a; Zhang et al., 2013). Three MicroRNAs that were identified to be regulated by p53 are forming a negative feedback loop vs p53, namely, MiR-141, MiR-92, and MiR-34a. MiR-141 and MiR-34a are upregulated, while MiR-92 is downregulated by p53 (Liu et al., 2017). MiR-141 and MiR-92 are targeting directly the 3′UTR of p53 (Neveu et al., 2010), while MiR-34a is binding to the 5′UTR and coding sequence of p53 (Navarro and Lieberman, 2015).

2.5 p53 Is an Indirect Target of MicroRNAs

As expected, p53 is not only directly under extensive control of MicroRNAs but also indirectly via factors which either influence the protein stability or the activity of p53 (Fig. 5). MDM2 is the most critical regulator of p53 protein stability and therefore necessitates to be under tight control of a variety of MicroRNAs like MiR-192, MiR-194, MiR-215, MiR-221, MiR-605, MiR-17-3p, MiR-25, MiR-32, MiR-143, MiR-145, MiR-18b, MiR-193a, MiR-339-5p, MiR-509-5p, MiR-660 MiR-661, MiR-410, and MiR-1827 (Shen et al., 2014; Vijayakumaran et al., 2015; Zhang et al., 2016a). Some of the MicroRNAs that are targeting MDM2 are itself

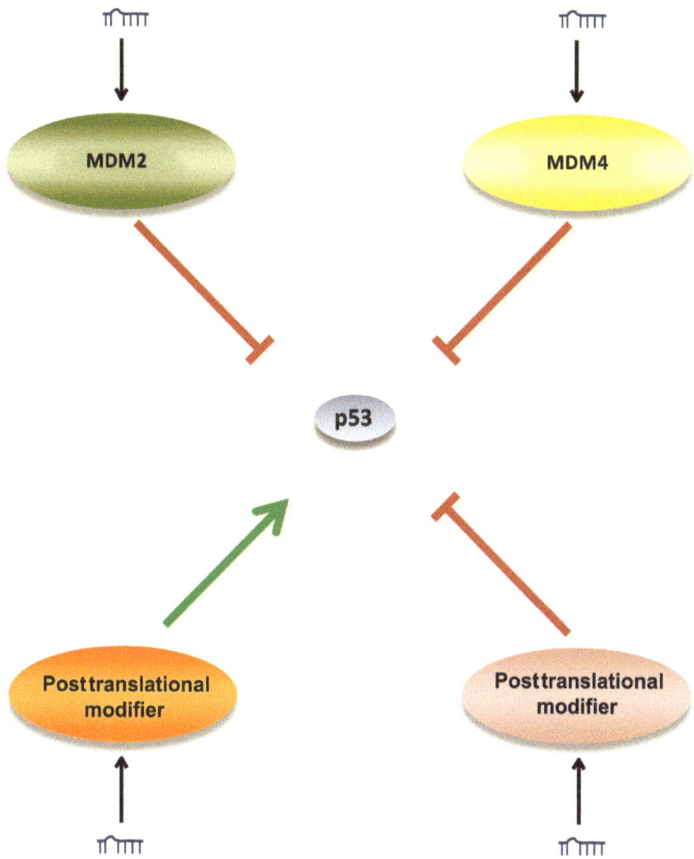

Fig. 5 Schematic representation of indirect regulation of p53 protein level or activity by MicroRNAs. p53 can also be indirectly regulated by MicroRNAs that target its major regulators like MDM2 (or HDM2) or MDM4 (HDMX) or by MicroRNAs that are downregulating factors that are posttranscriptional modifying p53 like SIRT1 or HDAC1.

regulated by p53, forming therefore a positive feedback loop and enhancing the p53 function in tumor suppression. MDM4, a homologue of MDM2, is another major negative regulator of p53. It mainly represses p53 transcriptional activity but was also reported to stimulate MDM2 in degrading p53 (Linares et al., 2003; Shadfan et al., 2012). Its protein level is reduced by MicroRNAs like MiR-191, MiR-10a, MiR-661, MiR-887, and let-7 (Vijayakumaran et al., 2015). Interestingly, MiR-34a was also demonstrated to target MDM4, reinforcing p53 in this indirect feedforward loop (Mandke et al., 2012; Okada et al., 2014). In fact, MiR-34 is participating in a rather extensive cross-talk to p53. Apart from targeting p53 itself and MDM4 it is also targeting additional factors that are influencing p53 activity. SIRT1 is one of those, a factor that deacetylates p53 decreasing therefore its transcriptional activity. MiR-34 is inhibiting the translation of SIRT1 by binding to its 3′UTR leading to an increase of p53 acetylation which finally gives rise to higher levels of p53 transcriptional targets like PUMA and p21 but also MiR-34a itself. The biological readout of SIRT1 downregulation in colon cancer cells is a rise in apoptosis (Yamakuchi et al., 2008). YY1 is another negative regulator of p53 that is targeted by MiR-34a. YY1 inhibits p53 by stimulating its ubiquitination and degradation (Chen et al., 2011). MTA2 and HDAC1, components of a complex that functions also in p53 deacetylation, are further targets of MiR-34a (Kaller et al., 2011). Therefore, MiR-34 is not only a critical downstream mediator of p53 functions but is forming an intricate network for regulating and reinforcing the p53 response.

3. MicroRNAs ENGAGED IN FULFILLING GOF ACTIVITIES OF MUTANT p53

3.1 MicroRNAs Transcriptionally Regulated by Mutant p53

As mentioned earlier, p53 is mutated in approximately 50%–70% of all human cancers. Most mutations are missense mutations and occur at its DNA-binding domain which abolishes the ability of p53 to bind to its cognate DNA elements. The mutations are divided roughly into two categories, the DNA contact mutants which affect amino acids crucial for DNA binding and the structural mutants which cause important conformational alterations (Cho et al., 1994; Muller and Vousden, 2013). Notably, p53 mutations are not only associated with a mere loss of tumor-suppressive activity but in certain cases with a gain of oncogenic properties (Strano et al., 2007). Several hot spot mutations were shown to turn p53 into an oncogenic factor that is changing the metabolism and augmenting cell proliferation, genomic instability, resistance to apoptotic stimuli, and metastasis (Freed-Pastor and Prives, 2012; Oren and Rotter, 2010; Sorrentino et al., 2014). Although GOF mutant p53 is usually not able to bind to its wild-type consensus motif, it is characterized to effectuate transcriptional regulation. A part of this shift in transcription can be attributed to the fact that mutant p53 is binding to p73 and p63 and therefore functionally inhibiting them by either sequestration or prevention of their transcriptional activation activity (Ferraiuolo et al., 2016). Other GOF activities are mediated via an interaction with non-transcriptional factors like TopB1, Pin1, MRE11, PML (Freed-Pastor and Prives, 2012; Haupt et al., 2009; Mantovani et al., 2016). But mutant p53 was also demonstrated to be recruited to DNA by binding to a subset of transcription factors like NF-Y, SREBP, SP1, ETS protein family, NFκB, and E2F1 and regulates therefore directly transcription. Di Agostino et al. originally reported for instance that mutant p53 is able to interact with NF-Y, binding therefore NF-Y promoters and recruiting after DNA damage p300 to target genes like cyclin A, cyclin B1, cdk1, and cdc25C. This recruitment leads to an activation of those genes after DNA damage instead of a downregulation. Furthermore, this interaction was demonstrated to be crucial for an increase in S-phase after DNA damage (Di Agostino et al., 2006). ChIP-on-chip experiments revealed the vitamin D receptor response element as prevalent in promoter sequences bound by mutant p53. In fact, mutant p53 was shown to bind to the vitamin D receptor and modulate its target gene expression, turning vitamin D into an antiapoptotic agent (Stambolsky et al., 2010). A combined approach of ChIP-on-chip and ChIP-Seq for mutant p53 exhibited also the ETS-binding site motif as overrepresented within predicted mutant p53-binding sites. ETS2 was shown to interact and cooperate with mutant p53 to promote etoposide resistance (Do et al., 2012). Instead, interaction between ETS1 and mutant p53 was reported to cause an upregulation of the drug resistance gene MDR1 (Sampath et al., 2001). But also DNA-bound SP1 was revealed to be associated with mutated p53 which was shown to drive an activation of HIV-LTR-driven transcription (Chicas et al., 2000). An association with SREBP proteins unveiled to be crucial for the mutant p53 mediated upregulation of the mevalonate pathway which is associated with tumor invasiveness (Freed-Pastor et al., 2012). Cooks et al. illustrated a cross-talk between mutant p53 and NFκB. Mutant p53 provokes a robust prolongation of NFκB activation after TNFα treatment and therefore an extended expression of proinflammatory genes. Of note, mice carrying mutant p53 are prone to inflammation-associated colon cancer (Cooks et al., 2013). Finally, mutant p53 interacts with E2F1, augmenting thereby inhibitor of DNA binding 4 which is linking mutant p53 with tumor neoangiogenesis (Fontemaggi et al., 2009).

Thus, mutant p53 is able to regulate transcription despite the fact that it is not able to bind to wild-type p53-binding sites. As one would expect, mutant p53 is not only regulating the expression of protein-coding genes but also of noncoding MicroRNAs that are mediating in part the GOF activities of mutant p53 (Fig. 6). In nonsmall cell lung cancer cells MiR-128-2 is upregulated by the hotspot mutant p53R175H together with its host gene ARPP21. Mutant p53 was shown to bind to the promoter region of the MiR-128-2 host gene ARPP21, leading to a transactivation. MiR-128-2 is posttranscriptionally repressing E2F5 which results in an increase of p21 that is usually repressed by E2F5. p21 localizes to the cytoplasm and exerts antiapoptotic functions by reducing procaspase 3 cleavage. Ectopic expression of MiR-128-2 confers increased resistance to chemotherapeutics like doxorubicin, cisplatin, and 5-fluorouracil (Donzelli et al., 2012). MiR-155 was shown in breast cancer cells to be upregulated by p53 mutant R248Q or R282W. In this case the upregulation is accomplished by inhibiting p63 which is directly repressing the MiR-155 host gene.

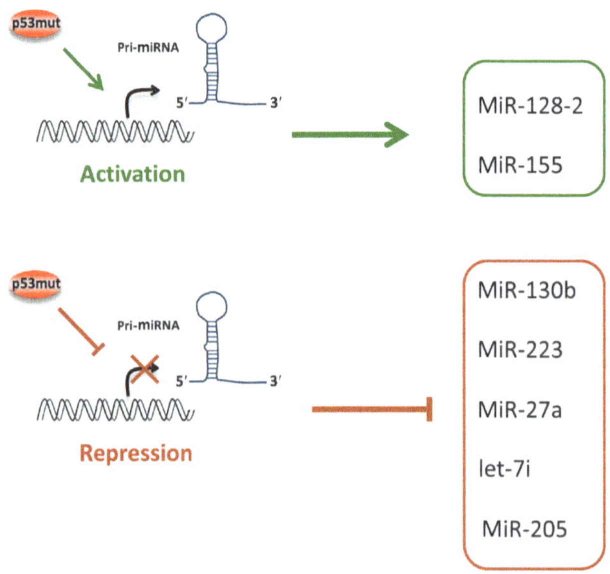

Fig. 6 Overview of MicroRNAs directly up- or downregulated by mutant p53.

The zinc-finger transcriptional repressor ZNF652 that is inhibiting various genes that are driving metastasis was revealed to be a crucial target of MiR-155. Upregulation of MiR-155 and subsequently downregulation of ZNF652 promote invasion of breast cancer cells (Neilsen et al., 2013). In endometrial cancer cells MiR-130b is directly repressed by p53 mutants, and this downregulation was identified to be crucial for EMT and invasion. MiR-130b expression is reduced in endometrial cancer, and lower expression correlates with shorter survival. ZEB1 was demonstrated to be a direct target of MiR-130b. Therefore, mutant p53 downregulates MiR-130b which results in an upregulation of ZEB1, driving subsequently EMT and invasion (Dong et al., 2013). Another MicroRNA that is involved in the increased chemoresistance conferred by mutant p53 is MiR-223. Diverse GOF p53 mutants bind to its promoter in conjunction with the transcriptional repressor ZEB1 and downregulate the expression of MiR-223. Stathmin-1 was identified as a MiR-223 target. It is overexpressed in various cancers and known to be crucial for the cell cycle. Correspondingly, silencing of Stathmin-1 or exogenous expression of MiR-223 augments apoptosis and renders cells more sensitive to DNA-damaging treatments (Masciarelli et al., 2014). MiR-27a displays another MicroRNA that is downregulated by mutant p53R273H, R175H, and G279E via binding to its promoter. EGFR was discovered to be downregulated by MiR-27a, being targeted at its 3′UTR which leads to translational inhibition. Downregulation of MiR-27a by mutant p53 provokes an increase of EGFR and a sustained EGFR-ERK pathway activation. The mutant p53–miR27a–EGFR axis contributes to the GOF activity of mutant p53 by increasing colony formation, proliferation, and tumor formation of breast cancer cells in xenograft mouse models (Wang et al., 2013b). Subramanian et al. reported let-7i to be downregulated by mutant GOF p53 proteins. The p53 family member p63 was shown to upregulate let-7i, while mutant p53 variants R273H, R280K, and R248W triggered a downregulation of the tumor-suppressor MicroRNA. This repression is executed by physically binding to p63 in the promoter region and antagonizing its transcriptional activation. let-7i was shown to inhibit migration, invasion, and metastasis formation by targeting the 3′UTR of oncogenic drivers like E2F5, LIN28B, MYC, and several RNA posttranscriptional modification genes. Downregulation of let-7i by mutant p53 leads to a derepression of these oncogenic target genes, giving rise to cell migration, invasion, and metastasis (Subramanian et al., 2015). In a microarray analysis also the p53 mutant R282W was shown to upregulate let-7i (Zhang et al., 2016b). Whether wild-type p53 is regulating the expression of let-7i is still controversially discussed (Subramanian et al.,

2015; Zhang et al., 2016b). MiR-205 is described as a crucial factor for inhibiting cell migration and metastasis in prostate cancer. It is upregulated by both p63 isoforms TAp63 and ΔNp63 that are binding to two putative p53 consensus response elements in the upstream region of MiR-205. ΔNp63 was shown to downregulate ZEB1 via the induction of MiR-205 that was previously described to target EMT master regulators like ZEB1 (Gregory et al., 2008) and to reduce migration and invasion of prostate cancer cells (Gandellini et al., 2009). Interestingly, ectopic expression of the mutants R175H, R248W, R249S, and R273H led to a strong downregulation of MiR-205. The p53 mutants could counteract p63 by binding to it and decreasing its protein stability. Thus, mutant p53 inhibited p63-mediated MiR-205 upregulation, enhancing thereby motility and migration. Furthermore, expression of ΔNp63 or MiR-205 reduced the incidence of lung metastasis in vivo after injecting prostate cancer cells in the tail vein of nude mice. Importantly, loss of the p63/MiR-205 axis was associated with an increased metastatic potential of prostate cancer and predictive of poor survival (Tucci et al., 2012).

Thus, mutant p53 is transcriptionally regulating specific MicroRNAs by binding to DNA through other transcription factors or by inhibiting the p53 family member p63. These mutant p53-regulated MicroRNAs participate in the GOF activities such as chemoresistance and metastatic behavior.

3.2 Mutant p53 Regulates MicroRNA Biogenesis at Posttranscriptional Level

As already mentioned, human tumors are often characterized by a global decrease in MicroRNA expression (Lu et al., 2005). Many tumor-suppressor MicroRNAs are silenced by epigenetic modifications. But also the posttranscriptional biogenesis of MicroRNAs is often perturbed in cancer. In fact, wild-type p53 was proven to influence the maturation of MicroRNAs by a direct association with the Microprocessor complex and to modulate the accessibility of the RISC complex loaded with certain MicroRNAs to their target mRNAs (Gurtner et al., 2016; Krell et al., 2016; Leveille et al., 2011; Suzuki et al., 2009). In line with that, mutations or decreased expression of MiR biogenesis enzymes are associated with increased malignancy, underlining the importance of MicroRNAs in tumor suppression (Hata and Kashima, 2016). Notably, not only wild-type p53 but also the mutant p53 versions are interfering at various steps in the biogenesis of MicroRNAs (Fig. 7). Like for the transcriptional regulation, they are not only losing but changing their activity in influencing the MicroRNA

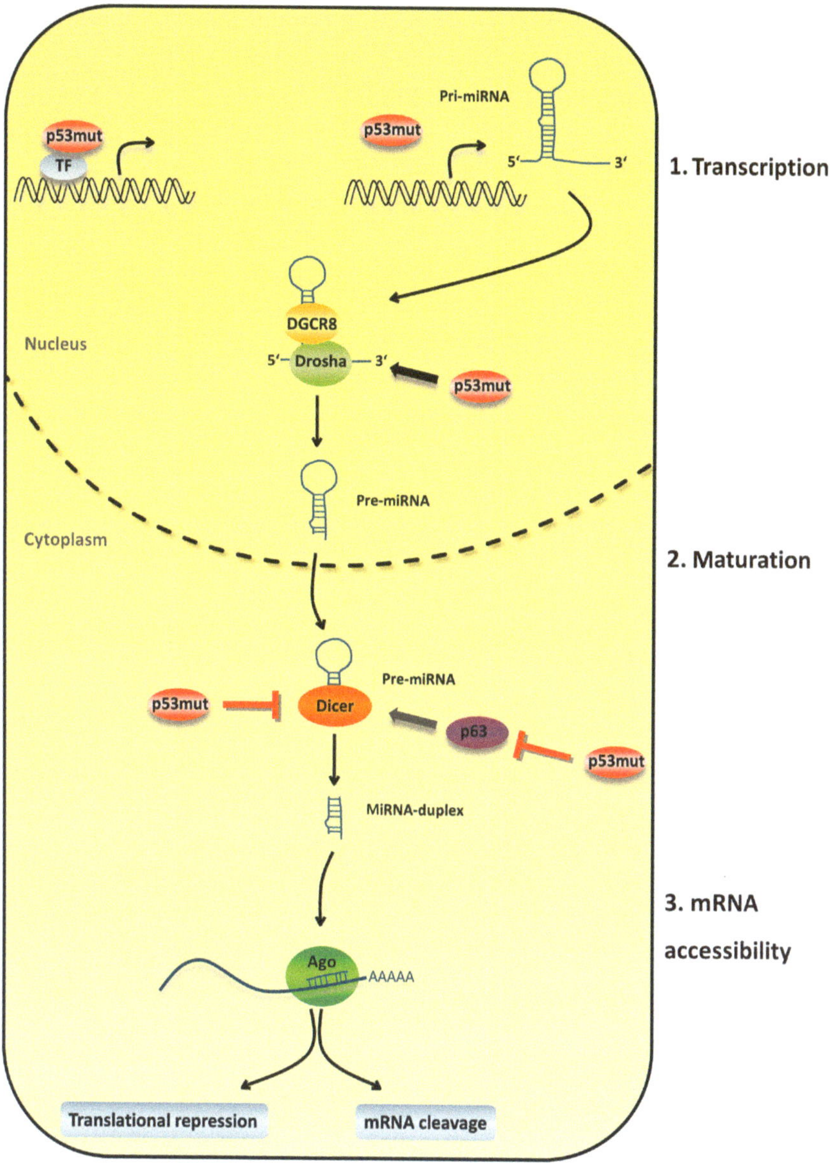

Fig. 7 Schematic representation of the various MicroRNA processing steps modulated by mutant p53. *1. Transcription*: mutant p53 proteins are up- or downregulating MicroRNAs by binding to DNA-bound transcription factors or by inhibiting p63. *2. Maturation*: p53 gain-of-function mutants modulate MicroRNA biogenesis by interacting with auxiliary proteins of the Drosha Microprocessor complex and by downregulating DICER. *3. mRNA accessibility for MicroRNAs*: mutant p53 in contrast to the wild-type protein is not increasing the association of AGO2 with certain MicroRNAs and their corresponding targets. Some mutants are rather decreasing the association.

processing, gaining therefore oncogenic properties. Mutant p53R175H was reported to be able to bind to the p68 auxiliary subunit of the Microprocessor complex like wild-type p53 but decreased the interaction of p68 with Drosha and the association of the Microprocessor complex with certain MicroRNAs like pri-miR-16-1 and pri-miR-143 (Fig. 3). The p53 mutants C135Y, R175H, and R273 reduced the processing and therefore the mature levels of MiR-16-1, MiR-143, and MiR-206 (Suzuki et al., 2009). Also the biogenesis of MiR-26a was reported to be modulated by mutant p53 variants. In endometrial cancer cells p53 mutants C135Y, R175H, R248Q, and R273H decreased pre- and mature MiR-26a without affecting the pri-microRNA levels. This is thought to be accomplished by reducing the association of pri-miR-26a with the Drosha/p68 complex. Mutant p53 R248Q was shown to associate with p68 and was suggested to disrupt in this way the Drosha–p68 complex necessary for MiR-26a maturation (Jiang et al., 2015). EZH2 was identified as a direct MiR-26a target, and its expression was found to be elevated in the more aggressive endometrial carcinoma type II. Inversely, ectopic expression of MiR-26a in endometrial cancer cells reduced EMT markers like N-cadherin, Vimentin, and SNAIL and inhibited proliferation, migration, and invasion. Furthermore, in nude mice MiR-26a expression decreased the tumor volume in a xenograft model. High expression of MiR-26a is correlated with longer survival in endometrial carcinoma. Thus, mutant p53 was proposed to increase EMT by downregulating the maturation of MiR-26a followed by an increase of the MiR-26a target EZH2 (Jiang et al., 2015). DEAD-box RNA helicase p72/p82 displays another auxiliary factor of the Microprocessor complex. Garibaldi et al. demonstrated that mutant p53 was able to bind to p72, inhibiting thereby the recruitment of Drosha to certain pri-miRNAs. This association was shown for the endogenous mutants R273H/P309S and R175H in colon and breast cancer cells, respectively. Therefore, mutant p53 was suggested to sequester p72/p82 which results in a downregulation of the processing and therefore mature levels of MicroRNAs like MiR-517a, MiR-519a, MiR-218, and MiR-105. MiR-517a, MiR-519a, and MiR-105 decreased the viability of colon cancer cells upon ectopic expression. MiR-519a and MiR-218 were shown to inhibit the migration of colon cancer cells and to downregulate ZEB1 followed by an increase of E-cadherin protein levels upon exogenous expression, indicating that they might promote MET. Of note, some of these MicroRNAs whose maturation is inhibited by mutant p53 were shown to be upregulated by the wild-type p53 protein (Fig. 3) (Garibaldi et al., 2016). In female HCC the processing of pri- to pre-miR-18a was shown to be enhanced by various mutant p53 forms as well as wild type. Increased levels of MiR-18a were reported to decrease the MiR-18a target ERα, promoting thereby proliferation of hepatoma cells. Overexpression of the p53/MiR-18a pathway was claimed to associate with a trend of worse overall survival in hepatitis B virus-related female HCC by disrupting the tumor-protective estrogen pathway (Li et al., 2015a).

But mutant p53 is not only decreasing the activity of the Microprocessor complex, it is also intervening at the next endonucleolytic step. Mutant p53 drives the downregulation of the endonuclease Dicer which is responsible for the cleavage of the pre-microRNAs into the mature MicroRNAs (Fig. 7). The downregulation occurs at the transcriptional level by inhibiting TAp63-mediated transcriptional activation and at the posttranscriptional level in a TAp63-independent manner. Mutant p53-mediated downregulation of Dicer is accompanied by enhanced recycling and activation of growth factor receptors MET and EGFR and an increase of invasion and cell scattering after growth factor stimulation (HGF, EGF). The decrease of Dicer led to a downregulation of mature MiR-31, MiR-130b, and MiR-206 (Muller et al., 2014). Recently Walerych et al. revealed also an additional effect of mutant p53 on MicroRNA maturation. Five different p53 missense variants were shown to increase the proteasome machinery as common oncogenic signature (Walerych et al., 2016). The enhanced proteasome activity led to a decrease of the KH-type splicing regulatory protein KSRP which is an RNA-binding protein involved in the maturation of MicroRNAs at the Drosha and Dicer level (Trabucchi et al., 2009). KSRP is binding specifically to the terminal loop of certain pri- and pre-microRNAs, favoring therefore their association with the Drosha and Dicer complex. Downregulation of the KSRP protein by mutant p53 via proteasomal degradation causes the decrease of the tumor-suppressive mature MicroRNAs let-7a and MiR-30c (Walerych et al., 2016).

As mentioned before, wild-type p53 was demonstrated to modulate the MiRNA-targetome network by increasing the association of AGO2 with almost the entire let-7 family. Mutant p53 variants were not able to increase this interaction. In more detail, mutant R273H did not influence the association, while mutants R175H and R248W elicited even a reduction (Krell et al., 2016).

4. CONCLUSIONS

p53 is not only orchestrating protein-coding genes to inhibit initiation and progression of cancer but also noncoding RNAs like MicroRNAs.

MicroRNAs tightly control a network fine-tuning upstream and downstream the p53 pathway, forming an intricate network with several positive feedforward and negative feedback loops. In general, wild-type p53 regulates the transcription of MicroRNAs by binding directly to DNA via its consensus sequence. In contrast, mutant p53 proteins usually lost their ability to bind to the standard p53 response elements and regulate transcriptional expression of MicroRNAs by interacting with other transcription factors. Of note, whole genome gene expression data in combination with genomic DNA-binding analysis revealed that wild-type p53 predominantly activates genes if it was bound nearby (Botcheva et al., 2011; Hunten et al., 2015; Menendez et al., 2013; Nikulenkov et al., 2012; Schlereth et al., 2013). MicroRNAs upregulated by p53 might therefore represent a bridge, enabling the repression of a plethora of genes in an indirect and posttranscriptional manner. In the wild-type p53 context MicroRNAs are key effectors in inhibiting cell proliferation, in apoptosis, stress response, maintenance of stem cell potency, and metabolism. On the contrary, in the case of mutant GOF p53 they are crucial in conferring oncogenic properties like chemoresistance and the capability to drive metastasis. Both wild-type and mutant p53 were also shown to regulate the processing of MicroRNAs. So far, wild-type p53 increased the maturation of specific MicroRNAs, while mutant GOF p53 was mainly reducing it (Fig. 3). This is in line with the observation that the overall amount of MicroRNAs is generally downregulated in tumors compared with normal tissues, underling the importance of these small RNAs in controlling various steps in the development of cancer with p53 at the center of this molecular network (Lu et al., 2005).

ACKNOWLEDGMENTS

Contribution of AIRC (AIRC 14455), EPIGEN Flagship Project (13/05/R/42) to G.B. and Aboca to S.S. was greatly appreciated. We would also like to thank all members of the groups from G.B. and S.S.

REFERENCES

Ahuja, D., Goyal, A., Ray, P.S., 2016. Interplay between RNA-binding protein HuR and microRNA-125b regulates p53 mRNA translation in response to genotoxic stress. RNA Biol. 13, 1152–1165.

Au Yeung, C.L., Tsang, T.Y., Yau, P.L., Kwok, T.T., 2011. Human papillomavirus type 16 E6 induces cervical cancer cell migration through the p53/microRNA-23b/urokinase-type plasminogen activator pathway. Oncogene 30, 2401–2410.

Barsotti, A.M., Beckerman, R., Laptenko, O., Huppi, K., Caplen, N.J., Prives, C., 2012. p53-dependent induction of PVT1 and miR-1204. J. Biol. Chem. 287, 2509–2519.

Bates, G.J., Nicol, S.M., Wilson, B.J., Jacobs, A.M., Bourdon, J.C., Wardrop, J., Gregory, D.J., Lane, D.P., Perkins, N.D., Fuller-Pace, F.V., 2005. The DEAD box protein p68: a novel transcriptional coactivator of the p53 tumour suppressor. EMBO J. 24, 543–553.

Beard, J.A., Tenga, A., Hills, J., Hoyer, J.D., Cherian, M.T., Wang, Y.D., Chen, T., 2016. The orphan nuclear receptor NR4A2 is part of a p53-microRNA-34 network. Sci. Rep. 6, 25108.

Becker, L.E., Lu, Z., Chen, W., Xiong, W., Kong, M., Li, Y., 2012. A systematic screen reveals microRNA clusters that significantly regulate four major signaling pathways. PLoS One 7. e48474.

Bersani, F., Lingua, M.F., Morena, D., Foglizzo, V., Miretti, S., Lanzetti, L., Carra, G., Morotti, A., Ala, U., Provero, P., Chiarle, R., Singer, S., Ladanyi, M., Tuschl, T., Ponzetto, C., Taulli, R., 2016. Deep sequencing reveals a novel miR-22 regulatory network with therapeutic potential in rhabdomyosarcoma. Cancer Res. 76, 6095–6106.

Bieging, K.T., Mello, S.S., Attardi, L.D., 2014. Unravelling mechanisms of p53-mediated tumour suppression. Nat. Rev. Cancer 14, 359–370.

Bohlig, L., Friedrich, M., Engeland, K., 2011. p53 activates the PANK1/miRNA-107 gene leading to downregulation of CDK6 and p130 cell cycle proteins. Nucleic Acids Res. 39, 440–453.

Bommer, G.T., Gerin, I., Feng, Y., Kaczorowski, A.J., Kuick, R., Love, R.E., Zhai, Y., Giordano, T.J., Qin, Z.S., Moore, B.B., MacDougald, O.A., Cho, K.R., Fearon, E.R., 2007. p53-mediated activation of miRNA34 candidate tumor-suppressor genes. Curr. Biol. 17, 1298–1307.

Borchert, G.M., Lanier, W., Davidson, B.L., 2006. RNA polymerase III transcribes human microRNAs. Nat. Struct. Mol. Biol. 13, 1097–1101.

Botcheva, K., McCorkle, S.R., McCombie, W.R., Dunn, J.J., Anderson, C.W., 2011. Distinct p53 genomic binding patterns in normal and cancer-derived human cells. Cell Cycle 10, 4237–4249.

Braun, C.J., Zhang, X., Savelyeva, I., Wolff, S., Moll, U.M., Schepeler, T., Orntoft, T.F., Andersen, C.L., Dobbelstein, M., 2008. p53-responsive microRNAs 192 and 215 are capable of inducing cell cycle arrest. Cancer Res. 68, 10094–10104.

Bublik, D.R., Bursac, S., Sheffer, M., Orsolic, I., Shalit, T., Tarcic, O., Kotler, E., Mouhadeb, O., Hoffman, Y., Fuchs, G., Levin, Y., Volarevic, S., Oren, M., 2017. Regulatory module involving FGF13, miR-504, and p53 regulates ribosomal biogenesis and supports cancer cell survival. Proc. Natl. Acad. Sci. U.S.A. 114, E496–E505.

Calin, G.A., Dumitru, C.D., Shimizu, M., Bichi, R., Zupo, S., Noch, E., Aldler, H., Rattan, S., Keating, M., Rai, K., Rassenti, L., Kipps, T., Negrini, M., Bullrich, F., Croce, C.M., 2002. Frequent deletions and down-regulation of micro- RNA genes miR15 and miR16 at 13q14 in chronic lymphocytic leukemia. Proc. Natl. Acad. Sci. U.S.A. 99, 15524–15529.

Calin, G.A., Cimmino, A., Fabbri, M., Ferracin, M., Wojcik, S.E., Shimizu, M., Taccioli, C., Zanesi, N., Garzon, R., Aqeilan, R.I., Alder, H., Volinia, S., Rassenti, L., Liu, X., Liu, C.G., Kipps, T.J., Negrini, M., Croce, C.M., 2008. MiR-15a and miR-16-1 cluster functions in human leukemia. Proc. Natl. Acad. Sci. U.S.A. 105, 5166–5171.

Chang, T.C., Wentzel, E.A., Kent, O.A., Ramachandran, K., Mullendore, M., Lee, K.H., Feldmann, G., Yamakuchi, M., Ferlito, M., Lowenstein, C.J., Arking, D.E., Beer, M.A., Maitra, A., Mendell, J.T., 2007. Transactivation of miR-34a by p53 broadly influences gene expression and promotes apoptosis. Mol. Cell 26, 745–752.

Chang, J., Davis-Dusenbery, B.N., Kashima, R., Jiang, X., Marathe, N., Sessa, R., Louie, J., Gu, W., Lagna, G., Hata, A., 2013. Acetylation of p53 stimulates miRNA processing and determines cell survival following genotoxic stress. EMBO J. 32, 3192–3205.

Chang, S., He, S., Qiu, G., Lu, J., Wang, J., Liu, J., Fan, L., Zhao, W., Che, X., 2016. MicroRNA-125b promotes invasion and metastasis of gastric cancer by targeting STARD13 and NEU1. Tumour Biol. 37, 12141–12151.

Chen, Q.R., Yu, L.R., Tsang, P., Wei, J.S., Song, Y.K., Cheuk, A., Chung, J.Y., Hewitt, S.M., Veenstra, T.D., Khan, J., 2011. Systematic proteome analysis identifies transcription factor YY1 as a direct target of miR-34a. J. Proteome Res. 10, 479–487.

Chen, L., Chen, X.R., Zhang, R., Li, P., Liu, Y., Yan, K., Jiang, X.D., 2013. MicroRNA-107 inhibits glioma cell migration and invasion by modulating Notch2 expression. J. Neurooncol 112, 59–66.

Chen, W.Y., Liu, S.Y., Chang, Y.S., Yin, J.J., Yeh, H.L., Mouhieddine, T.H., Hadadeh, O., Abou-Kheir, W., Liu, Y.N., 2015. MicroRNA-34a regulates WNT/TCF7 signaling and inhibits bone metastasis in Ras-activated prostate cancer. Oncotarget 6, 441–457.

Chicas, A., Molina, P., Bargonetti, J., 2000. Mutant p53 forms a complex with Sp1 on HIV-LTR DNA. Biochem. Biophys. Res. Commun. 279, 383–390.

Cho, Y., Gorina, S., Jeffrey, P.D., Pavletich, N.P., 1994. Crystal structure of a p53 tumor suppressor-DNA complex: understanding tumorigenic mutations. Science 265, 346–355.

Choi, Y.J., Lin, C.P., Ho, J.J., He, X., Okada, N., Bu, P., Zhong, Y., Kim, S.Y., Bennett, M.J., Chen, C., Ozturk, A., Hicks, G.G., Hannon, G.J., He, L., 2011. miR-34 miRNAs provide a barrier for somatic cell reprogramming. Nat. Cell Biol. 13, 1353–1360.

Cimmino, A., Calin, G.A., Fabbri, M., Iorio, M.V., Ferracin, M., Shimizu, M., Wojcik, S.E., Aqeilan, R.I., Zupo, S., Dono, M., Rassenti, L., Alder, H., Volinia, S., Liu, C.G., Kipps, T.J., Negrini, M., Croce, C.M., 2005. miR-15 and miR-16 induce apoptosis by targeting BCL2. Proc. Natl. Acad. Sci. U.S.A. 102, 13944–13949.

Comel, A., Sorrentino, G., Capaci, V., Del Sal, G., 2014. The cytoplasmic side of p53's oncosuppressive activities. FEBS Lett. 588, 2600–2609.

Cooks, T., Pateras, I.S., Tarcic, O., Solomon, H., Schetter, A.J., Wilder, S., Lozano, G., Pikarsky, E., Forshew, T., Rosenfeld, N., Harpaz, N., Itzkowitz, S., Harris, C.C., Rotter, V., Gorgoulis, V.G., Oren, M., 2013. Mutant p53 prolongs NF-kappaB activation and promotes chronic inflammation and inflammation-associated colorectal cancer. Cancer Cell 23, 634–646.

Corney, D.C., Flesken-Nikitin, A., Godwin, A.K., Wang, W., Nikitin, A.Y., 2007. MicroRNA-34b and MicroRNA-34c are targets of p53 and cooperate in control of cell proliferation and adhesion-independent growth. Cancer Res. 67, 8433–8438.

Cortez, M.A., Ivan, C., Valdecanas, D., Wang, X., Peltier, H.J., Ye, Y., Araujo, L., Carbone, D.P., Shilo, K., Giri, D.K., Kelnar, K., Martin, D., Komaki, R., Gomez, D.R., Krishnan, S., Calin, G.A., Bader, A.G., Welsh, J.W., 2016. PDL1 regulation by p53 via miR-34. J. Natl. Cancer Inst. 108.

Cui, S.Y., Wang, R., Chen, L.B., 2014. MicroRNA-145: a potent tumour suppressor that regulates multiple cellular pathways. J. Cell. Mol. Med. 18, 1913–1926.

Dar, A.A., Majid, S., de Semir, D., Nosrati, M., Bezrookove, V., Kashani-Sabet, M., 2011. miRNA-205 suppresses melanoma cell proliferation and induces senescence via regulation of E2F1 protein. J. Biol. Chem. 286, 16606–16614.

De Cola, A., Volpe, S., Budani, M.C., Ferracin, M., Lattanzio, R., Turdo, A., D'Agostino, D., Capone, E., Stassi, G., Todaro, M., Di Ilio, C., Sala, G., Piantelli, M., Negrini, M., Veronese, A., De Laurenzi, V., 2015. miR-205-5p-mediated downregulation of ErbB/HER receptors in breast cancer stem cells results in targeted therapy resistance. Cell Death Dis. 6. e1823.

Di Agostino, S., Strano, S., Emiliozzi, V., Zerbini, V., Mottolese, M., Sacchi, A., Blandino, G., Piaggio, G., 2006. Gain of function of mutant p53: the mutant

p53/NF-Y protein complex reveals an aberrant transcriptional mechanism of cell cycle regulation. Cancer Cell 10, 191–202.
Do, P.M., Varanasi, L., Fan, S., Li, C., Kubacka, I., Newman, V., Chauhan, K., Daniels, S.R., Boccetta, M., Garrett, M.R., Li, R., Martinez, L.A., 2012. Mutant p53 cooperates with ETS2 to promote etoposide resistance. Genes Dev. 26, 830–845.
Dong, P., Karaayvaz, M., Jia, N., Kaneuchi, M., Hamada, J., Watari, H., Sudo, S., Ju, J., Sakuragi, N., 2013. Mutant p53 gain-of-function induces epithelial-mesenchymal transition through modulation of the miR-130b-ZEB1 axis. Oncogene 32, 3286–3295.
Donzelli, S., Fontemaggi, G., Fazi, F., Di Agostino, S., Padula, F., Biagioni, F., Muti, P., Strano, S., Blandino, G., 2012. MicroRNA-128-2 targets the transcriptional repressor E2F5 enhancing mutant p53 gain of function. Cell Death Differ. 19, 1038–1048.
Donzelli, S., Mori, F., Bellissimo, T., Sacconi, A., Casini, B., Frixa, T., Roscilli, G., Aurisicchio, L., Facciolo, F., Pompili, A., Carosi, M.A., Pescarmona, E., Segatto, O., Pond, G., Muti, P., Telera, S., Strano, S., Yarden, Y., Blandino, G., 2015. Epigenetic silencing of miR-145-5p contributes to brain metastasis. Oncotarget 6, 35183–35201.
Fabbri, M., Bottoni, A., Shimizu, M., Spizzo, R., Nicoloso, M.S., Rossi, S., Barbarotto, E., Cimmino, A., Adair, B., Wojcik, S.E., Valeri, N., Calore, F., Sampath, D., Fanini, F., Vannini, I., Musuraca, G., Dell'Aquila, M., Alder, H., Davuluri, R.V., Rassenti, L.Z., Negrini, M., Nakamura, T., Amadori, D., Kay, N.E., Rai, K.R., Keating, M.J., Kipps, T.J., Calin, G.A., Croce, C.M., 2011. Association of a microRNA/TP53 feedback circuitry with pathogenesis and outcome of B-cell chronic lymphocytic leukemia. JAMA 305, 59–67.
Fan, Y., Yin, S., Hao, Y., Yang, J., Zhang, H., Sun, C., Ma, M., Chang, Q., Xi, J.J., 2014. miR-19b promotes tumor growth and metastasis via targeting TP53. RNA 20, 765–772.
Ferraiuolo, M., Di Agostino, S., Blandino, G., Strano, S., 2016. Oncogenic intra-p53 family member interactions in human cancers. Front. Oncol. 6, 77.
Fontemaggi, G., Dell'Orso, S., Trisciuoglio, D., Shay, T., Melucci, E., Fazi, F., Terrenato, I., Mottolese, M., Muti, P., Domany, E., Del Bufalo, D., Strano, S., Blandino, G., 2009. The execution of the transcriptional axis mutant p53, E2F1 and ID4 promotes tumor neo-angiogenesis. Nat. Struct. Mol. Biol. 16, 1086–1093.
Freed-Pastor, W.A., Prives, C., 2012. Mutant p53: one name, many proteins. Genes Dev. 26, 1268–1286.
Freed-Pastor, W.A., Mizuno, H., Zhao, X., Langerod, A., Moon, S.H., Rodriguez-Barrueco, R., Barsotti, A., Chicas, A., Li, W., Polotskaia, A., Bissell, M.J., Osborne, T.F., Tian, B., Lowe, S.W., Silva, J.M., Borresen-Dale, A.L., Levine, A.J., Bargonetti, J., Prives, C., 2012. Mutant p53 disrupts mammary tissue architecture via the mevalonate pathway. Cell 148, 244–258.
Friedman, R.C., Farh, K.K., Burge, C.B., Bartel, D.P., 2009. Most mammalian mRNAs are conserved targets of microRNAs. Genome Res. 19, 92–105.
Galluzzi, L., Morselli, E., Kepp, O., Vitale, I., Pinti, M., Kroemer, G., 2011. Mitochondrial liaisons of p53. Antioxid. Redox Signal. 15, 1691–1714.
Gandellini, P., Folini, M., Longoni, N., Pennati, M., Binda, M., Colecchia, M., Salvioni, R., Supino, R., Moretti, R., Limonta, P., Valdagni, R., Daidone, M.G., Zaffaroni, N., 2009. miR-205 exerts tumor-suppressive functions in human prostate through down-regulation of protein kinase cepsilon. Cancer Res. 69, 2287–2295.
Garibaldi, F., Falcone, E., Trisciuoglio, D., Colombo, T., Lisek, K., Walerych, D., Del Sal, G., Paci, P., Bossi, G., Piaggio, G., Gurtner, A., 2016. Mutant p53 inhibits miRNA biogenesis by interfering with the microprocessor complex. Oncogene 35, 3760–3770.
Geng, L., Chaudhuri, A., Talmon, G., Wisecarver, J.L., Are, C., Brattain, M., Wang, J., 2014. MicroRNA-192 suppresses liver metastasis of colon cancer. Oncogene 33, 5332–5340.

Georges, S.A., Biery, M.C., Kim, S.Y., Schelter, J.M., Guo, J., Chang, A.N., Jackson, A.L., Carleton, M.O., Linsley, P.S., Cleary, M.A., Chau, B.N., 2008. Coordinated regulation of cell cycle transcripts by p53-inducible microRNAs, miR-192 and miR-215. Cancer Res. 68, 10105–10112.

Green, D.R., Kroemer, G., 2009. Cytoplasmic functions of the tumour suppressor p53. Nature 458, 1127–1130.

Greene, S.B., Gunaratne, P.H., Hammond, S.M., Rosen, J.M., 2010. A putative role for microRNA-205 in mammary epithelial cell progenitors. J. Cell Sci. 123, 606–618.

Gregory, R.I., Yan, K.P., Amuthan, G., Chendrimada, T., Doratotaj, B., Cooch, N., Shiekhattar, R., 2004. The Microprocessor complex mediates the genesis of microRNAs. Nature 432, 235–240.

Gregory, P.A., Bert, A.G., Paterson, E.L., Barry, S.C., Tsykin, A., Farshid, G., Vadas, M.A., Khew-Goodall, Y., Goodall, G.J., 2008. The miR-200 family and miR-205 regulate epithelial to mesenchymal transition by targeting ZEB1 and SIP1. Nat. Cell Biol. 10, 593–601.

Gulyaeva, L.F., Kushlinskiy, N.E., 2016. Regulatory mechanisms of microRNA expression. J. Transl. Med. 14, 143.

Gurtner, A., Falcone, E., Garibaldi, F., Piaggio, G., 2016. Dysregulation of microRNA biogenesis in cancer: the impact of mutant p53 on Drosha complex activity. J. Exp. Clin. Cancer Res. 35, 45.

Ha, M., Kim, V.N., 2014. Regulation of microRNA biogenesis. Nat. Rev. Mol. Cell Biol. 15, 509–524.

Hata, A., Kashima, R., 2016. Dysregulation of microRNA biogenesis machinery in cancer. Crit. Rev. Biochem. Mol. Biol. 51, 121–134.

Haupt, S., di Agostino, S., Mizrahi, I., Alsheich-Bartok, O., Voorhoeve, M., Damalas, A., Blandino, G., Haupt, Y., 2009. Promyelocytic leukemia protein is required for gain of function by mutant p53. Cancer Res. 69, 4818–4826.

He, L., He, X., Lim, L.P., de Stanchina, E., Xuan, Z., Liang, Y., Xue, W., Zender, L., Magnus, J., Ridzon, D., Jackson, A.L., Linsley, P.S., Chen, C., Lowe, S.W., Cleary, M.A., Hannon, G.J., 2007. A microRNA component of the p53 tumour suppressor network. Nature 447, 1130–1134.

Hermeking, H., 2012. MicroRNAs in the p53 network: micromanagement of tumour suppression. Nat. Rev. Cancer 12, 613–626.

Herrera-Merchan, A., Cerrato, C., Luengo, G., Dominguez, O., Piris, M.A., Serrano, M., Gonzalez, S., 2010. miR-33-mediated downregulation of p53 controls hematopoietic stem cell self-renewal. Cell Cycle 9, 3277–3285.

Hill, L., Browne, G., Tulchinsky, E., 2013. ZEB/miR-200 feedback loop: at the crossroads of signal transduction in cancer. Int. J. Cancer 132, 745–754.

Hollstein, M., Sidransky, D., Vogelstein, B., Harris, C.C., 1991. p53 mutations in human cancers. Science 253, 49–53.

Hu, W., Chan, C.S., Wu, R., Zhang, C., Sun, Y., Song, J.S., Tang, L.H., Levine, A.J., Feng, Z., 2010. Negative regulation of tumor suppressor p53 by microRNA miR-504. Mol. Cell 38, 689–699.

Humphries, B., Yang, C., 2015. The microRNA-200 family: small molecules with novel roles in cancer development, progression and therapy. Oncotarget 6, 6472–6498.

Hunten, S., Kaller, M., Drepper, F., Oeljeklaus, S., Bonfert, T., Erhard, F., Dueck, A., Eichner, N., Friedel, C.C., Meister, G., Zimmer, R., Warscheid, B., Hermeking, H., 2015. p53-regulated networks of protein, mRNA, miRNA, and lncRNA expression revealed by integrated pulsed stable isotope labeling with amino acids in cell culture (pSILAC) and next generation sequencing (NGS) analyses. Mol. Cell. Proteomics 14, 2609–2629.

Hwang, C.I., Matoso, A., Corney, D.C., Flesken-Nikitin, A., Korner, S., Wang, W., Boccaccio, C., Thorgeirsson, S.S., Comoglio, P.M., Hermeking, H., Nikitin, A.Y., 2011. Wild-type p53 controls cell motility and invasion by dual regulation of MET expression. Proc. Natl. Acad. Sci. U.S.A. 108, 14240–14245.

Ito, T., Yagi, S., Yamakuchi, M., 2010. MicroRNA-34a regulation of endothelial senescence. Biochem. Biophys. Res. Commun. 398, 735–740.

Jiang, F.Z., He, Y.Y., Wang, H.H., Zhang, H.L., Zhang, J., Yan, X.F., Wang, X.J., Che, Q., Ke, J.Q., Chen, Z., Tong, H., Zhang, Y.L., Wang, F.Y., Li, Y.R., Wan, X.P., 2015. Mutant p53 induces EZH2 expression and promotes epithelial-mesenchymal transition by disrupting p68-Drosha complex assembly and attenuating miR-26a processing. Oncotarget 6, 44660–44674.

Jiang, X., Hu, C., Arnovitz, S., Bugno, J., Yu, M., Zuo, Z., Chen, P., Huang, H., Ulrich, B., Gurbuxani, S., Weng, H., Strong, J., Wang, Y., Li, Y., Salat, J., Li, S., Elkahloun, A.G., Yang, Y., Neilly, M.B., Larson, R.A., Le Beau, M.M., Herold, T., Bohlander, S.K., Liu, P.P., Zhang, J., Li, Z., He, C., Jin, J., Hong, S., Chen, J., 2016. miR-22 has a potent anti-tumour role with therapeutic potential in acute myeloid leukaemia. Nat. Commun. 7, 11452.

Jin, L., Hu, W.L., Jiang, C.C., Wang, J.X., Han, C.C., Chu, P., Zhang, L.J., Thorne, R.F., Wilmott, J., Scolyer, R.A., Hersey, P., Zhang, X.D., Wu, M., 2011. MicroRNA-149*, a p53-responsive microRNA, functions as an oncogenic regulator in human melanoma. Proc. Natl. Acad. Sci. U.S.A. 108, 15840–15845.

Kaller, M., Liffers, S.T., Oeljeklaus, S., Kuhlmann, K., Roh, S., Hoffmann, R., Warscheid, B., Hermeking, H., 2011. Genome-wide characterization of miR-34a induced changes in protein and mRNA expression by a combined pulsed SILAC and microarray analysis. Mol. Cell. Proteomics 10. M111.010462.

Kandoth, C., McLellan, M.D., Vandin, F., Ye, K., Niu, B., Lu, C., Xie, M., Zhang, Q., McMichael, J.F., Wyczalkowski, M.A., Leiserson, M.D., Miller, C.A., Welch, J.S., Walter, M.J., Wendl, M.C., Ley, T.J., Wilson, R.K., Raphael, B.J., Ding, L., 2013. Mutational landscape and significance across 12 major cancer types. Nature 502, 333–339.

Kang, L., Mao, J., Tao, Y., Song, B., Ma, W., Lu, Y., Zhao, L., Li, J., Yang, B., Li, L., 2015. MicroRNA-34a suppresses the breast cancer stem cell-like characteristics by downregulating Notch1 pathway. Cancer Sci. 106, 700–708.

Khella, H.W., Bakhet, M., Allo, G., Jewett, M.A., Girgis, A.H., Latif, A., Girgis, H., Von Both, I., Bjarnason, G.A., Yousef, G.M., 2013. miR-192, miR-194 and miR-215: a convergent microRNA network suppressing tumor progression in renal cell carcinoma. Carcinogenesis 34, 2231–2239.

Korpal, M., Ell, B.J., Buffa, F.M., Ibrahim, T., Blanco, M.A., Celia-Terrassa, T., Mercatali, L., Khan, Z., Goodarzi, H., Hua, Y., Wei, Y., Hu, G., Garcia, B.A., Ragoussis, J., Amadori, D., Harris, A.L., Kang, Y., 2011. Direct targeting of Sec23a by miR-200s influences cancer cell secretome and promotes metastatic colonization. Nat. Med. 17, 1101–1108.

Krell, J., Stebbing, J., Carissimi, C., Dabrowska, A.F., de Giorgio, A., Frampton, A.E., Harding, V., Fulci, V., Macino, G., Colombo, T., Castellano, L., 2016. TP53 regulates miRNA association with AGO2 to remodel the miRNA-mRNA interaction network. Genome Res. 26, 331–341.

Kruiswijk, F., Labuschagne, C.F., Vousden, K.H., 2015. p53 in survival, death and metabolic health: a lifeguard with a licence to kill. Nat. Rev. Mol. Cell Biol. 16, 393–405.

Kumar, M.S., Lu, J., Mercer, K.L., Golub, T.R., Jacks, T., 2007. Impaired microRNA processing enhances cellular transformation and tumorigenesis. Nat. Genet. 39, 673–677.

Kumar, M., Lu, Z., Takwi, A.A., Chen, W., Callander, N.S., Ramos, K.S., Young, K.H., Li, Y., 2011. Negative regulation of the tumor suppressor p53 gene by microRNAs. Oncogene 30, 843–853.
Lane, D.P., 1992. Cancer. p53, guardian of the genome. Nature 358, 15–16.
Lane, D.P., Crawford, L.V., 1979. T antigen is bound to a host protein in SV40-transformed cells. Nature 278, 261–263.
Lankenau, M.A., Patel, R., Liyanarachchi, S., Maharry, S.E., Hoag, K.W., Duggan, M., Walker, C.J., Markowitz, J., Carson 3rd, W.E., Eisfeld, A.K., de la Chapelle, A., 2015. MicroRNA-3151 inactivates TP53 in BRAF-mutated human malignancies. Proc. Natl. Acad. Sci. U.S.A. 112, E6744–E6751.
Le, M.T., Teh, C., Shyh-Chang, N., Xie, H., Zhou, B., Korzh, V., Lodish, H.F., Lim, B., 2009. MicroRNA-125b is a novel negative regulator of p53. Genes Dev. 23, 862–876.
Leveille, N., Elkon, R., Davalos, V., Manoharan, V., Hollingworth, D., Oude Vrielink, J., le Sage, C., Melo, C.A., Horlings, H.M., Wesseling, J., Ule, J., Esteller, M., Ramos, A., Agami, R., 2011. Selective inhibition of microRNA accessibility by RBM38 is required for p53 activity. Nat. Commun. 2, 513.
Levine, A.J., Oren, M., 2009. The first 30 years of p53: growing ever more complex. Nat. Rev. Cancer 9, 749–758.
Lezina, L., Purmessur, N., Antonov, A.V., Ivanova, T., Karpova, E., Krishan, K., Ivan, M., Aksenova, V., Tentler, D., Garabadgiu, A.V., Melino, G., Barlev, N.A., 2013. miR-16 and miR-26a target checkpoint kinases Wee1 and Chk1 in response to p53 activation by genotoxic stress. Cell Death Dis. 4. e953.
Li, X., Zhang, Y., Shi, Y., Dong, G., Liang, J., Han, Y., Wang, X., Zhao, Q., Ding, J., Wu, K., Fan, D., 2011. MicroRNA-107, an oncogene microRNA that regulates tumour invasion and metastasis by targeting DICER1 in gastric cancer. J. Cell. Mol. Med. 15, 1887–1895.
Li, C.L., Yeh, K.H., Liu, W.H., Chen, C.L., Chen, D.S., Chen, P.J., Yeh, S.H., 2015a. Elevated p53 promotes the processing of miR-18a to decrease estrogen receptor-alpha in female hepatocellular carcinoma. Int. J. Cancer 136, 761–770.
Li, J., Aung, L.H., Long, B., Qin, D., An, S., Li, P., 2015b. miR-23a binds to p53 and enhances its association with miR-128 promoter. Sci. Rep. 5, 16422.
Lian, J., Jing, Y., Dong, Q., Huan, L., Chen, D., Bao, C., Wang, Q., Zhao, F., Li, J., Yao, M., Qin, L., Liang, L., He, X., 2016. miR-192, a prognostic indicator, targets the SLC39A6/SNAIL pathway to reduce tumor metastasis in human hepatocellular carcinoma. Oncotarget 7, 2672–2683.
Liang, M., Yao, G., Yin, M., Lu, M., Tian, H., Liu, L., Lian, J., Huang, X., Sun, F., 2013. Transcriptional cooperation between p53 and NF-kappaB p65 regulates microRNA-224 transcription in mouse ovarian granulosa cells. Mol. Cell. Endocrinol. 370, 119–129.
Linares, L.K., Hengstermann, A., Ciechanover, A., Muller, S., Scheffner, M., 2003. HdmX stimulates Hdm2-mediated ubiquitination and degradation of p53. Proc. Natl. Acad. Sci. U.S.A. 100, 12009–12014.
Linzer, D.I., Levine, A.J., 1979. Characterization of a 54K dalton cellular SV40 tumor antigen present in SV40-transformed cells and uninfected embryonal carcinoma cells. Cell 17, 43–52.
Liu, C., Kelnar, K., Liu, B., Chen, X., Calhoun-Davis, T., Li, H., Patrawala, L., Yan, H., Jeter, C., Honorio, S., Wiggins, J.F., Bader, A.G., Fagin, R., Brown, D., Tang, D.G., 2011. The microRNA miR-34a inhibits prostate cancer stem cells and metastasis by directly repressing CD44. Nat. Med. 17, 211–215.
Liu, Y., Xing, R., Zhang, X., Dong, W., Zhang, J., Yan, Z., Li, W., Cui, J., Lu, Y., 2013. miR-375 targets the p53 gene to regulate cellular response to ionizing radiation and etoposide in gastric cancer cells. DNA Repair (Amst) 12, 741–750.

Liu, J., Zhang, C., Zhao, Y., Feng, Z., 2017. MicroRNA control of p53. J. Cell Biochem. 118, 7–14.
Lu, J., Getz, G., Miska, E.A., Alvarez-Saavedra, E., Lamb, J., Peck, D., Sweet-Cordero, A., Ebert, B.L., Mak, R.H., Ferrando, A.A., Downing, J.R., Jacks, T., Horvitz, H.R., Golub, T.R., 2005. MicroRNA expression profiles classify human cancers. Nature 435, 834–838.
Majid, S., Saini, S., Dar, A.A., Hirata, H., Shahryari, V., Tanaka, Y., Yamamura, S., Ueno, K., Zaman, M.S., Singh, K., Chang, I., Deng, G., Dahiya, R., 2011. MicroRNA-205 inhibits Src-mediated oncogenic pathways in renal cancer. Cancer Res. 71, 2611–2621.
Mandke, P., Wyatt, N., Fraser, J., Bates, B., Berberich, S.J., Markey, M.P., 2012. MicroRNA-34a modulates MDM4 expression via a target site in the open reading frame. PLoS One 7. e42034.
Mantovani, F., Walerych, D., Sal, G.D., 2017. Targeting mutant p53 in cancer: a long road to precision therapy. FEBS J. 284, 837–850.
Martello, G., Rosato, A., Ferrari, F., Manfrin, A., Cordenonsi, M., Dupont, S., Enzo, E., Guzzardo, V., Rondina, M., Spruce, T., Parenti, A.R., Daidone, M.G., Bicciato, S., Piccolo, S., 2010. A microRNA targeting dicer for metastasis control. Cell 141, 1195–1207.
Masciarelli, S., Fontemaggi, G., Di Agostino, S., Donzelli, S., Carcarino, E., Strano, S., Blandino, G., 2014. Gain-of-function mutant p53 downregulates miR-223 contributing to chemoresistance of cultured tumor cells. Oncogene 33, 1601–1608.
Menendez, D., Nguyen, T.A., Freudenberg, J.M., Mathew, V.J., Anderson, C.W., Jothi, R., Resnick, M.A., 2013. Diverse stresses dramatically alter genome-wide p53 binding and transactivation landscape in human cancer cells. Nucleic Acids Res. 41, 7286–7301.
Migliore, C., Petrelli, A., Ghiso, E., Corso, S., Capparuccia, L., Eramo, A., Comoglio, P.M., Giordano, S., 2008. MicroRNAs impair MET-mediated invasive growth. Cancer Res. 68, 10128–10136.
Mori, F., Ferraiuolo, M., Santoro, R., Sacconi, A., Goeman, F., Pallocca, M., Pulito, C., Korita, E., Fanciulli, M., Muti, P., Blandino, G., Strano, S., 2016. Multitargeting activity of miR-24 inhibits long-term melatonin anticancer effects. Oncotarget 7, 20532–20548.
Mudhasani, R., Zhu, Z., Hutvagner, G., Eischen, C.M., Lyle, S., Hall, L.L., Lawrence, J.B., Imbalzano, A.N., Jones, S.N., 2008. Loss of miRNA biogenesis induces p19Arf-p53 signaling and senescence in primary cells. J. Cell Biol. 181, 1055–1063.
Muller, P.A., Vousden, K.H., 2013. p53 mutations in cancer. Nat. Cell Biol. 15, 2–8.
Muller, P.A., Trinidad, A.G., Caswell, P.T., Norman, J.C., Vousden, K.H., 2014. Mutant p53 regulates Dicer through p63-dependent and -independent mechanisms to promote an invasive phenotype. J. Biol. Chem. 289, 122–132.
Nakazawa, K., Dashzeveg, N., Yoshida, K., 2014. Tumor suppressor p53 induces miR-1915 processing to inhibit Bcl-2 in the apoptotic response to DNA damage. FEBS J. 281, 2937–2944.
Navarro, F., Lieberman, J., 2015. miR-34 and p53: new insights into a complex functional relationship. PLoS One 10. e0132767.
Neilsen, P.M., Noll, J.E., Mattiske, S., Bracken, C.P., Gregory, P.A., Schulz, R.B., Lim, S.P., Kumar, R., Suetani, R.J., Goodall, G.J., Callen, D.F., 2013. Mutant p53 drives invasion in breast tumors through up-regulation of miR-155. Oncogene 32, 2992–3000.
Neveu, P., Kye, M.J., Qi, S., Buchholz, D.E., Clegg, D.O., Sahin, M., Park, I.H., Kim, K.S., Daley, G.Q., Kornblum, H.I., Shraiman, B.I., Kosik, K.S., 2010. MicroRNA profiling reveals two distinct p53-related human pluripotent stem cell states. Cell Stem Cell 7, 671–681.

Nikulenkov, F., Spinnler, C., Li, H., Tonelli, C., Shi, Y., Turunen, M., Kivioja, T., Ignatiev, I., Kel, A., Taipale, J., Selivanova, G., 2012. Insights into p53 transcriptional function via genome-wide chromatin occupancy and gene expression analysis. Cell Death Differ. 19, 1992–2002.

Nishida, N., Yokobori, T., Mimori, K., Sudo, T., Tanaka, F., Shibata, K., Ishii, H., Doki, Y., Kuwano, H., Mori, M., 2011. MicroRNA miR-125b is a prognostic marker in human colorectal cancer. Int. J. Oncol. 38, 1437–1443.

Okada, N., Lin, C.P., Ribeiro, M.C., Biton, A., Lai, G., He, X., Bu, P., Vogel, H., Jablons, D.M., Keller, A.C., Wilkinson, J.E., He, B., Speed, T.P., He, L., 2014. A positive feedback between p53 and miR-34 miRNAs mediates tumor suppression. Genes Dev. 28, 438–450.

Oren, M., Rotter, V., 2010. Mutant p53 gain-of-function in cancer. Cold Spring Harb. Perspect. Biol. 2. a001107.

Ozsolak, F., Poling, L.L., Wang, Z., Liu, H., Liu, X.S., Roeder, R.G., Zhang, X., Song, J.S., Fisher, D.E., 2008. Chromatin structure analyses identify miRNA promoters. Genes Dev. 22, 3172–3183.

Pandey, A.K., Zhang, Y., Zhang, S., Li, Y., Tucker-Kellogg, G., Yang, H., Jha, S., 2015. TIP60-miR-22 axis as a prognostic marker of breast cancer progression. Oncotarget 6, 41290–41306.

Park, S.M., Gaur, A.B., Lengyel, E., Peter, M.E., 2008. The miR-200 family determines the epithelial phenotype of cancer cells by targeting the E-cadherin repressors ZEB1 and ZEB2. Genes Dev. 22, 894–907.

Park, E.Y., Chang, E., Lee, E.J., Lee, H.W., Kang, H.G., Chun, K.H., Woo, Y.M., Kong, H.K., Ko, J.Y., Suzuki, H., Song, E., Park, J.H., 2014. Targeting of miR34a-NOTCH1 axis reduced breast cancer stemness and chemoresistance. Cancer Res. 74, 7573–7582.

Peng, X., Guo, W., Liu, T., Wang, X., Tu, X., Xiong, D., Chen, S., Lai, Y., Du, H., Chen, G., Liu, G., Tang, Y., Huang, S., Zou, X., 2011. Identification of miRs-143 and -145 that is associated with bone metastasis of prostate cancer and involved in the regulation of EMT. PLoS One 6. e20341.

Pichiorri, F., Suh, S.S., Rocci, A., De Luca, L., Taccioli, C., Santhanam, R., Zhou, W., Benson Jr., D.M., Hofmainster, C., Alder, H., Garofalo, M., Di Leva, G., Volinia, S., Lin, H.J., Perrotti, D., Kuehl, M., Aqeilan, R.I., Palumbo, A., Croce, C.M., 2010. Downregulation of p53-inducible microRNAs 192, 194, and 215 impairs the p53/MDM2 autoregulatory loop in multiple myeloma development. Cancer Cell 18, 367–381.

Piovan, C., Palmieri, D., Di Leva, G., Braccioli, L., Casalini, P., Nuovo, G., Tortoreto, M., Sasso, M., Plantamura, I., Triulzi, T., Taccioli, C., Tagliabue, E., Iorio, M.V., Croce, C.M., 2012. Oncosuppressive role of p53-induced miR-205 in triple negative breast cancer. Mol. Oncol. 6, 458–472.

Raver-Shapira, N., Marciano, E., Meiri, E., Spector, Y., Rosenfeld, N., Moskovits, N., Bentwich, Z., Oren, M., 2007. Transcriptional activation of miR-34a contributes to p53-mediated apoptosis. Mol. Cell 26, 731–743.

Ren, Z.J., Nong, X.Y., Lv, Y.R., Sun, H.H., An, P.P., Wang, F., Li, X., Liu, M., Tang, H., 2014. Mir-509-5p joins the Mdm2/p53 feedback loop and regulates cancer cell growth. Cell Death Dis. 5. e1387.

Rokavec, M., Oner, M.G., Li, H., Jackstadt, R., Jiang, L., Lodygin, D., Kaller, M., Horst, D., Ziegler, P.K., Schwitalla, S., Slotta-Huspenina, J., Bader, F.G., Greten, F.R., Hermeking, H., 2014. IL-6R/STAT3/miR-34a feedback loop promotes EMT-mediated colorectal cancer invasion and metastasis. J. Clin. Invest. 124, 1853–1867.

Sachdeva, M., Zhu, S., Wu, F., Wu, H., Walia, V., Kumar, S., Elble, R., Watabe, K., Mo, Y.Y., 2009. p53 represses c-Myc through induction of the tumor suppressor miR-145. Proc. Natl. Acad. Sci. U.S.A. 106, 3207–3212.

Saleh, A.D., Savage, J.E., Cao, L., Soule, B.P., Ly, D., DeGraff, W., Harris, C.C., Mitchell, J.B., Simone, N.L., 2011. Cellular stress induced alterations in microRNA let-7a and let-7b expression are dependent on p53. PLoS One 6. e24429.

Salzman, D.W., Nakamura, K., Nallur, S., Dookwah, M.T., Metheetrairut, C., Slack, F.J., Weidhaas, J.B., 2016. miR-34 activity is modulated through 5′-end phosphorylation in response to DNA damage. Nat. Commun. 7, 10954.

Sampath, J., Sun, D., Kidd, V.J., Grenet, J., Gandhi, A., Shapiro, L.H., Wang, Q., Zambetti, G.P., Schuetz, J.D., 2001. Mutant p53 cooperates with ETS and selectively up-regulates human MDR1 not MRP1. J. Biol. Chem. 276, 39359–39367.

Schlereth, K., Heyl, C., Krampitz, A.M., Mernberger, M., Finkernagel, F., Scharfe, M., Jarek, M., Leich, E., Rosenwald, A., Stiewe, T., 2013. Characterization of the p53 cistrome–DNA binding cooperativity dissects p53's tumor suppressor functions. PLoS Genet. 9. e1003726.

Shadfan, M., Lopez-Pajares, V., Yuan, Z.M., 2012. MDM2 and MDMX: alone and together in regulation of p53. Transl. Cancer Res. 1, 88–89.

Shao, Y., Qu, Y., Dang, S., Yao, B., Ji, M., 2013. MiR-145 inhibits oral squamous cell carcinoma (OSCC) cell growth by targeting c-Myc and Cdk6. Cancer Cell Int. 13, 51.

Shen, J., Niu, W., Zhou, M., Zhang, H., Ma, J., Wang, L., 2014. MicroRNA-410 suppresses migration and invasion by targeting MDM2 in gastric cancer. PLoS One 9. e104510.

Shi, L., Jackstadt, R., Siemens, H., Li, H., Kirchner, T., Hermeking, H., 2014. p53-induced miR-15a/16-1 and AP4 form a double-negative feedback loop to regulate epithelial-mesenchymal transition and metastasis in colorectal cancer. Cancer Res. 74, 532–542.

Shu, L., Yan, W., Chen, X., 2006. RNPC1, an RNA-binding protein and a target of the p53 family, is required for maintaining the stability of the basal and stress-induced p21 transcript. Genes Dev. 20, 2961–2972.

Siemens, H., Jackstadt, R., Hunten, S., Kaller, M., Menssen, A., Gotz, U., Hermeking, H., 2011. miR-34 and SNAIL form a double-negative feedback loop to regulate epithelial-mesenchymal transitions. Cell Cycle 10, 4256–4271.

Song, S.J., Ito, K., Ala, U., Kats, L., Webster, K., Sun, S.M., Jongen-Lavrencic, M., Manova-Todorova, K., Teruya-Feldstein, J., Avigan, D.E., Delwel, R., Pandolfi, P.P., 2013a. The oncogenic microRNA miR-22 targets the TET2 tumor suppressor to promote hematopoietic stem cell self-renewal and transformation. Cell Stem Cell 13, 87–101.

Song, S.J., Poliseno, L., Song, M.S., Ala, U., Webster, K., Ng, C., Beringer, G., Brikbak, N.J., Yuan, X., Cantley, L.C., Richardson, A.L., Pandolfi, P.P., 2013b. MicroRNA-antagonism regulates breast cancer stemness and metastasis via TET-family-dependent chromatin remodeling. Cell 154, 311–324.

Sorrentino, G., Ruggeri, N., Specchia, V., Cordenonsi, M., Mano, M., Dupont, S., Manfrin, A., Ingallina, E., Sommaggio, R., Piazza, S., Rosato, A., Piccolo, S., Del Sal, G., 2014. Metabolic control of YAP and TAZ by the mevalonate pathway. Nat. Cell Biol. 16, 357–366.

Spike, B.T., Wahl, G.M., 2011. p53, stem cells, and reprogramming: tumor suppression beyond guarding the genome. Genes Cancer 2, 404–419.

Stambolsky, P., Tabach, Y., Fontemaggi, G., Weisz, L., Maor-Aloni, R., Siegfried, Z., Shiff, I., Kogan, I., Shay, M., Kalo, E., Blandino, G., Simon, I., Oren, M., Rotter, V., 2010. Modulation of the vitamin D3 response by cancer-associated mutant p53. Cancer Cell 17, 273–285.

Strano, S., Dell'Orso, S., Di Agostino, S., Fontemaggi, G., Sacchi, A., Blandino, G., 2007. Mutant p53: an oncogenic transcription factor. Oncogene 26, 2212–2219.

Subramanian, M., Francis, P., Bilke, S., Li, X.L., Hara, T., Lu, X., Jones, M.F., Walker, R.L., Zhu, Y., Pineda, M., Lee, C., Varanasi, L., Yang, Y., Martinez, L.A., Luo, J., Ambs, S., Sharma, S., Wakefield, L.M., Meltzer, P.S., Lal, A., 2015. A mutant p53/let-7i-axis-regulated gene network drives cell migration, invasion and metastasis. Oncogene 34, 1094–1104.

Suh, S.O., Chen, Y., Zaman, M.S., Hirata, H., Yamamura, S., Shahryari, V., Liu, J., Tabatabai, Z.L., Kakar, S., Deng, G., Tanaka, Y., Dahiya, R., 2011. MicroRNA-145 is regulated by DNA methylation and p53 gene mutation in prostate cancer. Carcinogenesis 32, 772–778.

Sun, Y., Pan, J., Mao, S., Jin, J., 2014. IL-17/miR-192/IL-17Rs regulatory feedback loop facilitates multiple myeloma progression. PLoS One 9. e114647.

Suzuki, H.I., Yamagata, K., Sugimoto, K., Iwamoto, T., Kato, S., Miyazono, K., 2009. Modulation of microRNA processing by p53. Nature 460, 529–533.

Swarbrick, A., Woods, S.L., Shaw, A., Balakrishnan, A., Phua, Y., Nguyen, A., Chanthery, Y., Lim, L., Ashton, L.J., Judson, R.L., Huskey, N., Blelloch, R., Haber, M., Norris, M.D., Lengyel, P., Hackett, C.S., Preiss, T., Chetcuti, A., Sullivan, C.S., Marcusson, E.G., Weiss, W., L'Etoile, N., Goga, A., 2010. miR-380-5p represses p53 to control cellular survival and is associated with poor outcome in MYCN-amplified neuroblastoma. Nat. Med. 16, 1134–1140.

Takahashi, Y., Forrest, A.R., Maeno, E., Hashimoto, T., Daub, C.O., Yasuda, J., 2009. MiR-107 and MiR-185 can induce cell cycle arrest in human non small cell lung cancer cell lines. PLoS One 4. e6677.

Tarasov, V., Jung, P., Verdoodt, B., Lodygin, D., Epanchintsev, A., Menssen, A., Meister, G., Hermeking, H., 2007. Differential regulation of microRNAs by p53 revealed by massively parallel sequencing: miR-34a is a p53 target that induces apoptosis and G1-arrest. Cell Cycle 6, 1586–1593.

Tay, Y., Tan, S.M., Karreth, F.A., Lieberman, J., Pandolfi, P.P., 2014. Characterization of dual PTEN and p53-targeting microRNAs identifies microRNA-638/Dnm2 as a two-hit oncogenic locus. Cell Rep. 8, 714–722.

Tazawa, H., Tsuchiya, N., Izumiya, M., Nakagama, H., 2007. Tumor-suppressive miR-34a induces senescence-like growth arrest through modulation of the E2F pathway in human colon cancer cells. Proc. Natl. Acad. Sci. U.S.A. 104, 15472–15477.

Tian, S., Huang, S., Wu, S., Guo, W., Li, J., He, X., 2010. MicroRNA-1285 inhibits the expression of p53 by directly targeting its 3′ untranslated region. Biochem. Biophys. Res. Commun. 396, 435–439.

Trabucchi, M., Briata, P., Garcia-Mayoral, M., Haase, A.D., Filipowicz, W., Ramos, A., Gherzi, R., Rosenfeld, M.G., 2009. The RNA-binding protein KSRP promotes the biogenesis of a subset of microRNAs. Nature 459, 1010–1014.

Tsuchiya, N., Izumiya, M., Ogata-Kawata, H., Okamoto, K., Fujiwara, Y., Nakai, M., Okabe, A., Schetter, A.J., Bowman, E.D., Midorikawa, Y., Sugiyama, Y., Aburatani, H., Harris, C.C., Nakagama, H., 2011. Tumor suppressor miR-22 determines p53-dependent cellular fate through post-transcriptional regulation of p21. Cancer Res. 71, 4628–4639.

Tucci, P., Agostini, M., Grespi, F., Markert, E.K., Terrinoni, A., Vousden, K.H., Muller, P.A., Dotsch, V., Kehrloesser, S., Sayan, B.S., Giaccone, G., Lowe, S.W., Takahashi, N., Vandenabeele, P., Knight, R.A., Levine, A.J., Melino, G., 2012. Loss of p63 and its microRNA-205 target results in enhanced cell migration and metastasis in prostate cancer. Proc. Natl. Acad. Sci. U.S.A. 109, 15312–15317.

Vijayakumaran, R., Tan, K.H., Miranda, P.J., Haupt, S., Haupt, Y., 2015. Regulation of mutant p53 protein expression. Front. Oncol. 5, 284.

Vousden, K.H., Prives, C., 2009. Blinded by the light: the growing complexity of p53. Cell 137, 413–431.

Walerych, D., Lisek, K., Sommaggio, R., Piazza, S., Ciani, Y., Dalla, E., Rajkowska, K., Gaweda-Walerych, K., Ingallina, E., Tonelli, C., Morelli, M.J., Amato, A., Eterno, V., Zambelli, A., Rosato, A., Amati, B., Wisniewski, J.R., Del Sal, G., 2016. Proteasome machinery is instrumental in a common gain-of-function program of the p53 missense mutants in cancer. Nat. Cell Biol. 18, 897–909.

Wang, D.T., Ma, Z.L., Li, Y.L., Wang, Y.Q., Zhao, B.T., Wei, J.L., Qi, X., Zhao, X.T., Jin, Y.X., 2013a. miR-150, p53 protein and relevant miRNAs consist of a regulatory network in NSCLC tumorigenesis. Oncol. Rep. 30, 492–498.

Wang, W., Cheng, B., Miao, L., Mei, Y., Wu, M., 2013b. Mutant p53-R273H gains new function in sustained activation of EGFR signaling via suppressing miR-27a expression. Cell Death Dis. 4. e574.

Wang, Y., Ren, J., Gao, Y., Ma, J.Z., Toh, H.C., Chow, P., Chung, A.Y., Ooi, L.L., Lee, C.G., 2013c. MicroRNA-224 targets SMAD family member 4 to promote cell proliferation and negatively influence patient survival. PLoS One 8. e68744.

Wang, X., Li, J., Dong, K., Lin, F., Long, M., Ouyang, Y., Wei, J., Chen, X., Weng, Y., He, T., Zhang, H., 2015. Tumor suppressor miR-34a targets PD-L1 and functions as a potential immunotherapeutic target in acute myeloid leukemia. Cell. Signal. 27, 443–452.

Wu, H., Zhu, S., Mo, Y.Y., 2009. Suppression of cell growth and invasion by miR-205 in breast cancer. Cell Res. 19, 439–448.

Wu, S.Y., Rupaimoole, R., Shen, F., Pradeep, S., Pecot, C.V., Ivan, C., Nagaraja, A.S., Gharpure, K.M., Pham, E., Hatakeyama, H., McGuire, M.H., Haemmerle, M., Vidal-Anaya, V., Olsen, C., Rodriguez-Aguayo, C., Filant, J., Ehsanipour, E.A., Herbrich, S.M., Maiti, S.N., Huang, L., Kim, J.H., Zhang, X., Han, H.D., Armaiz-Pena, G.N., Seviour, E.G., Tucker, S., Zhang, M., Yang, D., Cooper, L.J., Ali-Fehmi, R., Bar-Eli, M., Lee, J.S., Ram, P.T., Baggerly, K.A., Lopez-Berestein, G., Hung, M.C., Sood, A.K., 2016. A miR-192-EGR1-HOXB9 regulatory network controls the angiogenic switch in cancer. Nat. Commun. 7, 11169.

Xiao, J., Lin, H., Luo, X., Wang, Z., 2011a. miR-605 joins p53 network to form a p53:miR-605:Mdm2 positive feedback loop in response to stress. EMBO J. 30, 5021.

Xiao, J., Lin, H., Luo, X., Wang, Z., 2011b. miR-605 joins p53 network to form a p53: miR-605:Mdm2 positive feedback loop in response to stress. EMBO J. 30, 524–532.

Xiao, X., Huang, X., Ye, F., Chen, B., Song, C., Wen, J., Zhang, Z., Zheng, G., Tang, H., Xie, X., 2016. The miR-34a-LDHA axis regulates glucose metabolism and tumor growth in breast cancer. Sci. Rep. 6, 21735.

Xin, M., Qiao, Z., Li, J., Liu, J., Song, S., Zhao, X., Miao, P., Tang, T., Wang, L., Liu, W., Yang, X., Dai, K., Huang, G., 2016. miR-22 inhibits tumor growth and metastasis by targeting ATP citrate lyase: evidence in osteosarcoma, prostate cancer, cervical cancer and lung cancer. Oncotarget 7, 44252–44265.

Xu, N., Papagiannakopoulos, T., Pan, G., Thomson, J.A., Kosik, K.S., 2009. MicroRNA-145 regulates OCT4, SOX2, and KLF4 and represses pluripotency in human embryonic stem cells. Cell 137, 647–658.

Xu, Q., Liu, L.Z., Qian, X., Chen, Q., Jiang, Y., Li, D., Lai, L., Jiang, B.H., 2012. MiR-145 directly targets p70S6K1 in cancer cells to inhibit tumor growth and angiogenesis. Nucleic Acids Res. 40, 761–774.

Xu, K., Liang, X., Cui, D., Wu, Y., Shi, W., Liu, J., 2013. miR-1915 inhibits Bcl-2 to modulate multidrug resistance by increasing drug-sensitivity in human colorectal carcinoma cells. Mol. Carcinog. 52, 70–78.

Yamakuchi, M., Ferlito, M., Lowenstein, C.J., 2008. miR-34a repression of SIRT1 regulates apoptosis. Proc. Natl. Acad. Sci. U.S.A. 105, 13421–13426.

Yamakuchi, M., Lotterman, C.D., Bao, C., Hruban, R.H., Karim, B., Mendell, J.T., Huso, D., Lowenstein, C.J., 2010. P53-induced microRNA-107 inhibits HIF-1 and tumor angiogenesis. Proc. Natl. Acad. Sci. U.S.A. 107, 6334–6339.

Yan, H.L., Xue, G., Mei, Q., Wang, Y.Z., Ding, F.X., Liu, M.F., Lu, M.H., Tang, Y., Yu, H.Y., Sun, S.H., 2009. Repression of the miR-17-92 cluster by p53 has an important function in hypoxia-induced apoptosis. EMBO J. 28, 2719–2732.

Yin, H., Sun, Y., Wang, X., Park, J., Zhang, Y., Li, M., Yin, J., Liu, Q., Wei, M., 2015. Progress on the relationship between miR-125 family and tumorigenesis. Exp. Cell Res. 339, 252–260.

Zhang, S., Zhang, C., Li, Y., Wang, P., Yue, Z., Xie, S., 2011a. miR-98 regulates cisplatin-induced A549 cell death by inhibiting TP53 pathway. Biomed. Pharmacother. 65, 436–442.

Zhang, X., Yan, Z., Zhang, J., Gong, L., Li, W., Cui, J., Liu, Y., Gao, Z., Li, J., Shen, L., Lu, Y., 2011b. Combination of hsa-miR-375 and hsa-miR-142-5p as a predictor for recurrence risk in gastric cancer patients following surgical resection. Ann. Oncol. 22, 2257–2266.

Zhang, Y., Liao, J.M., Zeng, S.X., Lu, H., 2011c. p53 downregulates down syndrome-associated DYRK1A through miR-1246. EMBO Rep. 12, 811–817.

Zhang, N., Wei, X., Xu, L., 2013. miR-150 promotes the proliferation of lung cancer cells by targeting P53. FEBS Lett. 587, 2346–2351.

Zhang, Z., Zhang, L., Yin, Z.Y., Fan, X.L., Hu, B., Wang, L.Q., Zhang, D., 2014. miR-107 regulates cisplatin chemosensitivity of A549 non small cell lung cancer cell line by targeting cyclin dependent kinase 8. Int. J. Clin. Exp. Pathol. 7, 7236–7241.

Zhang, C., Liu, J., Tan, C., Yue, X., Zhao, Y., Peng, J., Wang, X., Laddha, S.V., Chan, C.S., Zheng, S., Hu, W., Feng, Z., 2016a. microRNA-1827 represses MDM2 to positively regulate tumor suppressor p53 and suppress tumorigenesis. Oncotarget 7, 8783–8796.

Zhang, Y., Hu, Y., Fang, J.Y., Xu, J., 2016b. Gain-of-function miRNA signature by mutant p53 associates with poor cancer outcome. Oncotarget 7, 11056–11066.

Zou, C., Xu, Q., Mao, F., Li, D., Bian, C., Liu, L.Z., Jiang, Y., Chen, X., Qi, Y., Zhang, X., Wang, X., Sun, Q., Kung, H.F., Lin, M.C., Dress, A., Wardle, F., Jiang, B.H., Lai, L., 2012. MiR-145 inhibits tumor angiogenesis and growth by N-RAS and VEGF. Cell Cycle 11, 2137–2145.

CHAPTER THREE

Impact of MicroRNAs in the Cellular Response to Hypoxia

Thomas Bertero*, Roger Rezzonico†, Nicolas Pottier‡, Bernard Mari†,1
*Université Côte d'Azur, CNRS, INSERM, IRCAN, FHU-OncoAge, Nice, France
†Université Côte d'Azur, CNRS, IPMC, FHU-OncoAge, Sophia-Antipolis, France
‡Faculté de Médecine de Lille, Pole Recherche, Lille, France
[1]Corresponding author: e-mail address: mari@unice.fr

Contents

Abstract

In mammalian cells, hypoxia, or inadequate oxygen availability, regulates the expression of a specific set of MicroRNAs (MiRNAs), termed "hypoxamiRs." Over the past 10 years, the appreciation of the importance of hypoxamiRs in regulating the cellular adaptation to hypoxia has grown dramatically. At the cellular level, each hypoxamiR, including the master hypoxamiR MiR-210, can simultaneously regulate expression of multiple target genes in order to fine-tune the adaptive response of cells to hypoxia. This review addresses the complex molecular regulation of MiRNAs in both physiological and

International Review of Cell and Molecular Biology, Volume 333
ISSN 1937-6448
http://dx.doi.org/10.1016/bs.ircmb.2017.03.006

pathological conditions of low oxygen adaptation and the multiple functions of hypoxamiRs in various hypoxia-associated biological processes, including apoptosis, survival, proliferation, angiogenesis, inflammation, and metabolism. From a clinical perspective, we also discuss the potential use of hypoxamiRs as new biomarkers and/or therapeutic targets in cancer and aging-associated diseases including cardiovascular and fibroproliferative disorders.

1. INTRODUCTION

Inadequate oxygen availability or hypoxia induces a complex and still incompletely understood set of adaptations that influence cellular survival and function (Semenza, 2012, 2014). Hypoxia results from physiological or pathological conditions such as high altitude, anemia, or abnormal and insufficient vasculature (Semenza, 2011). In the short term, adaptation to hypoxia is important to maintain homeostasis (Michiels, 2004), whereas chronic exposure to insufficient oxygen levels can lead to cellular dysfunction in particular for typically aerobic mammalian cells. Indeed, hypoxia is causatively linked to various pathological conditions such as tissue ischemia, inflammation, and solid tumorigenesis (Harris, 2002).

The molecular mechanisms underlying response to hypoxia are extremely complex, involving specific transcriptional and posttranscriptional control of gene expression. In recent years, significant progress has been made in our understanding of the molecular changes that dictate hypoxic adaptation and the recognition of the hypoxia-inducible factors (HIFs) as central elements of this process. Indeed, although additional non-HIF-dependent pathways have been described, the primary response to hypoxia in metazoan cells relies on HIF activity. The transcription factor (TF) HIF, the main driver of the cellular response to hypoxia (Iyer et al., 1998; Wang and Semenza, 1995), is a heterodimeric TF consisting of two basic helix-loop–helix protein subunits, a hypoxia-regulated and unstable α subunit (HIF-1α or HIF-2α) associated with a constitutive and stable β-subunit (HIF-1β or ANRT). Under normal oxygen tension (i.e., normoxia), HIF-α subunits possess a very short half life and are rapidly degraded by the ubiquitin–proteasome pathway (Salceda and Caro, 1997). This mechanism is mediated by a set of oxygen-sensing enzymes, the prolylhydroxylase domain-containing protein (PHD1–4), which by hydroxylating two specific proline amino acid residues (Ivan et al., 2001; Jaakkola et al., 2001), leads to recognition by the von Hippel–Lindau tumor suppressor (pVHL) protein and recruitment of the E3 ubiquitin ligase complex. Under hypoxic conditions, HIF-α protein remains unhydroxylated and, being stabilized, readily forms a heterodimeric complex with HIF-β. This heterodimer recognizes specific binding sites in the promoter region of target genes termed hypoxia response elements (HREs) to induce a transcriptional program ensuring optimal functional, metabolic, and vascular adaptation to hypoxia (Semenza, 2000). Currently, three HIF-α proteins have been described in higher metazoans: HIF-1α and HIF-2α, which regulate the expression of overlapping and unique transcriptional target genes, and HIF-3α whose alternative splice variants act as dominant inhibitors of HIF-dependent gene transcription (Maynard et al., 2005).

In addition, several complex posttranscriptional mechanisms also regulate mRNA turnover and translational control during hypoxia. Recent reports have in particular deciphered the critical role of MicroRNAs (MiRNAs) in the regulation of the molecular and cellular changes occurring during hypoxia (Cottrill et al., 2014; Gee et al., 2014; Greco et al., 2014).

MiRNAs are small, noncoding RNA molecules approximately 22 nucleotides in length that are largely conserved through evolution (Bartel, 2004). Since the discovery of the first MicroRNA in *Caenorhabditis elegans* in 1993 (Lee et al., 1993; Wightman et al., 1993), more than 2500 mature MiRNAs have been annotated in humans (miRBase v. 21, September 2016). Transcription of MiRNA genes leads to the formation of hairpin-looped primary MiRNA molecules (pri-miRNA), which are then processed in the nucleus into smaller precursors, named pre-miRNA (Fig. 1). Following export to the cytoplasm, the pre-miRNA hairpin loop is processed by the RNA endonuclease Dicer, yielding to a double-stranded MiRNA duplex (Gregory et al., 2004), which then congregates with the RNA-induced silencing complex (RISC). The duplex is loaded onto the ARGONAUTE (AGO1–4) proteins, and one strand of the duplex remains as mature MiRNA, whereas the other strand is discarded from AGO. MiRNAs originating from the 5′ and 3′ strands of pre-miRNA are referred to as the 5p and 3p MiRNAs, respectively. Although this aforementioned canonical pathway accounts for the production of most MiRNAs (Ha and Kim, 2014), several alternative pathways bypassing a part of the biogenesis steps mentioned earlier have been described (Kim et al., 2016b; Yang and Lai, 2011). Moreover, direct sequencing of MiRNAs has also revealed a significant level of diversity in MiRNA sequences including end sequence variations resulting in new sequences different from the annotated MiRNA, named "isomiRs" (Morin et al., 2008). These variations are mainly generated not only by a

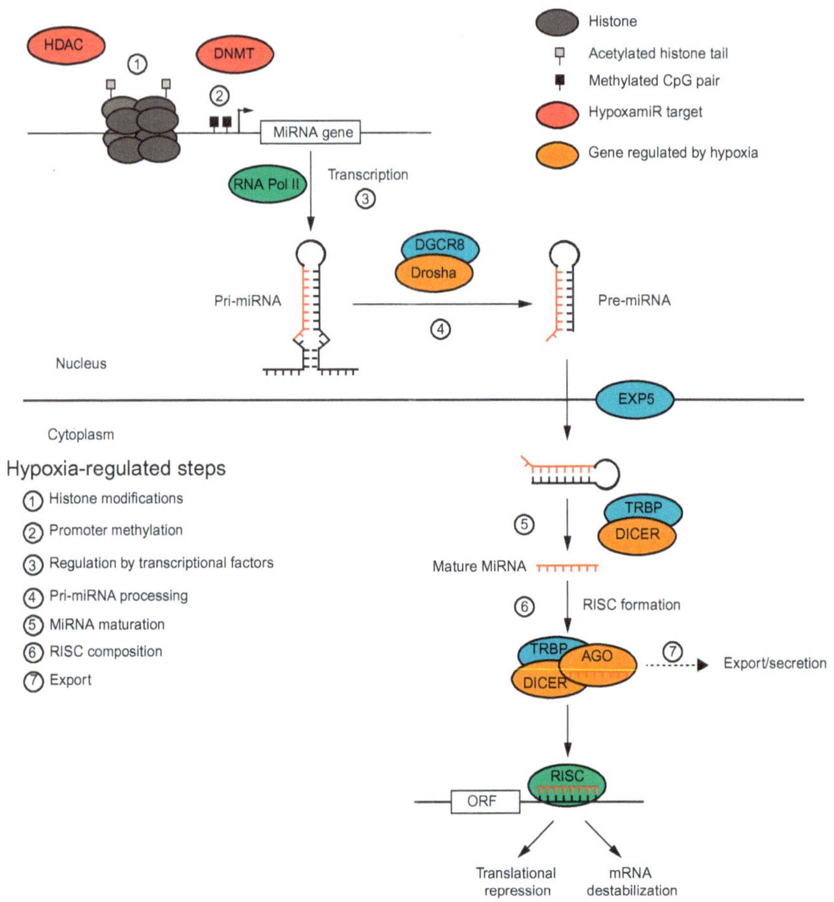

Fig. 1 hypoxamiR biogenesis and regulation by hypoxic pathways. MiRNA genes are transcribed by RNA polymerase II (RNA pol II) into long primary MiRNA transcripts (pri-miRNA) either from introns of protein-coding gene transcripts, or as intergenic MiRNAs under the control of their own promoters. The mature MiRNA is embedded in one of the two arms of the stem of a stem-loop structure within the pri-miRNA and must be excised to elicit gene silencing through a two stages process involving the RNase III family enzymes, Drosha and Dicer. The nuclear endoribonuclease Drosha, in complex with the dsRNA-binding protein DGCR8 (DiGeorge critical region 8, also known as Pasha), catalyzes the first endonucleolytic reaction by releasing an ~70 nt stem-loop precursor MiRNA (pre-miRNA) possessing a 3′-overhang. The pre-miRNA is then exported to the cytoplasm via Exportin-5 (EXP5) in complex with Ran-GTP. A second endoribonucleolytic reaction, catalyzed by Dicer, in concert with the trans-activating response RNA-binding protein (TRBP), occurs to generate an ~22 nt imperfectly base-paired RNA duplex with protruding 3′-overhangs at both ends. This duplex comprises the mature MiRNA (depicted in *red*) paired to the MiRNA* strand, which derives from the other arm of the stem of the pre-miRNA. The mature MiRNA is loaded

shift of Drosha and Dicer in the cleavage site, but also by nucleotide additions at the 3′-end. Therefore, in several instances, the most abundant sequence for individual MiRNA species may differ from the mature sequence listed in miRBase which may imply distinct functional roles for some isomiR sequences (Lee et al., 2010; Mercey et al., 2017).

The primary interaction between the MiRNA and its targets stems from a short stretch of 6–8-nt located at 5′ end of the MiRNA (termed the "seed sequence") that forms perfect Watson–Crick pairings with a MiRNA response element, often located in the 3′-UTR of the mRNA target, leading to either translation blockade or mRNA decay via interaction of AGO with deadenylation factors and translational machinery (Doench and Sharp, 2004; Lewis et al., 2005; Nam et al., 2014; Shin et al., 2010). As such 6-mers sequences are present every 4 kb on average, it can be deduced that MiRNAs are predicted to regulate a broad range of targets. In addition to these canonical sites, several noncanonical sites have been shown to play relevant biological roles, expanding the catalog of functional site types and in turn the complexity of the gene regulatory network controlled by MiRNAs (Kim et al., 2016a). Finally, the joint action of several MiRNAs on a same transcript adds an additional level of complexity to the MiRNA regulatory network.

During the past 10 years, more than 50 hypoxia-regulated MiRNAs (hypoxamiRs) have been identified including the "master hypoxamiR" MiR-210, which shows a high and consistent upregulation under hypoxia in virtually all cell types (Chan and Loscalzo, 2010; Chan et al., 2012) in contrast to most hypoxamiRs that are rather cell type specific. Identification of hypoxamiRs has initially focused on culturing a variety of cell lines and

---

into the RNA-induced silencing complex (RISC) and guides RISC to target transcripts, while the remaining MiRNA* strand is released and degraded. In animals, the vast majority of MiRNAs only form partial duplexes with their targets, typically in their 3′-untranslated region (UTR), which usually results in reduced protein expression through a variety of mechanisms that involve mRNA degradation and translational repression. Different steps in this biogenesis pathway can be affected during hypoxia and are indicated (1–6). For instance, epigenetic mechanisms including histone modifications (1) and DNA methylation (2) can regulate their expression. hypoxamiRs-associated pri-miRNAs can be also directly transcribed by HIF (MiR-210) or modulated by a cascade of hypoxia-associated molecular events (3). Finally, hypoxia has been shown to influence several steps of MiRNA biogenesis depending on the cell type and the hypoxic context including downregulation of Drosha (4), DICER suppression through posttranslational or epigenetic mechanisms (5), AGO2 posttranslational modifications (6), and MiRNA export (7).

primary cells of different tissue origin under hypoxic conditions (as reviewed in Chan and Loscalzo, 2010; Cottrill et al., 2014; Gee et al., 2014; Greco et al., 2014; Nallamshetty et al., 2013). Of note, the hypoxamiRs identified in each study are not completely identical, which likely reflects a combination of technical differences in sensitivities of MiRNA detection, the range of MiRNA screened, the duration and severity of oxygen deprivation and the cellular context. Consequently, the comprehensive list of MiRNAs that are modulated by low oxygen exposure in various biological contexts is still debated and deep-sequencing techniques in various models, such as the full characterization of endothelial cells (Voellenkle et al., 2012) or glioblastoma (Agrawal et al., 2014) miRNome in hypoxia will be required to identify the complex repertoire of hypoxamiRs, including isomiRs variants.

This review will address the complex molecular regulation of MiRNAs in both physiological and pathological conditions of low oxygen adaptation and the multiple functions of hypoxamiRs in fine-tuning the adaptive response of cells to hypoxia. From a clinical perspective, we also discuss the potential use of hypoxamiRs as new biomarkers and/or therapeutic targets in cancer and aging-associated diseases including cardiovascular and fibroproliferative disorders.

2. REGULATION OF MiRNA EXPRESSION BY HYPOXIA

Few hypoxamiRs, such as MiR-210, possess a functional HRE in their promoter regions and have been unambiguously identified as direct targets of HIF (Chan et al., 2009; Giannakakis et al., 2008; Huang et al., 2009; Puissegur et al., 2011). Therefore, most hypoxamiRs are modulated by a cascade of hypoxia-associated molecular events that indirectly contribute to their differential expression including inflammation and modulation of numerous signaling pathways such as BMP and MYC signaling. In addition to these transcriptional regulations, other hypoxia-mediated mechanisms including epigenetic modifications and posttranscriptional regulations of several enzymes involved in MiRNA biogenesis have been described (Fig. 1) and are detailed below.

2.1 Control of hypoxamiRs Expression Through HIFs-Dependent Mechanisms

2.1.1 MiR-210

In contrast to most hypoxamiRs, MicroRNA-210 (MiR-210) is a unique hypoxamiR that is robustly and ubiquitously induced in both transformed

(Kulshreshtha et al., 2007b; reviewed in Gee et al., 2014; Shen et al., 2013a) and primary cell types (as reviewed in Greco et al., 2014) regardless of their tissue origin. While some heterogeneity exists in MiR-210 baseline level of expression and relative degree of upregulation, its consistent overexpression during hypoxia likely reflects a fundamental and highly conserved role for this MiRNA.

An interesting observation has revealed a dose-dependent upregulation of MiR-210 as oxygen levels decline, suggesting that MiR-210 expression directly relies on oxygen concentration (Chan et al., 2009). In line with this, mimicking hypoxia-associated processes (i.e., exposure of cultured cells to low pH, oxidant stress, growth factor deprivation, and osmotic stress) did not affect MiR-210 expression (Chan et al., 2009; Fasanaro et al., 2008). Furthermore, the genomic location of the MiR-210 gene is also consistent with its regulation by low oxygen concentrations (Camps et al., 2008; Kulshreshtha et al., 2007b). It is embedded in the intronic sequence of a noncoding transcript MiR-210HG that itself is hypoxia-inducible and MiR-210 gene is close to HRAS and RASSF7 genes, whose expression correlates with both MiR-210 levels and hypoxic exposure in vivo (Huang et al., 2009). Finally, independent biochemical and genetic data have confirmed that MiR-210 promoter region carries a functional HRE recognized by HIF-1α upon hypoxic exposure (Chan et al., 2009; Giannakakis et al., 2008; Huang et al., 2009; Puissegur et al., 2011) (Fig. 1). Of note, while multiple groups have shown that MiR-210 is specifically induced by HIF-1α, other investigators have found a specific HIF-2α-dependent regulation (Zhang et al., 2009).

2.1.2 Other hypoxamiRs

Some evidence indicates that HIF may at least contribute in part to the transcription of other hypoxamiRs through direct binding to HREs in their respective promoter regions (Agrawal et al., 2014; Kulshreshtha et al., 2007a,b). In silico analysis of the putative promoter region of more than 20 hypoxamiRs upregulated in multiple breast and colon cancer cell lines unveiled a significant enrichment of HREs compared to the promoter regions of randomly selected, unrelated MiRNAs. Experimental validation was further performed on several of these hypoxamiRs, including MiR-24–1, MiR-26, MiR-103/107, MiR-181, and MiR-213, using genetic approaches, promoter–reporter assays, and chromatin immunoprecipitation (ChIP) (Kulshreshtha et al., 2008). Two other MiRNAs, MiR-155 and MiR-373, have been shown to be potently induced in a HIF-1α-dependent manner in normal or tumoral epithelial cells (Bruning et al., 2011; Crosby et al., 2009).

Overall these data suggest that HIF may directly exert (i.e., by HRE recognition alone) multitiered effects on a substantial network of MiRNAs (Nallamshetty et al., 2013; White et al., 2012) although a direct involvement of HIF in the transcriptional regulation of MiRNAs under hypoxia has only been demonstrated in a limited number of cases and a specific cellular context. In particular, the presence of a consensus HRE sequence in the promoter region of hypoxamiRs should be interpreted with caution. For example, while MiR-21 is induced by hypoxia in several cell types and contains a consensus HRE in its promoter region (Kulshreshtha et al., 2007b; Parikh et al., 2012), its regulation is HIF independent under conditions of low oxygen tension (Gorospe et al., 2011; Parikh et al., 2012). Induction of MiR-21 occurs downstream of Akt2 activation and promotes resistance to hypoxia in primary immortalized lung fibroblasts and a number of mammary and ovarian cancer cell lines (Polytarchou et al., 2011). In fact, for a significant number of hypoxamiRs, data on HIF recruitment to and/or transactivation of respective MiRNA promoters have not been presented yet to support the direct involvement of HIF in their transcriptional regulation. It is thus likely that, as described for MiR-21 in certain specific contexts, numerous hypoxamiRs may be primarily regulated by HIF-independent pathways including inflammation or activation of BMP or MYC signaling (Boldin et al., 2011; Taganov et al., 2006; Wood et al., 1998). Deciphering the interconnections and crosstalks between these different pathways is challenging and development of bioinformatic approaches, notably network theory, may address these issues.

2.2 HIF-Independent Transcriptional Regulation of hypoxamiRs

Similarly to MiR-21 and MiR-155, several MiRNAs have been shown to be regulated by HIF-independent hypoxia-induced TFs. Whereas HIFs play a predominant role in the coordination of transcriptional changes during hypoxic stress, several other TFs, such as NF-κB, PU.1, and p53, also execute important regulatory roles in the context of oxygen deprivation (Cummins and Taylor, 2005) (Fig. 1).

For example, hypoxic stresses can potently induce expression of TP53, which encodes p53. Similar to HIF, p53 regulates the expression of both protein-coding genes as well as noncoding RNA transcripts (Hermeking, 2012) including several MiRNAs implicated in tumor suppression and cell cycle arrest such as the MiR-34 family or the MiR-15/16 cluster (Okada et al., 2014; Raver-Shapira et al., 2007; Suzuki et al., 2009; Tarasov et al., 2007). However, these MiRNAs have been most often found down-regulated in hypoxic conditions (Dejean et al., 2011; Du et al., 2012; Hua et al., 2006) suggesting a more complex regulation of the p53 pathway following hypoxic stress or transcriptional regulation of these MiRNAs by other regulators. In addition, p53 controls the transcription of the MiR-200 family that inhibit epithelial-to-mesenchymal transition (EMT), a phenotype associated to tumor metastasis and chemoresistance (Chang et al., 2011; Fischer et al., 2015). Although the p53–MiRNA network has been more extensively studied in the context of oncogenesis and tumor biology, a recent work indicates that p53 may also control transcriptional regulation of specific MiRNAs during hypoxia by a HIF-independent pathway (Mutharasan et al., 2011). Under hypoxic conditions, p53 accumulates in HIF-β knockout mouse embryonic fibroblasts (MEFs) that lack effective HIF signaling and can directly induce MiR-210. However, as mentioned in this study, a robust induction of MiR-210 by p53 required severe conditions of hypoxia. In addition, p53 has been involved in the regulation of other hypoxamiRs, such as MiR-107 and MiR-192, in the context of tumorigenesis (Hermeking, 2012). However, direct experimental evidence demonstrating a functional link between p53 and a hypoxic stress in the induction of these MiRNAs is currently lacking.

Both acute and chronic hypoxic conditions can efficiently activate nuclear factor-kappaB (NF-κB), a well-studied TF connected to inflammation that regulates an expanding network of MiRNAs (Ma et al., 2011; Yang and Wang, 2016). A systematic screening approach in human monocytes identified several NF-κB-regulated MiRNAs, including MiR-146a and MiR-155, which have subsequently been shown to exert opposite effects on inflammatory processes (Taganov et al., 2006). In addition, other NF-κB-induced MiRNAs such as MiR-21 (Sheedy et al., 2010) and the MiR-130 family (MiR-130a/b and MiR-301a/b) (Bertero et al., 2014a; Cui et al., 2016; Zhou et al., 2010) also possess NF-κB-binding sites in their promoter region and function to modulate inflammatory responses. Interestingly, the MiR-210 promoter also contains a conserved NF-κB-binding site approximately 200 bp upstream of the primary stem-loop structure in the pri-miRNA (Zhang et al., 2012b). ChIP as well as promoter–reporter assays and gene knockdown studies, indicate that NF-κB directly interacts with and transactivates the MiR-210 promoter under hypoxic conditions. A complex temporal regulation of HIF-1 and NF-κB components is taking place following hypoxia and recent evidence indicate that several NF-κB downstream MiRNAs including MiR-199a-5p and MiR-93 participates in the posttranscriptional control of the hypoxic response (Jiang et al., 2015).

The functional importance of HIF-independent transcriptional regulation of MiRNAs under conditions of low oxygen availability is also underscored by the emergence of MiR-424 as an important component of the response to hypoxic stress in endothelial cells (Ghosh et al., 2010). Hypoxic endothelial cells activate transcription of MiR-424 through HIF-independent mechanisms that involve the PU.1 runt-related TF1 (RUNX-1) and CCAAT/enhancer-binding protein α (C/EBPα).

Reactivation of embryonic TFs such as POU5F1 (also known as OCT4) under hypoxia has also been proposed (Covello et al., 2006). In this study, the authors identify POU5F1 as a HIF-2α-specific target gene and report that HIF-2α can regulate stem cell function and/or differentiation through activation of POU5F1. Interestingly, recent studies in the context of pulmonary hypertension (PH), including pulmonary arterial hypertension (PAH), have found that the MiR-130/301 was increased by hypoxia in both pulmonary arterial smooth muscle cells (PASMCs) and pulmonary artery endothelial cells (PAECs) (Bertero et al., 2014a, 2015a; Brock et al., 2015). It was notably found that this induction was mediated by activation of a HIF-2α-POU5F1/OCT4 regulatory axis via multiple approaches including the characterization of functional-binding sequence for POU5F1 in the promoter of all these MiRNA family members (Bertero et al., 2014a).

Finally, a huge amount of clinical data have been accumulated linking other MiRNAs including let-7 family, MiR-1, MiR-133a/b, MiR-19a/b, MiR-150, MiR-195, MiR-221, MiR-23a/b, MiR-29a/b, MiR-30 family, MiR-106b–25 polycistron, MiR-204, MiR-223, and MiR-320 to ischemic/hypoxic diseases (Courboulin et al., 2011; Lee et al., 2012a; Liang et al., 2014; Liu and Olson, 2010; Meloche et al., 2015a; Olson, 2014; Quiat and Olson, 2013; Shi et al., 2016; Small et al., 2010; Zeng et al., 2016). Interestingly, recent evidence indicates that a single MiRNA locus can orchestrate subordinate MiRNA and target gene networks to control multiple signaling pathways across disparate cell types. A recent example concerns the MiR-130 family in the context of hypoxia-driven PH (Bertero et al., 2014a). First, in PASMCs, MiR-130 family directly targets peroxisome proliferator-activated receptor gamma (PPARγ) that in turn controls the expression of signal transducer and activator of transcription 3 (STAT3), a known regulator of MiR-204 also involved in PH (Courboulin et al.,

2011). Second, in PAECs, MiR-130 family through its action on PPARγ controls the expression of another effector, apelin, a protein known to regulate the expression of MiR-424/503 also involved in the development of PH (Kim et al., 2013a). Thus, the hypoxia-induced MiR-130 family controls the expression of distinct hypoxamiRs (i.e., MiR-204 and MiR-424/503) depending on the cellular context. Further efforts should be done to decipher the complex interconnections between all of these actors in particular through systems biology approaches (Barabasi et al., 2011; Chan and Loscalzo, 2012).

2.3 Epigenetic Mechanisms

Given the high degree of structural similarity between promoters and enhancers of MiRNA loci and protein-coding regions, epigenetic modifications are also emerging as important transcriptional regulatory mechanisms in MiRNA biogenesis (Fig. 1). Silencing of MiRNAs through DNA methylation of CpG islands in promoter regions is already known to hold important pathobiological and prognostic implications in a wide spectrum of malignancies (Bueno et al., 2008; Garzon et al., 2009). Recent and provocative findings demonstrated that hypoxia-induced loss of oxygen-dependent ten–eleven translocation (TET) enzyme activity increases hypermethylation at gene promoters. As a consequence, up to half of hypermethylation events would be directly due to hypoxia in cancer patients, with these events conferring a selective tumor advantage (Thienpont et al., 2016). These data support a central role for hypoxia in DNA methylation events, at least in cancer. Although data on epigenetic regulation of genomic regions that encode MiRNAs in hypoxia is limited, a recent study demonstrated that hypoxic stress decreased DNA methyltransferase (DMNT) activity in neural progenitor cells (Xiong et al., 2012). This decline in DMNT activity correlated with demethylation of a CpG island associated with the MiR-210 coding region and a corresponding increase in MiR-210 expression. Conversely, upregulation of DNMT, previously associated with PAH pathogenesis (Kim et al., 2011a), has been recently involved in the downregulation of MiR-126 in decompensated RV hypertrophy-derived endothelial cells (Potus et al., 2015). It is interesting to note that mutual regulation of MiRNAs and DNA methylation has been also often observed in human cancers (Wang et al., 2017). The molecular pathways driving alterations in DMNT activity and DNA methylation status have not been fully elucidated and further work is needed to determine if hypoxia-regulated chromatin structure through other mechanisms such as histone modifications (acetylation/methylation) or modulation of expression of epigenetic erasers, writers, and readers such as bromodomain-containing proteins (Meloche et al., 2015b) can also affect MiRNAs expression.

2.4 Regulation of MiRNA Biogenesis by Hypoxia

While an initial study performed on human trophoblasts cultured under hypoxic conditions indicated that MiRNA biosynthetic pathway was functional and that hypoxia did not alter the expression of key MiRNA machinery proteins (Donker et al., 2007), several studies reported instead that hypoxia affected the MiRNA biosynthesis machinery in many types of cancer via distinct and original molecular mechanisms, thereby eliciting broad effects on the MiRNA cellular content (Fig. 1). The generation of mature MiRNAs is a complex, multistep process that can be regulated by hypoxic stress at virtually each level depending on the cell type and the hypoxic context (i.e., transient or chronic): (i) downregulation of Drosha via a HIF-independent mechanism (ETS1/ELK1) (Rupaimoole et al., 2014); (ii) DICER suppression through posttranslational or epigenetic mechanisms (Ho et al., 2012; van den Beucken et al., 2014); and (iii) AGO2 tyrosine-phosphorylation by EGFR which reduces the binding of Dicer to AGO2 and in turn inhibits MiRNA processing (Shen et al., 2013b). A recent study has described a temporal regulation of DICER1 in hypoxic hepatocarcinoma cells with first a decrease during acute hypoxia followed by a late increase via a HIF-1- and NF-κB-dependent mechanisms (Jiang et al., 2015). Surprisingly, in most cases, only a fraction of cellular MiRNAs was usually affected, possibly due to DICER-independent maturation or to the high stability of most mature MiRNAs. Of note, reduced expression of Dicer was also detected in a rat model of PAH, during the onset of the disease caused by chronic hypoxia and leading to the downregulation of a subset of MiRNAs including MiR-21, -22, -30c, and let-7f (Caruso et al., 2010). It is important to note that from one *DICER* gene, multiple Dicer transcript variants can be produced as a result of the initiation of transcription from alternative promoters and alternative splicing (Kurzynska-Kokorniak et al., 2015). Moreover, four antisense transcripts associated with DICER1 have also been identified, suggesting complex transcriptional and posttranscriptional mechanisms of regulation. Interestingly, one of these DICER1 antisense was found upregulated in hypoxic endothelial cells (Voellenkle et al., 2016) and may thus contribute to the repression of DICER protein in hypoxic conditions. In the same study, the authors identified the H19 LncRNA (hosting MiR-675) and MiR-503HG as two of the best hypoxia-regulated LncRNAs in endothelial cells (Voellenkle et al., 2016). Surprisingly, the two mature MiRNAs were not induced by hypoxia in the same model (Voellenkle et al., 2012), suggesting a specific blockade of their maturation in hypoxic condition. While most of these studies described an inhibition of MiRNA biogenesis by hypoxia, a recent study has shown that biogenesis of MiR-215 and other MiRNAs was accelerated posttranscriptionally by HIFs through HIF–Drosha interaction in glioblastoma (Hu et al., 2016).

2.5 MiRNA Export and Hypoxia

Complex organisms may coordinate molecular responses to hypoxia by specialized avenues of communication across multiple tissues. Recently, it has been discovered that MiRNAs are stably "secreted" into the bloodstream at rest and in response to tissue injury or other pathological conditions (Kosaka et al., 2010; Valadi et al., 2007, reviewed in Chen et al., 2012b). Increased levels of MiR-210 expression in plasma have been notably reported in the context of various cancers as well as other systemic diseases (Camps et al., 2008; Foekens et al., 2008; Gunel et al., 2011; Jung et al., 2012; Wulfken et al., 2011; Zeng et al., 2011; Zhang et al., 2013). Plasma-associated circulating MiRNAs may be protected from degradation by several complementary mechanisms including packaging in phospholipid bilayer-encapsulated microvesicles or exosomes (Valadi et al., 2007) and/or via the formation of MiRNA–protein complexes, such as those bound to AGO2 (Arroyo et al., 2011) or high-density lipoprotein (HDL) (Vickers et al., 2011). In the setting of microvesicle release, it has been reported that some MiRNAs complexed with AGO2 can be also directly taken up by recipient cells and mediate important cellular functions associated with diseases (Hergenreider et al., 2012; Kosaka et al., 2010; Mittelbrunn et al., 2011; Pegtel et al., 2010; Vickers et al., 2011; Zhang et al., 2010). Interestingly, several studies have reported that hypoxia promoted MiR-210 export via direct or indirect mechanisms (Cui et al., 2015; Hale et al., 2014; Tadokoro et al., 2013). In one of this report, Hale et al. have dissected a complex mechanism pointing MiR-210 extracellular release as a highly regulated process (Hale et al., 2014). Depending on the hypoxic status, prolylhydroxylation of AGO2 not only acts as a molecular switch that reciprocally modulates MiR-210 release but also as intracellular activity in recipient cells after

MiR-210 delivery. Therefore, AGO2-dependent control of released MiR-210 represents a unique communication system that integrates the hypoxic response across anatomically distinct cells, preventing unnecessary activity of delivered MiR-210 in normoxia while still preparing recipient tissues for incipient hypoxic stress and accelerating adaptation.

3. THE DIVERSE ROLES OF MiRNAs IN CELLULAR ADAPTATION TO HYPOXIA

Considering the multifaceted contribution of hypoxia in various pathophysiological conditions including cancer, neurological dysfunctions, myocardial infarction, and lung diseases, a better characterization of the roles played by these hypoxamiRs in these complex networks appears mandatory for any development of future therapeutic applications. Although individual MiRNAs mediate relatively modest inhibitory effects via 30%–50% reductions on target protein levels, they collectively represent an important component of the cellular response to hypoxia through combinatorial and coordinated regulation of key target genes. An exhaustive discussion of the constantly expanding portfolio of hypoxamiR functions is beyond the scope of this report. Yet, it is worth reviewing the broad themes that characterize their general actions in overall hypoxic reprogramming, including the reinforcement of HIF adaptive responses and the regulation of proteins levels involved in metabolic adaptation, cell cycle progression, DNA repair, apoptosis, angiogenesis, inflammation, and immune response (Fig. 2).

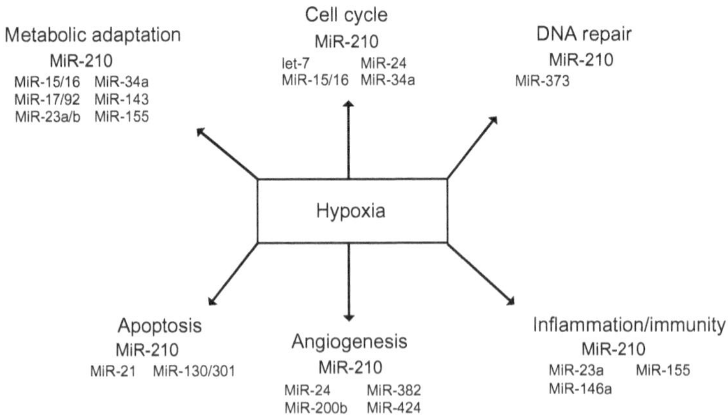

Fig. 2 The diverse functions of MiRNAs in cellular adaptation to hypoxia. The main hypoxamiRs and their associated functions are indicated.

3.1 The hypoxamiR Targetome: Present and Future

Both in silico and experimental approaches have been widely developed to identify MiRNA targets during the last 10 years and a recent review using hypoxamiRs as a paradigm for defining functional MiRNA targets and characterization of their targetome has detailed these different methodologies (Bertero et al., 2014b). We will briefly here summarize the main strategies available. Although many improvements have been performed on bioinformatics MiRNA-target prediction tools including improved quantitative models and use of experimental datasets (Agarwal et al., 2015; Vlachos et al., 2015), their single use has strong limitations, such as the lack of data on the set of transcripts in a particular cellular context and the diversity of targets in terms of MiRNA:mRNA Watson–Crick base-pairing requirements. This should justify to combine insofar as possible in silico target enrichment analysis (Backes et al., 2016; Corapcioglu and Ogul, 2015; Fan et al., 2016; Le Brigand et al., 2010; Steinfeld et al., 2013; Ulitsky et al., 2010; van Dongen et al., 2008) with open experimental approaches including quantitative proteomics, transcriptomics, AGO/MiRNA pulldown methodologies (Chi et al., 2009; Hafner et al., 2010; Helwak et al., 2013; Orom and Lund, 2007), or ribosome profiling (Ingolia et al., 2012; Popa et al., 2016). Given the context of MiRNA action in hypoxia models in which multiple levels of gene regulation occur, combining RNA-seq and ribo-profiling should more particularly enable the analysis of direct and indirect functions of hypoxamiRs at a genome-wide level. The recent development of single-cell RNA sequencing may also provide several interesting findings related to the role of MiRNAs in the variability of cells to hypoxic stresses. Based on the cell-to-cell variation in MiRNA activity observed by the recent work of Schmiedel et al. (2015), variability in MiRNA activity among a cell population, for instance among a tumor, would provide different cellular responses to hypoxia and promote distinct phenotypes.

The different approaches described in the following sections have shown that most of these hypoxamiRs can target a large number of transcripts. The MiR-210 example perfectly illustrates this complex situation with a large set of more than 30 confirmed targets, identified and validated using various methodologies in several different tissues (Table 1) (Bertero et al., 2012; Chan et al., 2009; Chen et al., 2010; Chio et al., 2013; Crosby et al., 2009; Fasanaro et al., 2008, 2009, 2012; Giannakakis et al., 2008; He et al., 2013; Huang et al., 2009; Ishibashi et al., 2012; Kai et al., 2016;

Table 1 Main Validated Targets of MiR-210

Function	Target Gene	Cell Type	Identification	Validation	References
Mitochondrial metabolism	*ALDH5A1*	Lung carcinoma	Microarray	Luciferase	Bertero et al. (2012)
	COX10	Colon cancer	Bioinformatics	Luciferase, WB	Chen et al. (2010)
	ISCU	Endothelial	Bioinformatics	Luciferase, WB	Chan et al. (2009)
	NDUFA4	Lung carcinoma	Microarray	Luciferase	Puissegur et al. (2011)
	SDHD	Lung carcinoma	Microarray	Luciferase	Puissegur et al. (2011)
	TfR	Breast cancer	Bioinformatics	Luciferase, WB	Yoshioka et al. (2012)
Cell cycle/DNA damage	*BUB1B*	Nasopharyngeal	Bioinformatics	Luciferase, qPCR, WB	He et al. (2013)
	CBX1	Endothelial	Microarray/proteomics	Ago IP, luciferase	Fasanaro et al. (2009)
	CCNF	Nasopharyngeal	Bioinformatics	Luciferase, qPCR, WB	He et al. (2013)
	CDC25B	Nasopharyngeal	Bioinformatics	Luciferase, qPCR, WB	He et al. (2013)
	E2F3	Embryonic kidney	Bioinformatics	Luciferase, WB	Giannakakis et al. (2008)
	FAM83D	Nasopharyngeal	Bioinformatics	Luciferase, qPCR, WB	He et al. (2013)
	PLK1	Nasopharyngeal	Bioinformatics	Luciferase, qPCR, WB	He et al. (2013)
	RAD52	Breast cancer	Bioinformatics	Luciferase, WB	Crosby et al. (2009)
	STMN1	Gastric epithelial	Microarray	Luciferase, qPCR	He et al. (2013) and Kiga et al. (2014)
	DIMT1	Gastric epithelial	Microarray	Luciferase, qPCR	Kiga et al. (2014)

Survival/ apoptosis	*ACVR1B*	Stromal	Bioinformatics	Luciferase	Mizuno et al. (2009)
	AIFM3	Hepatocarcinoma	Bioinformatics	Luciferase, WB	Yang et al. (2012)
	BCL2	Neuroblastoma	Bioinformatics	qPCR	Chio et al. (2013)
	CASP8AP2	Mesenchymal stem	Bioinformatics	Luciferase, qPCR, IF	Kim et al. (2009)
	EFNA3	Endothelial	Bioinformatics	Luciferase, IF	Fasanaro et al. (2008)
	FGFRL1	Pancreatic cancer	Ago IP	Luciferase	Huang et al. (2009)
	GPD1L	Endothelial	Microarray/proteomics	Ago IP, luciferase, qPCR	Fasanaro et al. (2009)
	HSD17B1	Trophoblastic	Bioinformatics	Luciferase, qPCR, ELISA	Ishibashi et al. (2012)
	MNT	Colon cancer	Microarray	Luciferase WB	Zhang et al. (2009)
	THBS1	Colon cancer	Microarray	Luciferase WB	Zhang et al. (2009)
	PTPN1	Endothelial	Microarray/proteomics	Ago IP, luciferase, WB	Fasanaro et al. (2009)
	PTPN2	Adipose-derived stem	Bioinformatics	Luciferase, qPCR, WB	Kim et al. (2009)
	ROD1	Endothelial	Ago IP	Luciferase, WB, qPCR	Fasanaro et al. (2012)
	SHIP-1	Myeloid	Bioinformatics	Luciferase, IHC	Lee et al. (2012b)
	TP53I11	Pancreatic cancer	Ago IP	Luciferase	Huang et al. (2009)
	VMP1	Hepatocarcinoma	Microarray	Luciferase, qPCR, WB	Ying et al. (2011)
	DR6	Chondrocytes	Bioinformatics	Luciferase, qPCR, WB	Zhang et al. (2015)

Continued

Table 1 Main Validated Targets of MiR-210—cont'd

Function	Target Gene	Cell Type	Identification	Validation	References
Development	*HOXA1*	Pancreatic cancer	Ago IP	Luciferase, WB	Huang et al. (2009)
	HOXA3	Embryonic kidney	Bioinformatics	Luciferase	Giannakakis et al. (2008)
	HOXA9	Pancreatic cancer	Ago IP	Luciferase	Huang et al. (2009)
HIF components	*HIF-1α*	T lymphocyte	Bioinformatics	Luciferase, qPCR, WB	Wang et al. (2014a)
	HIF-3α	Hepatocarcinoma	Bioinformatics	Luciferase	Kai et al. (2016)
Inflammation and immunity	*NDRG1*	MDSC	Microarray	Luciferase, qPCR, WB	Noman et al. (2015)
	IL16	MDSC	Microarray	Luciferase, qPCR, WB	Noman et al. (2015)
	CXCL12	MDSC	Microarray	Luciferase, qPCR, WB	Noman et al. (2015)
	ARG1[a]	MDSC	Microarray	Luciferase, qPCR, WB[a]	Noman et al. (2015)
Other	*KCMF1*	Trophoblastic	Bioinformatics	Luciferase, qPCR, WB	Luo et al. (2014)
	TWIST1	Renal cell carcinoma	Bioinformatics	Luciferase, qPCR	Yoshino et al. (2017)

[a]Upregulated by MiR-210 by a direct unknown mechanism.
The cell type as well as the identification and validation methods for the main MiR-210 validated targets are described. MDSC: myeloid-derived suppressor cells.

Kiga et al., 2014; Kim et al., 2009, 2013b; Lee et al., 2012b; Luo et al., 2014; Mizuno et al., 2009; Noman et al., 2012, 2015; Puissegur et al., 2011; Wang et al., 2014a; Yang et al., 2012; Ying et al., 2011; Yoshino et al., 2017; Yoshioka et al., 2012; Zhang et al., 2009, 2015). As detailed below, the consequence of this wide set of targets is a wealth of independent functional studies, mainly focused on one or a limited number of targets, that have dissected distinct mechanisms of action on multiple signaling pathways and different cell types, establishing MiR-210 as a master regulator of the hypoxic response (Chan and Loscalzo, 2010; Chan et al., 2012).

Many challenges are remaining in the field such as modeling the functional influence of the whole set of MiR-210 and other hypoxamiRs targets in a particular cellular context as well as understanding the influence of those that may act as competitive inhibitors of MiRNA function, preventing MiRNAs from binding their "effective" targets (Seitz, 2009). The search of additional natural MiRNA sponges or endogenous MiRNA-competing transcripts (ceRNAs), including pseudogenes (Poliseno et al., 2010), circular RNAs (Hansen et al., 2013; Memczak et al., 2013), or long noncoding RNAs (lncRNAs) (Das et al., 2014) represent an additional level of complexity.

Another challenge will be to explore the relevance of the best MiRNAs–mRNA target hits that could be facilitated by the use of CRISPR genome-editing approaches to specifically disrupt such interactions in vivo and in vitro. A very nice example of the importance of such interaction has been shown with the mutation of the MiR-155-binding site in the SOCS1 3′-UTR in the mouse germline, leading to a functional defect of T and NK cell function (Lu et al., 2015). The phenotypic consequences of only one MiRNA::mRNA interaction is likely to be subtle and may require complex experimental protocols focusing on specific conditions such as chronic hypoxia or reoxygenation.

The use of highly reproducible standardized culture conditions associated with computational network modeling (Barabasi et al., 2011; Chan and Loscalzo, 2012) should also largely improve the elucidation of these discrete interactions. Initial attempts to model computationally and interrogate MiRNA regulatory networks in diseases relative to hypoxia such as PH (Bertero et al., 2014a, 2015a; Parikh et al., 2012) and cancer (Kreeger and Lauffenburger, 2010; Yang et al., 2013a) have yielded positive results. Recently, Bertero and colleagues designed an in silico approach to rank MiRNAs with systems wide effects on known downstream PH relevant genes and their first degree interactors, based on several independent gene network architectural parameters (Bertero et al., 2014a). Not only was the MiR-130/301 family identified as the top ranked MiRNA family, but also such network analyses outlined a model of downstream actions of these MiRNAs to include two known proliferative pathways, a pathway influencing vasoconstriction, and a novel signaling mechanism controlling extracellular matrix deposition and remodeling. These findings were verified in vivo and in vitro (Bertero et al., 2014a, 2015a,c; Brock et al., 2015), thus demonstrating the power of combining computational bioinformatics with experimental biology and affording distinct advantages as compared with traditional scientific approaches. In the future, a particular attention should be given to the modeling of feedback regulatory loops between MiRNAs and TFs, often associated for fine-tuning of signaling pathways as detailed in the next paragraph below.

The use of wide-scale genomic analysis (Harper et al., 2015) should also be expanded to concentrate on the importance of genetic associations relevant to MiRNA biogenesis genes, MiRNA genes themselves, and target sequences recognized by those MiRNAs. Mapping of polymorphisms and cancer-related somatic mutations among the set of hypoxamiRs target sites may uncover functionally relevant interactions, as shown for a let-7 MicroRNA-binding site polymorphism in the KRAS 3′-UTR associated with increased risk for several cancers (Chin et al., 2008; Keane and Ratner, 2010; Paranjape et al., 2011; Pharoah et al., 2011; Ratner et al., 2010). At the same time, the vast information emerging regarding mammalian metabolic networks and metabolomics could lead to profound systems-based discoveries regarding the MiRNA-based control of the pervasive metabolic reprogramming central to hypoxia.

3.2 Feedback Regulatory Loops Between MiRNAs and Hypoxia

Particular attention has been paid to the observation that MiRNAs and TFs are often highly coordinated. From that perspective, MiRNAs may help confer robustness to biological processes by reinforcing transcriptional programs (Ebert and Sharp, 2012) through establishment of positive or negative feedback loops. This specific mode of regulation is likely important for the hypoxic response by fine-tuning HIF-mediated transcriptional program to the fluctuating oxygen concentrations during either acute or chronic hypoxia. Several factors have been involved in HIF negative feedback regulation such as the antisense HIF RNA, the oxygen-sensing enzymes PHDs, the tumor suppressors pVHL and REDD1, as well as nitric oxide synthases

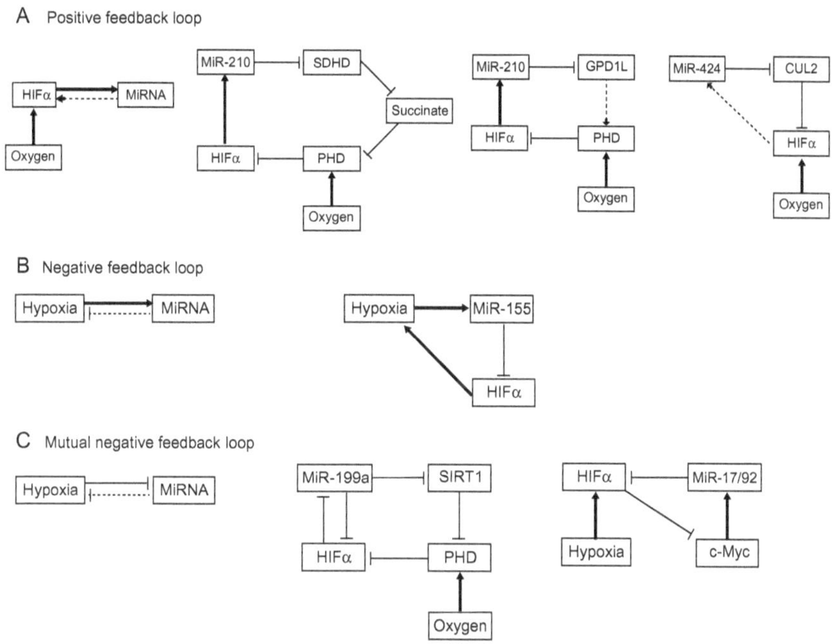

Fig. 3 Examples of MiRNAs–TFs regulatory loops involved in cell adaptation to hypoxia. (A) A positive feedback loop enforces hypoxia. Following oxygen downregulation PHDs are inhibited and HIF-1α is stabilized. Stabilization of HIF-a1 increases MiR-210 expression by directly binding the promoter of MiR-210. In turn MiR-210 downregulates SDHD and GPD1L, leading to direct or indirect PHD inhibition, and amplification of HIF-1α stabilization. A similar loop has been described with the induction of MiR-424 targeting CUL2. (B) A negative feedback loop contributes to resolution of HIF-1α. MiR-155 is induced in intestinal cell by prolonged hypoxia which in turn directly bound the 3′-UTR of HIF-1α and decreases its expression. (C) Mutual negative feedback loops involving MiR-199a and MiR-17–92, commonly downregulated by hypoxia can contribute to a full derepression of the hypoxic response through direct and indirect mechanisms. Direct interactions are represented by *solid lines*, indirect interactions are represented by *dotted lines*.

(Henze and Acker, 2010). Several hypoxamiRs have also been recently involved in HIF regulatory feedback regulations (Fig. 3).

3.2.1 MiR-210

There is evidence of two mechanisms by which MiR-210 triggers a positive feedback loop where HIF1A drives MiR-210 expression, which further induces HIF1A protein stability. The first mechanism is mediated through targeting the enzyme glycerol-3-phosphate dehydrogenase 1-like (GPD1L) that mediates hyperhydroxylation of HIF-1alpha via increasing PHD activity (Kelly et al., 2011). In normoxia, small amounts of MiR-210 result in high GPD1L protein levels and high activity of the PHDs, contributing to reduce HIF1A levels. When oxygen tension is lowered, decreased GPD1L protein potentiates the inhibition of PHD activity, leading to increased HIF1A stability (Fig. 3A). The second mechanism reported by our group involved the modulation of the oxidative phosphorylation by MiR-210 (Puissegur et al., 2011). Indeed, in lung adenocarcinoma cells, MiR-210 targets the subunit D of the succinate dehydrogenase complex (SDHD), resulting in increased succinate accumulation, a natural inhibitor of PHD activity, and subsequent increased HIF-1A stability (Fig. 3A).

Conversely, HIF1A has been identified in T cells, as a bona fide target of MiR-210 (Wang et al., 2014a). In this study, the authors identified MiR-210 as an important regulator of T-cell differentiation in hypoxia, which can limit immune-mediated inflammatory diseases. Indeed, in a mice model of experimental colitis, MiR-210 reduced the abundance of Hif1a transcripts and the proportion of cells that produced inflammatory cytokines and controlled disease severity. Therefore, the authors demonstrate that targeting HIF-1α by administrating MiR-210 mimics in vivo might be an effective therapeutic strategy for TH17 cell-dependent autoimmune diseases. Taken together, these results surprisingly describe a negative feedback mechanism mediated by MiR-210 on HIF-1α expression, suggesting that different and even opposite regulatory loops between HIF1 and MiR-210 can occur according to the cell type and the local environment.

3.2.2 Other hypoxamiRs

Positive and negative feedback loops between MiRNAs and hypoxia are not limited to MiR-210. Indeed, the induction of MiR-155 after prolonged hypoxia in intestinal epithelial cells has been associated with a negative feedback loop with HIF-1A (Bruning et al., 2011). Indeed, MiR-155 directly target HIF1A and overexpression of MiR-155 decreases the HIF1A protein and its associated transcriptional activity in hypoxia while the blockade of endogenous MiR-155 prevents the resolution of HIF1A stabilization and activity. Thus, MiR-155 induction contributes to a negative feedback loop for the resolution of HIF1A activity in cells exposed to prolonged hypoxia, leading to the oscillatory behavior of HIF1A-dependent transcription (Fig. 3B).

Another type of regulatory loops is exemplified by MiR-199a and MiR-17–92, commonly downregulated by hypoxia and which have been shown to negatively regulate HIF activity. This mutual negative feedback loop

(Fig. 3C) may thus result in a full derepression of the hypoxic response (Cascio et al., 2010; Rane et al., 2009; Taguchi et al., 2008). For instance, the rapid downregulation of MiR-199a in cardiac myocytes upon hypoxia is required to allow the upregulation of its target, HIF1A and the subsequent hypoxic transcriptional program (Rane et al., 2009). The override of MiR-199a decrease during hypoxia inhibits HIF1A expression and HIF1α-mediated stabilization of p53. As a consequence, apoptosis is attenuated as well. On the other hand, the blockade of MiR-199a during normoxia results in the upregulation of HIF1A, both directly and indirectly. Indeed, MiR-199a also targets sirtuin-1 that, in turn, negatively regulates PHD2, required for the destabilization of HIF1A. The relevance of this molecular circuitry is highlighted by the fact that MiR-199a inhibition reproduces hypoxia preconditioning in vitro. The MiR-17/92 cluster, a ubiquitously expressed set of MiRNA known to be induced by the proto-oncoprotein Myc, has been also shown to engage in an antagonistic relationship with HIF (Taguchi et al., 2008). Working with human lung cancer cells, Taguchi et al. found that members of the MiR-17/92 cluster target HIF-1A directly, and decrease its expression in a dose-dependent manner. Conversely, HIF has been shown to suppress MiR-17/92 expression via inhibition of Myc, a mechanism that is likely responsible for the marked downregulation of this MiRNA cluster under hypoxic conditions.

Another efficient molecular mechanism used by MiRNAs to regulate the hypoxic response is by targeting the HIF1α ubiquitination machinery. Ghosh et al. found that MiR-424 was induced in human endothelial cells exposed to hypoxia and in turn, targeted cullin 2, a scaffolding protein critical to the assembly of the ubiquitin ligase system, thereby stabilizing HIF1A (Ghosh et al., 2010) (Fig. 3A). The relevance of this pathway in postischemic vascular remodeling and angiogenesis was demonstrated by the fact that MiR-424 promoted angiogenesis in vivo in mice. In agreement with these data, mmu-MiR-322/rno-MiR-322, the mouse/rat homolog of human MiR-424, is also significantly upregulated along with HIF1A in experimental models of ischemia- (Ghosh et al., 2010) and hypoxia-induced PH (Kim et al., 2013a).

In summary, these studies demonstrate that hypoxamiRs represent an important and additional level of regulation of the HIF system through various negative and positive feedback loops, allowing the system to fine-tune the hypoxic response to notably adapt from acute to chronic hypoxia and rapidly respond to a second hypoxic event. Given the relative high stability of MiRNAs, these examples notably underline the role of these hypoxamiRs in modulating the dynamic of HIF-1α degradation after reoxygenation.

3.3 Metabolic Adaptation

The metabolic response to hypoxia is characterized by a shift in ATP production to glycolysis and lactic acid fermentation at the expense of oxidative phosphorylation (Warburg, 1956). This shift is associated with the suppression of apoptosis, as well as a reduction in oxygen-sensing and quenching of cytosolic reactive oxygen species (ROS) (Michelakis et al., 2002). Since anaerobic metabolism is inherently less efficient than glucose oxidation, such cells also show a concomitant increase in glucose transport and processing to compensate for the loss of ATP (as reviewed in Metallo and Vander Heiden, 2013). All metazoan cells display this so-called "glycolytic shift" when exposed to low levels of oxygen, known as the "Pasteur effect," and on short time scales, such adaptations serve to improve cell survival and function by striking an optimal balance between cellular energy production and oxidative stress (Aisenberg and Potter, 1957). During chronic or prolonged hypoxia, however, this phenomenon can result in persistent changes in cellular energy metabolism that do not resolve when oxygen supplies are restored. This "Warburg effect" is considered as a major component of many chronic diseases, including cancer (Kroemer and Pouyssegur, 2008; Vander Heiden et al., 2009), PH (Bertero et al., 2016; Ryan and Archer, 2015; Tuder et al., 2012), and fibrogenic diseases (Maher, 2015; Xie et al., 2015). Moreover, even when anaerobic metabolism does not persist, the long-term effects of mitochondrial ROS production during hypoxia can be seen in cases of stroke, hypoxic-ischemic injury, and diabetes mellitus (Petersen et al., 2004). In such cases, hypoxia has a profound effect on cellular metabolism, and these changes have clinical relevance to a wide range of seemingly disparate diseases.

HIF-1 plays a critical role in this switch, by upregulating the expression of most glycolytic enzymes and actively downregulating mitochondrial respiration and biogenesis (Semenza, 2013). For example, HIF is responsible for the induction of lactate dehydrogenase A (LDHA), a cytoplasmic enzyme that is responsible for the conversion of pyruvate into lactate, for the upregulation of glucose transporters (GLUT1, GLUT3) and the glycolytic enzyme hexokinase II (HK-II) as well as for the transcription of the PDK1 gene encoding a kinase that phosphorylates and inactivates pyruvate dehydrogenase, thereby shunting pyruvate from the mitochondria by preventing its oxidative decarboxylation to acetyl–CoA (Koppenol et al., 2011; Semenza, 2013).

In general, MiRNAs represent reasonable candidates for prolonged regulation of the glycolytic shift, as they possess a remarkably long half life. By examining MiRNA decay rates in MEFs after the silencing of Dicer, Gantier et al. demonstrated that MiRNA molecules may be approximately 10 times more stable than mRNA, with an average half life of 5 days (Gantier et al., 2011). This finding implicates hypoxamiRs as potential key contributors to the persistence of a Warburg phenotype that is characteristic of tumorigenic and hypertrophic cells in cancer and heart diseases, respectively, more particularly in cases in which HIF stabilization has been nominally resolved. As recently reviewed (Chen et al., 2012a; Singh et al., 2011), MiRNAs regulate cell metabolic processes through complex mechanisms, including directly targeting key enzymes or transporters of metabolic processes and regulating TFs, oncogenes/tumor suppressors as well as multiple oncogenic signaling pathways. We will focus below on the main examples related to hypoxamiRs (Fig. 4).

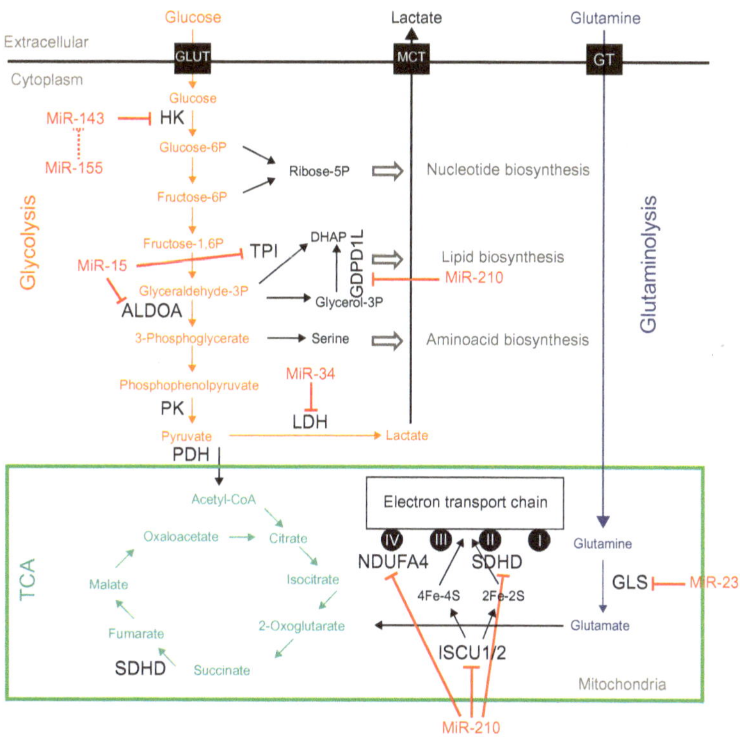

Fig. 4 hypoxamiRs and regulation of cell metabolism. The main hypoxamiRs and their validated targets in the different metabolic pathways are indicated.

3.3.1 MiR-210

MiR-210 is known to be induced in a variety of disease conditions associated with hypoxia, including preeclampsia, ischemia, PH, cardiovascular pathologies, and multiple cancers (Camps et al., 2008; Chan et al., 2009; Giannakakis et al., 2008; Greco et al., 2009; Hu et al., 2010; Pineles et al., 2007; Puissegur et al., 2011; Pulkkinen et al., 2008; Wigle et al., 2002; Zhu et al., 2009). At a metabolic level, Chan et al. found that MiR-210 directly targets the iron–sulfur cluster assembly proteins ISCU1 and ISCU2, members of a family of proteins that play a key role in the assembly of [4Fe-4S] and [2Fe-2S] iron–sulfur clusters (Chan et al., 2009) (Fig. 4). Serving essential roles in electron transport and cellular redox state, iron–sulfur clusters represent an important component of the mitochondrial respiratory complexes in the ETC and are incorporated into tricarboxylic acid (TCA) cycle enzymes, such as aconitase and succinate dehydrogenase (SDH) (Rouault and Tong, 2008). Upon identifying ISCU1/2 as a MiR-210 target, it has been shown that HIF-expressing tissues in vivo demonstrate increased levels of MiR-210, decreased expression of ISCU1/2, and consequent disruption of the integrity of iron–sulfur clusters, as assessed by electron paramagnetic resonance spectroscopy (Chan et al., 2009; White et al., 2015). In turn, through repressing ISCU1/2 during hypoxia, MiR-210 decreases the activity of prototypical iron–sulfur enzymes controlling mitochondrial metabolism, including complex I and aconitase leading to an inhibition of mitochondrial respiration and consequently an increased cell survival in the acute setting. These results are consistent across pulmonary arterial endothelial cells, tumorigenic cells of breast cancer, and trophoblasts in human placental tissue, reflecting the potential relevance of this finding to pathologies as disparate as PH, preeclampsia, and cancer (Chan et al., 2009; Chen et al., 2010; Favaro et al., 2010). In contrast, during hypoxia, when repression of electron transport corrects the mismatch with reduced oxygen tension, MiR-210 appears to increase ATP levels and optimize energy production in the hypoxic cell via the Pasteur effect (Semenza, 2007). In addition to ISCU, which affects mitochondrial function indirectly, several integral components of the mitochondrial ETC have been found to be MiR-210 targets: NADH dehydrogenase 1 alpha subcomplex 4 (NDUFA4) and SDHD, as well as cytochrome *c* oxidase assembly homolog 10 (COX10), thereby potentially producing a synergistic effect on inhibiting electron transport and mitochondrial respiration (Chen et al., 2010; Puissegur et al., 2011) (Fig. 4). Interestingly, SDHD, a subunit of complex II, is a well-documented tumor suppressor gene, suggesting a protumorigenic effect of MiR-210 (Puissegur et al., 2011).

An intriguing metabolic target experimentally confirmed is GPD1L (Kelly et al., 2011). GPD1L is highly homologous to glycerol-3-phosphate dehydrogenases that transfer electrons from the cytoplasmic NADH to the mitochondrial ETC and may be itself a key regulator of NAD+/NADH ratio. A study by Kelly et al. proposed a feedback mechanism based on GPD1L repression by MiR-210, which inactivates HIF prolylhydroxylase activity, leading to stabilization of HIF (Kelly et al., 2011) (Fig. 4). In this model, indirectly supported by our data from Puissegur et al. (2011), MiR-210 is both a downstream and upstream of HIF-1 (Fig. 3A).

3.3.2 Other hypoxamiRs

While HIF is primarily responsible for the induction of the glycolytic shift, many of its transcriptional targets will continue to maintain this cellular phenotype even in the absence of further stimulation from HIF itself. MiR-155 and MiR-23a/b clusters are two such examples—both acting directly on enzymes of the TCA cycle and glycolytic pathways (Gao et al., 2009; Jiang et al., 2012) (Fig. 4). Indeed, MiR-155 has been directly linked to the upregulation of glycolysis through a complex molecular mechanism involving the suppression of another MiRNA, the antiglycolytic MiR-143 (Fang et al., 2012; Jiang et al., 2012). While not itself a hypoxamiR, MiR-143 targets hexokinase II (HK-II) that catalyzes the first step of the glycolysis pathway. This relationship was demonstrated by Fang et al. in small cell lung cancer cells (Fang et al., 2012) and may be extended to other cell types as well. Working with human breast cancer cells, Jiang et al. demonstrated that MiR-155 suppresses MiR-143 expression via targeting its transcriptional activator, C/EBPβ, thus increasing HK-II expression at the posttranscriptional level. Supporting the relevance of this relationship, MiR-155 has been also shown to upregulate HK-II in this same cell type via the SOCS1/STAT3 axis, known to promote the transcription of HK-II (Jiang et al., 2012). Expressed in a variety of tissues, another MiRNA, MiR-23, is suppressed by the proto-oncoprotein c-Myc while upregulated by hypoxia (Gao et al., 2009). Interestingly, c-Myc expression was found correlated with high levels of glutaminase (GLS) in human prostate cancer cells. By screening predicted MiR-23 targets and using reporter gene construct expression in both gain- and loss-of-function assays, these authors demonstrated that GLS is a direct target of MiR-23 and can be rescued by overexpression of either c-Myc or anti-miR-23 agents. GLS catalyzes the hydrolysis of glutamine to glutamate, which generates α-ketoglutarate, a source of fuel for the TCA cycle (Hensley et al., 2013). In the context of hypoxia, induction of MiR-23a/b thus results in the suppression of oxidative metabolism by reducing α-ketoglutarate entry into the TCA cycle.

Many tumors contain genetic alterations, such as MYC activation, that can collaborate with HIF to confer metabolic advantages to tumor cells. A complex interplay occurs between MYC and HIF in cancer. Although HIF1 could counter normal levels of MYC by inhibiting its function (Gordan et al., 2007), HIF1 but also HIF2 are able to cooperate with overexpressed MYC to enhance aerobic glycolysis via multiple targets including the glucose transporter SLC2A1 (GLUT1), LDHA or PDK1 (Dang et al., 2008). Interestingly, the c-Myc-regulated oncomiR and hypoxamiR MiR-17–92 cluster may further contribute to influence the central carbon metabolism, the mammalian target of rapamycin complex 1 (mTORC1) signaling through negative regulation of the LKB1 tumor suppressor (Izreig et al., 2016). This key study raises the important question of the metabolic consequences of the complex regulation of the MiR-17–92 cluster following hypoxia in tumor cells.

Another example of hypoxamiR involved in the regulation of the hypoxia metabolism is MiR-15a, a hematopoietic MiRNA whose deletion is commonly associated with chronic lymphocytic leukemia (Calin et al., 2008). MiR-15a has been shown to control the expression level of the glycolytic enzymes aldolase A (ALDOA) and triose-phosphate isomerase I (TPI) (Fig. 4). As such, MiR-15a impairs rapid glycolysis which is necessary to sustain cells that have suppressed glucose oxidation and it is concordantly downregulated under hypoxic conditions (Calin et al., 2008). Interestingly, a recent study indicated that fatty acid synthase (FASN) is also a primary target of MiR-15a and MiR-16-1 in breast cancer (Wang et al., 2016). According to the Warburg effect, increased anaerobic glycolysis in cancer cells produces excess pyruvate, some of which can be converted to acetyl–CoA and used for de novo fatty acid synthesis via notably upregulation of FASN, promoting growth, and metastasis (Sounni et al., 2014). Overall these data suggest that MiR-15a–16-1 may thus suppress oncogenesis through blocking several downstream steps of the Warburg effect.

The recent identification of the channel responsible for mitochondrial Ca^{2+} entry, the mitochondrial Ca^{2+} uniporter (MCU) has opened a new era for the study of mitochondrial Ca^{2+} regulation and its key role in a variety of processes including cell death and metabolism (Rizzuto et al., 2012).

Mitochondrial Ca^{2+} overload is considered as a fundamental trigger of the apoptotic process, and several oncogenes and tumor suppressors modify the activity of the MCU complex to control apoptosis. Recent data have also shown that MiR-25, one of the MiRNAs transcribed from the MiR-106b–25 polycistron and upregulated in several cancers including colon cancer, decreases mitochondrial Ca^{2+} uptake through selective MCU downregulation, thus conferring resistance to apoptotic challenges (Marchi et al., 2013). Importantly, regulation of intracellular calcium and mitochondrial bioenergetics are also closely linked (McCormack and Denton, 1993). For instance, vascular cells in PAH exhibit a cancer-like phenotype, associated with altered mitochondrial metabolism and dynamics, as well as an increase in intracellular calcium and plasmalemmal membrane depolarization leading to excessive proliferation and apoptosis resistance (Sutendra and Michelakis, 2014). A recent report indicated that two MiRNAs, MiR-138 and MiR-25 are upregulated in PAH PASMCs and target the MCU subunit, causing a downregulation of the MCU complex, that accounts for the majority of the cancer-like mitochondrial-metabolic phenotype in PAH (Hong et al., 2017). In conclusion, the interplay between intracellular Ca^{2+} and MiRNAs might be a key aspect related to mitochondrial metabolism in several physiological and pathological conditions.

3.4 Cell Cycle

3.4.1 MiR-210

Control of proliferation of hypoxic cells has also been linked to MiR-210. In particular, in vivo tumor growth of several cell lines derived from various cancers in immunodeficient mice is substantially repressed after forced expression of MiR-210 (Huang et al., 2009; Tsuchiya et al., 2011). This phenotype correlates with the known induction of cell cycle arrest at the G1/S transition during hypoxia (Gordan et al., 2007; Hammer et al., 2007). However, Zhang and colleagues have found that this MiRNA appears to induce, rather than repress, the G1/S transition in other transformed cells, allowing the cells to bypass hypoxia-induced cell cycle arrest (Zhang et al., 2009). Therefore, it is unclear if MiR-210 acts primarily as a positive or negative regulator of proliferation, and if this action may vary depending upon contextual cues. A proliferative phenotype carries obvious therapeutic implications in hypoxic tumor growth; additionally, if present in primary cell types, such as in the vasculature, this phenotype may carry equally important ramifications in the progression of a variety of vascular diseases marked by dysregulated proliferation (e.g., atherosclerosis and PH).

A careful delineation of the mechanisms at play in cell cycle control and proliferation is required for exploitation of this MiRNA for therapeutic gain. As in the case of metabolic adaptation, the control of proliferation by MiR-210 appears likely mediated by repression of a significant number of direct targets. A few of these targets have been interrogated in sufficient details to establish plausible mechanisms by which proliferation is affected during hypoxia. First, Huang and colleagues have identified the novel targets FGFRL1 and HOXA3 as factors downregulated by MiR-210, which, in turn, appear to decrease tumor initiation and growth (Huang et al., 2007). FGFRL1 is the fifth and most recently discovered fibroblast growth factor receptor (FGFR), a family of receptors important in a wide range of complex biological actions involving growth, differentiation, and organogenesis (Wiedemann and Trueb, 2000). Recent work suggests that FGFRL1 may act as a negative regulator of FGF signaling by acting as a decoy receptor for FGF ligands (Steinberg et al., 2010). HOXA3 belongs to the HOX family of TFs with well-described roles in control of morphogenesis during embryonic development. Interestingly, HOXA3 has been implicated in both angiogenesis and wound repair via promotion of endothelial and epithelial migration (Mace et al., 2005). As a result, the independent identification of these genes as direct targets of MiR-210 correlates well with a function of this MiRNA as a critical regulator of growth and differentiation. Alternatively, Zhang and colleagues have confirmed MNT, a known antagonist of the proto-oncogene c-MYC, as a target of MiR-210 with profound effects on gene expression relevant to cell cycle and proliferation (Zhang et al., 2009). Interestingly, a significant overlap between the gene expression signatures of MiR-210 and c-MYC has been found using transcriptomic approaches in transformed cells, supporting the MiR-210-dependent downregulation of MNT. Notably in this setting, induction of MiR-210 and subsequent downregulation on MNT positively activates the G1/S transition, leading to cell cycle progression and tumor proliferation. Furthermore, in the context of c-MYC and HIF1 interplay (Dang et al., 2008; Gordan et al., 2007), a model has been proposed by which MiR-210 constitutes a negative feedback mechanism for the hypoxic response by reducing MNT and thereby increasing c-MYC activity. This interaction has also be investigated in the context of idiopathic pulmonary fibrosis (IPF), characterized by the relentless spread of fibroblasts from scarred alveoli into adjacent alveolar units, resulting in progressive hypoxia. Bodempudi et al. showed that hypoxia potently stimulates MiR-210 expression in IPF fibroblasts, and high MiR-210 expression drives fibroblast proliferation by repressing MNT (Bodempudi et al., 2014). It remains unclear how this model of

MiR-210/MNT axis-dependent initiation of cell cycling and growth may be integrated into the findings of MiR-210- and HOXA9/FGFRL1-dependent repression of tumor growth (Huang et al., 2009). Again, it is possible that some of these targets are not ubiquitously expressed in all cell types. For example, if activation of c-MYC by MiR-210 inhibits direct HIF1-dependent gene transcription in some cell types, this may not be the case in all cell types, as the effects of MiR-210 on such direct HIF-dependent gene expression appear minimal, at least in primary vascular endothelial cells (Chan et al., 2009). Thus, in case of proliferation, specific cellular context, notably the expression level of c-MYC, may again dictate the negative or positive regulatory effects of MiR-210.

The known functions of other confirmed targets of MiR-210 may also be predicted to influence the effects of MiR-210 on cellular proliferation. E2F3 is such a target that belongs to the E2F family of transcriptional activators and control cell cycle progression via influencing genes required for DNA synthesis at the G1/S phases (Dimova and Dyson, 2005). Alternative promoters of the E2F3 gene generate two forms (a and b) that are identical, apart from their first exon, and both carry the same 3′-UTR target sequence recognized by MiR-210 (Leone et al., 2000). E2F3a is an "activating" factor that can drive G0 cells to cycle and is induced when quiescent cells are stimulated to divide while E2F3b is a "repressive" factor that maintains G0 cells in quiescence. How MiR-210 regulates E2F3a and E2F3b to modulate the cell cycle and proliferation under hypoxia is still under investigation. ACVR1b, a member of the transforming growth factor-β (TGF-β) receptor superfamily, is another confirmed MiR-210 target that may influence proliferation and differentiation (Mizuno et al., 2009). In osteoblasts, MiR-210 appears to induce differentiation by inhibiting AVCR1b and consequently TGF/activin signaling. Because proliferation and differentiation can be inversely correlated, proliferation may also be affected through this mechanism. Again, more detailed and sophisticated studies will be required to determine how these targets and their downstream signals interact and influence with one another to converge ultimately upon the net proliferative capacity of the hypoxic cell.

3.4.2 Other hypoxamiRs

Since the seminal discovery of the frequent deletion and downregulation of MiRNA genes MiR-15/16 at 13q14 in chronic lymphocytic leukemia (Calin et al., 2002), many other MiRNAs have been involved in the regulation of the cell cycle and should also be considered as important players of the mammalian cell cycle (Bueno and Malumbres, 2011; Carleton et al., 2007). Oncogenic MiRNAs facilitate cell cycle entry and progression by targeting CDK inhibitors or transcriptional repressors of the retinoblastoma family while tumor suppressor MiRNAs induce cell cycle arrest by downregulating multiple components of the cell cycle machinery. Importantly, several of these MiRNAs are regulated by TFs involved in cell cycle regulation such as c-MYC, E2F, or p53, closely associated with HIF1/2 during the hypoxic response as previously underlined. These MiRNAs have been shown to participate collectively to the control of G1/S transition (MiR-15/16 and MiR-34 family) as well as entry and progression through mitosis (MiR-24). For instance, the MiR-15/16 cluster may induce cell cycle arrest at the G1 phase by targeting critical cell cycle regulators such as CDK1, CDK2, and CDK6 as well as cyclins (D1, D3, and E1) (Linsley et al., 2007; Liu et al., 2008; Takeshita et al., 2010; Wang et al., 2009). Of note, CDK4 or CDK6 mRNAs are also targeted by several other hypoxamiRs including MiR-24, MiR-34a, and let-7 family members (Johnson et al., 2007; Lal et al., 2008; Sun et al., 2008). Interestingly, MiR-24 and let-7 family are also able to regulate later phases of the mammalian cell cycle by targeting other cell cycle genes including cyclin A and cyclin B (Lal et al., 2009; Schultz et al., 2008).

While hypoxia has a profound impact on cell proliferation via a multitude of pathways, the relative contribution of these hypoxamiRs to the cell cycle extension in hypoxia is still unclear and should be precisely dissected using MiRNA knockout models.

3.5 DNA Repair

Genome integrity is challenged by diverse stresses, including mutagens, ROS, ultraviolet light, and chemo- or radiotherapeutic agents. Cellular responses to DNA damage involve a complex network of processes that detect and repair genomic lesions (Jackson and Bartek, 2009). MiRNAs have been demonstrated to participate in these processes (Wan et al., 2011).

3.5.1 MiR-210

While not standing out as robustly induced by irradiation, at least based on the available data, MiR-210, nevertheless, appears to have an impact on this complex process, as it downregulates RAD52 in breast cancer cells (Crosby et al., 2009), a key component in the homologous recombination-mediated repair of double-strand breaks (Van Dyck et al., 1999). Suppression of RAD52 by MiR-210 may provide an additional mechanism to help explain compromised homologous recombination repair in hypoxic cells. Conversely, high levels of MiR-210 may also be involved in lung (Grosso et al., 2013) and liver (Yang et al., 2012, 2013b) cancer cells radioresistance, suggesting that induction of this hypoxamiR both triggers mutation accumulation in the cancer cell genome and diminishes the effectiveness of radiotherapy.

3.5.2 Other hypoxamiRs

Correspondingly, another HIF-dependent MiRNA, MiR-373, can downregulate RAD23b (Crosby et al., 2009), a key component of the XPC/RAD23b complex which is involved in nucleotide excision repair and homology-dependent repair (Dantuma et al., 2009). Overall, MiR-373 and MiR-210 may thus participate in a hypoxamiR regulatory network modulating DNA repair during hypoxic stress. Direct data linking these MiRNAs to assays of DNA damage is limited. Furthermore, it is unclear if decreasing the DNA reparative capacity would act as an adaptive or maladaptive mechanism during hypoxia. Some theories suggest a beneficial impact of repressing DNA repair on ATP levels and/or autophagy (Eliopoulos et al., 2016), which could be important for survival in acute hypoxia; however, these putative mechanisms await more definitive corroboration.

A recent study proposed MiR-223, a hypoxia-downregulated MiRNA in PASMCs, as a crucial regulator of the DNA damage response via poly (ADP-ribose)polymerase 1 (PARP-1) targeting during the progression of PAH (Meloche et al., 2015a). The intensity of PARP-1/DNA damage signaling was shown to be finely controlled by this hypoxamiR both in vitro and in vivo with significant effects on PASMC apoptosis/proliferation balance and vascular remodeling. Interestingly, another hypoxamiR, MiR-204, was also implicated in the DNA damage response downstream of PARP-1 suggesting a complex network of MiRNA-regulated signaling cascade in PAH (Meloche et al., 2014).

3.6 Apoptosis

3.6.1 MiR-210

Hypoxic microenvironmental stresses, including hypoxia and nutrient deprivation, are well-known triggers of cell death, via both apoptosis and/or necrosis. Evasion from death responses is critical for diseases progression and represents a well-recognized hallmark of diseases. Thus, it is only fitting that MiR-210 has been investigated in the context of survival responses, and in particular, as a possible antiapoptotic component of the hypoxic response. Generally, the available evidence suggests a predominantly antideath role of MiR-210 in a variety of cell types, with overexpression protecting cells from apoptosis, and downregulation of MiR-210 during hypoxia promoting apoptosis (Bertero et al., 2012; Chan et al., 2009; Chen et al., 2010; Fasanaro et al., 2008; Favaro et al., 2010; Grosso et al., 2013; Hu et al., 2010; Kim et al., 2013b; Kulshreshtha et al., 2007b; Mutharasan et al., 2011; Puissegur et al., 2011).

While significant gaps remain in our understanding of this process, multiple targets have been identified to help explaining this effect: PTP1B (Hu et al., 2010; Kim et al., 2013b), caspase-8-associated protein-2 (CASP8AP2) (Kim et al., 2009), apoptosis-inducing factor, mitochondrion-associated, 3 (AIFM3) (Yang et al., 2012), the Bcl-2 adenovirus E1B 19 kDa-interacting protein 3 (BNIP3) (Wang et al., 2013) and SIN3 transcription regulator family member A (SIN3A) (Shang et al., 2014). In each case, MiR-210 downregulation of one of these different targets results in decreased apoptosis and increased cell proliferation. A recent investigation found notably that therapeutic inhibition of MiR-210 stimulated apoptosis and inhibited proliferation of a glioblastoma cell line (Wang et al., 2013). siRNA-mediated knockdown of SIN3A was sufficient to reestablish apoptosis evasion and unrestricted proliferation, highlighting SIN3A as another important MiR-210 target that regulates cellular decisions to proliferate or to undergo programmed death. Taken together, these studies support a model in which hypoxia-induced upregulation of MiR-210 results in the targeting and aberrant regulation of key members of different apoptosis-related cell signaling pathways in order to drive uncontrolled cell proliferation and carcinogenesis (Dang and Myers, 2015).

It is also important to note that the regulation of several mitochondrial components by the MiR-210 pathway, as described previously in the paragraph dedicated to metabolic adaptation, may have contrasting consequences on the regulation of cell death and survival under normoxic or hypoxic conditions. Indeed, clonogenic assays revealed that MiR-210 promoted clonogenic potential in hypoxia while forced MiR-210 overexpression in normoxia was rather proapoptotic (Puissegur et al., 2011). These apparently contradictory results may highlight the complex processes involved in the regulation of mitochondrial respiration and energy production in hypoxic cells. Following this model, the MiR-210-dependent repression of the ETC via SDHD, NDUFA4, or ISCU1/2 and other validated or yet unknown targets (Chen et al., 2010; Favaro et al., 2010;

Puissegur et al., 2011) would be protective during hypoxia by participating in the homeostatic downregulation of mitochondrial respiration. In contrast, MiR-210 overexpression in normoxia would create a mitochondrial dysfunction including a mismatch in electron transport that could lead notably to an increase in toxic ROS and increased apoptosis.

Finally, despite the predominant evidence for an antiapoptotic role, recent data suggested that MiR-210 may also exhibit a proapoptotic function, at least in neuroblastoma cells by targeting the antiapoptotic gene B-cell lymphoma 2 (BCL2) (Chio et al., 2013). Surprisingly, targeting of BCL2 by MiR-210 in endometriotic cells enhanced survival in hypoxia by promoting autophagy (Xu et al., 2016).

In summary, while information supporting a MiR-210-mediated blockade of apoptosis in hypoxia is accumulating for various cell types, it is still premature to state that this hypoxamiR represents a major protector against hypoxia-induced cell death in the tumor microenvironment.

3.6.2 Other hypoxamiRs

Persistent resistance to apoptosis could also be a direct result of the HIF-mediated glycolytic shift. Indeed, when glucose oxidation is suppressed, ROS production is low and relatively little H_2O_2 is present in the mitochondria and the cytosol (Solaini et al., 2010). A direct consequence is the hyperpolarization of the mitochondrial membrane, and the associated dysregulation of the mitochondrial permeability transition pore (MPTP), the large nonselective channel that allows the passage of proapoptotic mediators from the mitochondria into the cytoplasm (Tsujimoto and Shimizu, 2007). HIF-mediated upregulation of HK-II exacerbates this effect by binding to and inhibiting the voltage-dependent anion channel (VDAC), a key component of the MPTP (Michelakis et al., 2004). The proto-oncoprotein Akt contributes as well, by inducing the transport of HK-II across the mitochondrial membrane (Bonnet et al., 2007).

Several MiRNAs have been shown to play a role in this process. In addition to the MiR-155/MiR-143/HK-II signaling pathway previously described, MiR-21, a widely expressed hypoxamiR known to be upregulated in a variety of cancers (Asangani et al., 2008; Frankel et al., 2008; Garzon et al., 2010; Medina et al., 2010) as well as in tissue ischemia (Cheng and Zhang, 2010; Godwin et al., 2010), PH (Iannone et al., 2014; Parikh et al., 2012), and fibrosis (Chau et al., 2012; Liu et al., 2010; Thum et al., 2011), has been shown to upregulate Akt through suppression of its negative regulator phosphate and tensin homolog PTEN (Medina et al., 2010;

Sayed et al., 2010). The resulting inhibition of VDAC reduces ion flux between the mitochondria and the cytosol, inducing mitochondrial hyperpolarization and an increased threshold for the opening of the MPTP. Intriguingly, Zhang et al. have demonstrated that MiR-21 also modulates cellular ROS levels by targeting SOD2 and SOD3, the proteins which are responsible for reduction of more reactive free radical anion superoxide to the less reactive hydrogen peroxide (Zhang et al., 2012a). By promoting the accumulation of ROS, MiR-21 upregulation induces a cytosolic environment that mimics the one that would occur after a period of hypoxia, while the additional role of MiR-21 on MPTP regulation would prepare the cell to survive to such cytotoxic effects.

In addition to vascular cell hyperproliferation, resistance to apoptosis is also a key feature of PH. Indeed, several studies have demonstrated a spatiotemporal balance of PAECs apoptosis and proliferation (Bertero et al., 2016; Michelakis, 2006; Sakao et al., 2005). Following vascular injury, such as hypoxia, initial PAEC death will cause loss of small capillaries (which are essentially PAEC tubes), increasing the flow and shear stress in the remaining vessels, amplifying the effect. The emergence of apoptosis-resistant PAECs will lead to the proliferation in the intima and in plexogenic lesions. In this context, MiR-130/301 family, a master regulator of PH gene network induced by multiple triggers of PH, including hypoxia, has been found to regulate apoptosis and proliferation in PAECs (Bertero et al., 2014a). Forced expression of MiR-130a increased apoptotic caspase signaling in serum-starved cultured PAECs, while inhibition of MiR-130/301 protected against these effects. Notably, however, this MiRNA family did not exert similar control over apoptotic activity in serum-starved PASMCs, nor were these actions evident in the presence of serum. Thus, this dual nature of MiR-130/301 promoting context-specific apoptotic signaling and proliferation in PAECs indicates both the complexity of, and the substantially broad control over, multiple disparate pathways that converge to promote PH.

3.7 Angiogenesis

3.7.1 MiR-210

While most of the earlier functions are fundamental to the cellular hypoxic response, certain actions of MiR-210 are specialized to some specific cell or tissues. Relevant for tumor biology as well as for peripheral and pulmonary vascular diseases, hypoxic exposure significantly increases angiogenic potential, thereby offsetting limited oxygen and nutrient access and increasing cellular survival and growth. Angiogenesis is a physiological process involving the growth of new blood vessels from preexisting vessels. The regulation of angiogenesis by hypoxia is an important component of homeostatic mechanisms that link vascular oxygen supply to metabolic demand (Krock et al., 2011).

Notably, Fasanaro and colleagues have demonstrated that forced expression of MiR-210 in cultured HUVECs stimulates chemotaxis in response to vascular endothelial growth factor (VEGF) as well as the formation of capillary-like tubular structures in Matrigel (Fasanaro et al., 2008). At least in part, these processes are controlled by the downregulation of the receptor tyrosine kinase ligand ephrin-A3 (EFNA3), a direct target of MiR-210. Ephrin ligands and their receptors have prominent roles in the development of the cardiovascular system, vascular remodeling, and angiogenesis with known interactions of ephrin-A3 with such pathways (Kuijper et al., 2007; Pandey et al., 1995). However, MiR-210/EFNA3 relationships seem to be rather complex: hypoxia stimulates EFNA3 mRNA expression in endothelial cells, while MiR-210 decreases EFNA3 protein levels. In this respect, suppression by MiR-210 may represent a feedback loop to keep EFNA3 levels under control.

It is also likely that MiR-210 may act through other effectors as well, both directly and indirectly. Accordingly, increased angiogenic factor signaling has been reported in MiR-210-expressing cells (Foekens et al., 2008; Hu et al., 2010). Furthermore, as capillary tube formation and chemotaxis in cell culture are surrogate measures of true angiogenic potential, corroboration that MiR-210 can direct angiogenesis in vivo will be important in settings of tumor growth as well as primary vascular disease. It is also possible that such downstream, specialized phenotypes of cellular migration, and angiogenesis are subject to control by multiple MiR-210 targets and associated signaling pathways. For example, thrombospondin (THBS1), a c-MYC regulated gene involved in angiogenesis is also downregulated by MiR-210 and theoretically may also affect an angiogenic phenotype via control of MNT (a target mentioned previously) (Zhang et al., 2009). Finally, given the robust upregulation of MiR-210 in endothelium and the dependence of endothelial function on several MiR-210-regulated processes, such as mitochondrial metabolism (see earlier), it is tempting to speculate that multiple specialized endothelial processes beyond angiogenesis (such as thrombotic potential, solute permeability, and vascular smooth muscle phenotype modulation, among others) may also be sensitive to the effects of this MiRNA.

3.7.2 Other hypoxamiRs

Hypoxia negatively regulates MiR-200b, a MiRNA extensively studied in molecular oncology for its ability to regulate EMT (Brabletz and Brabletz, 2010). Overexpression of MiR-200b in human microvascular endothelial cells inhibits angiogenesis. Conversely, MiR-200b blockade enhances angiogenesis in vitro, as evidenced by tube formation on Matrigel and cell migration (Chan et al., 2011). These events are mediated by MiR-200b targeting of v-ets erythroblastosis virus E26 oncogene homolog 1 (ETS1) a crucial angiogenesis-related TF: knocking down endogenous ETS1 is sufficient to inhibit angiogenesis, whereas overexpression of ETS1 rescues MiR-200b-dependent impairment in angiogenic response and suppression of ETS1-associated gene expression. Interestingly, Magenta et al. found that MiR-200c overexpression induces endothelial cell growth arrest, apoptosis, and senescence, indicating that the whole MiR-200 family has an anti-angiogenic activity (Magenta et al., 2011).

Another hypoxia-induced MiRNA is MiR-24 (Kulshreshtha et al., 2007b), which acts as a negative regulator of angiogenesis. Fiedler et al. found that MiR-24 expression induces endothelial cell apoptosis, decreases endothelial capillary-like network formation on Matrigel, and inhibits cell sprouting from endothelial spheroids (Fiedler et al., 2011). These effects are mediated through direct targeting of the endothelium-enriched TF GATA-binding protein 2 (GATA2) and the p21-activated kinase PAK4. These targets, in turn, impinge on BCL2-associated agonist of cell death and sirtuin-1 signaling cascades. The relevance of MiR-24 modulation of angiogenesis is demonstrated by the impairment of zebrafish embryos angiogenesis upon overexpression of MiR-24 or silencing of its targets. Moreover, the blockade of endothelial MiR-24 limits mouse myocardial infarct size by preventing endothelial apoptosis and enhancing vascularity. These beneficial events, in turn, lead to preserved cardiac function and increased survival.

Conversely to MiR-200 and MiR-24, a recent study indicated that MiR-382 is induced by hypoxia in human gastric cancer cells and acts as an angiogenic oncogene by repressing PTEN (Seok et al., 2014). As discussed earlier, Ghosh et al. found that MiR-424 was induced in human endothelial cells exposed to hypoxia and promoted angiogenesis in vitro and in mice via targeting CUL2 (Ghosh et al., 2010), (Fig. 3A). Finally, given the complex TFs regulatory network induced by hypoxia, it is tempting to speculate that multiples MiRNAs involved in angiogenesis and known to be regulated by HIF-dependent and HIF-independent signals, such as MiR-130/301, MiR-126, MiR-27b, let-7, and MiR-1 may also play a role in these specialized phenotypes of cellular migration and angiogenesis. Again, more detailed and sophisticated studies will be required to determine how these MiRNAs, their targets and their downstream signals interact and influence with one another to converge ultimately to angiogenesis upon a hypoxic microenvironment.

3.8 Inflammation and Immunity

Inflammation is a primary response to injury or infection, allowing the body to eliminate pathogens and/or damaged tissue and to initiate repair processes. Low oxygen levels, or hypoxia, is a key feature of inflamed tissue and is due to damage to the local vasculature and increased oxygen consumption by pathogens and infiltrating immune cells. In addition to being a feature of inflammation, hypoxia also induces and regulates the inflammatory response by inducing the release of inflammatory cytokines, directing immune cell infiltration, and tuning the responses of the immune cells themselves (Imtiyaz and Simon, 2010; Taylor et al., 2016). At the molecular level, NF-κB is a central actor of the hypoxic response and can be activated by several mechanisms including the same oxygen-sensing hydroxylases responsible for conferring hypoxic sensitivity to the HIF pathway (Taylor, 2008). In this context, several hypoxamiRs, including MiR-210 have been involved in the modulation of several inflammatory and immune cells.

3.8.1 MiR-210

As described earlier, MiR-210 appears notably as a key modulator of TH17 polarization of helper T cells which limits inflammatory cytokines production and limits disease severity (Wang et al., 2014a). In the context of cancer, MiR-210 has been shown to be linked to hypoxia-dependent immune escape (Noman et al., 2012, 2015). On one hand, MiR-210 was shown to enhance myeloid-derived suppressor cells (MDSC)-mediated T-cell suppression by increasing arginase activity and nitric oxide production, resulting in increased tumor growth (Noman et al., 2015). On the other hand, upregulation of MiR-210 in tumor cells also decreases their susceptibility to lysis by antigen-specific cytotoxic T lymphocytes (Noman et al., 2012). These findings underline a potential immunosuppressive role for MiR-210 in oxygen-deprived regions of tumors where cancer stem-like cells and metastatic cellular behaviors are known to evolve. Surprisingly, germline deletion of MiR-210 results in the development of autoantibodies while mice overexpressing MiR-210 exhibit impaired class-switched Ab responses, expanding the immune regulatory function of MiR-210 to B cells (Mok et al., 2013).

3.8.2 Other hypoxamiRs

Two NF-κB-regulated MiRNAs, MiR-146a and MiR-155, have been shown to play important and opposite functions in several cell types including monocytes (O'Connell et al., 2010). Indeed, while MiR-146a was implicated in a negative feedback loop that limits inflammation by targeting TNF receptor-associated factor 6 (TRAF6) and interleukin-1 receptor-associated kinase 1 (IRAK1) (Boldin et al., 2011; Jurkin et al., 2010; Nahid et al., 2011), MiR-155 facilitates downstream proinflammatory signaling via suppression of multiple genes including phosphatidylinositol-3,4,5-trisphosphate 5-phosphatase 1 (INPP5D/SHIP1) required for enhancement pathways such as AKT2 and NOTCH1 signaling (Jablonski et al., 2016; O'Connell et al., 2009; Rodriguez et al., 2007; Virtue et al., 2012). Importantly, these two MiRNAs can also undergo a functional transfer between immune cells and constitute a mechanism of regulating the inflammatory response. Alexander et al. nicely showed that MiR-155 and MiR-146a are present in exosomes released from dendritic cells and pass between immune cells in vivo, as well as demonstrated that exosomal MiR-146a inhibits while MiR-155 promotes endotoxin-induced inflammation in mice (Alexander et al., 2015). Other examples of function in cell communication include the transfer of MiR-23a from hypoxic tumor-derived microvesicles (TD-MVs) to natural killer (NK) cells operating as a contribution in the various mechanisms used by cancer cells to block the antitumor immune response in an hypoxic environment (Berchem et al., 2016).

4. POTENTIAL CLINICAL APPLICATIONS

4.1 Diagnosis/Prognosis

The ideal biomarker should fit a number of stringent criteria depending on how the biomarker has to be used. It should ideally be accessible through noninvasive methods, specific to the disease or pathology of interest, represents a reliable indication of disease before clinical symptoms appear (early detection), sensitive to changes during disease course (disease progression or therapeutic response), and easily translatable from model systems to humans. Most of the current blood biomarkers are based on the levels of specific proteins in serum or plasma. Challenges for developing new protein-based biomarkers include the complexity of protein composition in blood, the diversity of posttranslational modifications, the low relative abundance of many proteins of interest, the sequence variations among different clinically relevant species, and the difficulties in developing suitable high-affinity detection agents (Ludwig and Weinstein, 2005; Mayeux, 2004). While proteins are more diverse and therefore potentially more informative, the challenges listed earlier have made the discovery and development of new protein-based biomarkers an expensive and time-consuming task. In this context, significant number of MiRNAs have been observed outside of cells (Arroyo et al., 2011; Chim et al., 2008; Valadi et al., 2007; Vickers et al., 2011), including various body fluids and their use as circulating biomarkers has indicated great promise (Reid et al., 2011). Indeed, these MiRNAs are stable and show distinct expression profiles among different fluid types, the sequences of most MiRNAs are conserved among different species, the expression of some MiRNAs is specific to tissues or biological stages and the level of MiRNAs can be easily assessed by various methods such as quantitative polymerase chain reaction (PCR). Of great interest, secreted MiRNAs fulfill many prerequisite required for clinical biomarker development (as reviewed in Etheridge et al., 2011). However, to further develop MiRNA-based biomarkers, several fundamental issues associated with MiRNA measurements still need to be addressed (Cookson et al., 2012; Jarry et al., 2014; Tiberio et al., 2015; Witwer, 2015). Although circulating extracellular MiRNAs clearly possess many desirable properties when compared to protein biomarkers, they require the same stringent validation processes as for protein-based biomarkers to demonstrate their specificity and selectivity. The most fundamental challenges in MiRNA-based biomarker development are the issues associated with accurate measurement. Setting up a standardization process for sample preparation and developing a more accurate method to assess the quality and quantity of MiRNA are urgently needed. Finally, finding informative biomarkers is not only key to understanding physiopathological processes of diseases, but is also critical for therapeutic development. We present below some examples of the promising use of hypoxamiRs as noninvasive biomarkers in various diseases.

4.1.1 MiR-210

Expression of MiR-210 has been consistently associated with poor clinical outcome in several cardiovascular pathologies such as PH (Lee et al., 2014), coronary artery disease, hypertension, heart failure (as reviewed in Greco et al., 2014) as well as solid tumor types, including soft-tissue sarcoma, breast, head and neck, lung and pancreatic tumors (Camps et al., 2008; Foekens et al., 2008; Gee et al., 2010; Wang et al., 2014b, reviewed in Gee et al., 2014). Of particular interest, expression levels of MiR-210 in plasma, serum, urine, and saliva have already been associated with different diseases (Chen et al., 2008; Gill et al., 2003; Humeau et al., 2015; Lawrie et al., 2008; Lorenzen et al., 2011; Mitchell et al., 2008). For instance, MiR-210 expression level was not only found upregulated in plasma and/or serum of several cancers and cardiovascular diseases, but also associated with disease progression. In addition, MiR-210 expression level was also increased in the saliva of patients with pancreatic cancer (Humeau et al., 2015), as well as differentially expressed in urine of patients with acute renal allograft rejection compared to stable transplant patients (Lorenzen et al., 2011).

Circulating MiR-210 levels have also been correlated with sensitivity to trastuzumab (a human epidermal growth factor receptor monoclonal antibody), tumor presence, and lymph node metastases in breast cancer patients (Cortez et al., 2011; Jung et al., 2012). This provides proof of concept that plasma MiR-210 level may also be used to monitor the response to anticancer therapies.

4.1.2 Other hypoxamiRs

Expression of multiple additional hypoxamiRs may be a viable alternative as a clinical biomarker of hypoxia. For example, a signature of hypoxia-related MiRNAs was shown to be a significant independent prognostic factor in breast cancer in multivariate analysis (Buffa et al., 2011). It has also been reported that several hypoxamiRs display the potential to act as biomarkers for a wide range of diseases, including myocardial infarction (MI), coronary artery diseases, hypertension, heart failure, type-2 diabetes mellitus, and several cancers (Chen et al., 2014; Creemers et al., 2012; Greco et al., 2014; Guay and Regazzi, 2013; Gupta et al., 2010; Hayes et al., 2014; Karakas et al., 2017; Li et al., 2016; McManus and Ambros, 2011; Navickas et al., 2016). For instance, several studies indicated an increase of circulating MiR-1, MiR-133a/b, and MiR-499-5p following acute MI in both humans and mice (D'Alessandra et al., 2010). Moreover, it has also been reported that circulating MiR-208 is enhanced in some, but not all patients with acute MI.

Several MiRNAs including MiR-223 and the MiR-130/301 family have also been proposed as biomarker of PH and cancers by several studies (Bertero et al., 2014a, 2015a; Wei et al., 2013; Zeng et al., 2016). Indeed, members of the MiR-130/301 family were significantly elevated in plasma sampled adjacent to the main pulmonary artery from PH individuals as compared with non-PH individuals. Further stratification of PH individuals based on pulmonary arterial pressures revealed an increasing level of MiRNA family member expression with hemodynamic severity. Moreover, a positive linear correlation was observed between circulating levels of EDN1 and individual MiR-130/301 family members. Circulating EDN1 levels were also correlated with the hemodynamic severity of disease. Thus, in animal models and human examples of PH in vivo, expression of the MiR-130/301 family is consistently elevated in pulmonary tissue and small diseased pulmonary vessels, a pattern that positively correlates with histologic and hemodynamic severity of disease (Bertero et al., 2015a). Coupled with the related functions of these MiRNAs in PH, such results provide a foundation for a larger study aimed at characterizing this MiRNA family as a profile of PH biomarkers that could offer more specificity as compared with a single MiRNA alone.

4.2 Therapeutic Perspectives

Understanding the pathogenic consequences of MiRNA alterations in diseases represents a new frontier in biomedical research. In particular, a better knowledge of hypoxamiR function in disease conditions for which hypoxia plays a major deleterious role may offer new innovative therapeutic strategies. Indeed, the development of MiRNA-based therapeutics is currently the subject of intensive research efforts especially in lethal and/or chronic diseases (Broderick and Zamore, 2011; Kim et al., 2011b; Pottier et al., 2014; Rupaimoole et al., 2011). MiRNAs have many advantages as a therapeutic modality. The mature MiRNA sequences are short and often completely conserved across multiple vertebrate species. These characteristics make MiRNAs relatively easy to target therapeutically and allows for using the same MiRNA-modulating compound in preclinical efficacy and safety studies as well as in clinical trials. Moreover, MiRNAs have typically many targets within cellular networks, which, in turn, enable modulation of entire pathways in a disease state via therapeutic targeting of disease-associated MiRNAs. Currently, three approaches are employed to modulate MiRNA activity: (i) restoring the function of a MiRNA using either synthetic double-stranded MiRNAs or viral vector-based overexpression; (ii) inhibiting the function of a MiRNA using chemically modified anti-miR oligonucleotides (ASOs) (Stenvang et al., 2012); and (iii) interfering between the MiRNA and a specific target using a modified ASO named target site blocker (TSB) or target protector (Choi et al., 2007).

4.2.1 MiR-210

Despite a large number of studies, few have reported a therapeutic benefit of MiR-210 modulation in a preclinical disease model. Based on its prosurvival and proangiogenic roles, it may be particularly relevant to therapeutically increase MiR-210 expression level, notably in cardiovascular diseases. Preclinical work showed that intracardiac injections with a vector carrying MiR-210 in a mouse model of myocardial infarction induced significant improvement of left ventricular fractional shortening, decreased cellular apoptosis, and increased neovascularisation (Hu et al., 2010). Recent data suggest that both chronic obstructive pulmonary disease and IPF share transcriptional mRNA–MicroRNA hypoxia/p53 regulation (Kusko et al., 2016). As previously presented, the potential fibroproliferative action of MiR-210 in IPF (Bodempudi et al., 2014) may also support the development of pharmacological treatments targeting the hypoxia-mediated MiR-210 proliferation pathway as a new therapeutic strategy for slowing the progression of IPF.

4.2.2 Other hypoxamiRs

Conversely, therapeutical effects of hypoxamiRs have been largely demonstrated in preclinical disease models. Examples of strategies blocking oncogenic MiRNAs or restoring tumor suppressor MiRNA expression have been extensively reviewed and include hypoxamiRs such as MiR-21, MiR-155, and MiR-34 (Cheng et al., 2015; Garzon et al., 2010; Hayes et al., 2014; Rothschild, 2014). Among these candidates, MiR-21 emerges as an interesting candidate for the development of therapeutic strategies against many forms of cancer as well as heart diseases (Bonci, 2010) and pulmonary fibrosis (Pottier et al., 2014; Rajasekaran et al., 2015). Additional attractive candidates for IPF include the hypoxamiR MiR-199a-5p associated with the control of TGF-β signaling in lung fibroblasts (Lino Cardenas et al., 2013). Other examples of potential clinical application are related to MiR-223 (Meloche et al., 2015a; Zeng et al., 2016), MiR-204 (Courboulin et al., 2011; Lee et al., 2012a; Ruffenach et al., 2016), MiR-138, MiR-25 (Hong et al., 2017), and the MiR-130/301 family in PH. In particular, inhibition of MiR-130/301 family has been reported to prevent and ameliorate hemodynamic as well as histological manifestation of PH in rodent models of PH (Bertero et al., 2014a, 2015a, 2016; Brock et al., 2015). Importantly, while these reports demonstrate the efficacy of the inhibition of

MiR-130/301 in preclinical PH models, inhibition of MiR-130/301 has been achieved by different methods. While Brock and colleagues used intraperitoneal delivery of LNA anti-miRNA (Brock et al., 2015), Bertero et al. used intrapharyngal or intravenous (Bertero et al., 2014a, 2015a) or intraperitoneal injection (Bertero et al., 2015b,c) of short anti-miRNA named "Short-mers." Short-mer is a tiny oligonucleotide (eight nucleotides) designed as fully modified antisense oligonucleotides complementary to the seed sequence of the MiR-130/301 MiRNA family (Stenvang et al., 2012). Specifically, the MiR-130/301 short-mer carried a sequence complementary to the active site of the MiR-130/301 MiRNA family, containing a phosphorothioate backbone and modifications (fluoro-, methoxyethyl, and bicyclic sugar) at the sugar 2′ position. Importantly, short-mer oligonucleotides were demonstrated to be nontoxic, lipid-permeable, and high-affinity oligonucleotides. What is remarkable is that each of these delivery routes has demonstrated efficacy in: (1) reaching the lungs, as determined by changes in the MiRNA and respective target expression levels and (2) altering the severity of PH. Although most of these delivery techniques demonstrate efficacy in ameliorating the severity of PH in the experimental models, a number of key limitations remain inadequately addressed. One of the major obstacles remains characterization of their potential off-target effects. While most of the published studies have confirmed delivery to the lungs, either via conducting quantitative PCR for the MiRNA itself or the proposed targets, only a limited number of these studies have evaluated the off-target effects, including effects on the liver, which is widely known to take up oligonucleotides from the circulation (Moschos et al., 2011). Moreover, because current studies are based on short-term therapeutic interventions, long-term toxicities, and putative rebound effects remain to be established. The stability of the oligonucleotide-based therapies also remains to be optimized. Dosage and frequency of delivery have also not been fully optimized, and converting effective doses in rodents to humans remains a major hurdle. Lastly, the financial costs of translating such findings to humans represent a key challenge, especially for a disease such as PH where lifelong therapies are warranted.

Oligonucleotide therapies, including MiRNA and short interfering RNA (siRNA), have entered the clinical stage for a number of diseases beyond hypoxia-related disease. These include mipomersen, a siRNA targeting apolipoprotein B and FDA-approved for treatment of homozygous familial hypercholesterolemia (Raal et al., 2010), as well as the number of other siRNAs and MiRNAs for conditions including hepatitis C (Janssen et al., 2013), hemophilia, tissue fibrosis, and hematological malignancies, that are at various stages of clinical trials. While the signaling axes implicated in MiRNA biology have been robustly validated in many cases, replication of the direct therapeutic efficacy of MiRNA mimetics or inhibitors in vivo have often been varied or have been attempted only by the discovering lab alone (as reviewed in Chun et al., 2017). Thus, the choice of which of several MiRNAs to develop as therapeutic targets is not a straightforward decision. There exist numerous variables to consider regarding the inherent biology of any given MiRNA including its breadth (narrow vs broad) of target gene engagement, efficacy of repression of any single target gene, and landscape of activity across single or different tissue/cell types. Tailoring each variable may offer advantages and compromises in terms of potency, pleiotropy, and selectivity/side effect profile, but it is currently not clear whether a systematic method to address or prioritize such biologic activity may facilitate for more efficient MiRNA-based drug discovery.

In summary, scientific endeavors have been accelerating over the recent years to identify the fundamental and robust actions of MiRNAs in controlling hypoxia and hypoxia-related diseases. Attendant with those discoveries, the potential has been growing for clinical application in the development of new RNA-based biomarkers of disease and in the advent of a new generation of medications based on RNA-specific biology (Janssen et al., 2013). Yet, in spite of these promising results, challenges still remain with regards to deciphering the expansive complexity of MiRNA-target gene networks in hypoxia and hypoxia-related diseases, demonstrating the feasibility and utility of quantifying plasma MiRNAs for diagnosis or prognosis, and proving the efficacy and specificity of MiRNA-based therapies.

ACKNOWLEDGMENTS

We thank people of the different teams involved for interesting discussion. Team of Physiological Genomics at IPMC is supported by CNRS, FRM, ARC, Cancéropole PACA, Fondation Unice, Fondation du Souffle, and labex Signalife [ANR-11-LABX-0028-01]. Team of Tumor and Stroma Interaction at IRCAN is supported by ARC (grant PJA20131200325), Cancéropole PACA, and the French National Research Agency (ANR-11-LABX-0028-01).

REFERENCES

Agarwal, V., Bell, G.W., Nam, J.W., Bartel, D.P., 2015. Predicting effective microRNA target sites in mammalian mRNAs. Elife 4, e05005.

Agrawal, R., Pandey, P., Jha, P., Dwivedi, V., Sarkar, C., Kulshreshtha, R., 2014. Hypoxic signature of microRNAs in glioblastoma: insights from small RNA deep sequencing. BMC Genomics 15, 686.

Aisenberg, A.C., Potter, V.R., 1957. Studies on the Pasteur effect. II. Specific mechanisms. J. Biol. Chem. 224, 1115–1127.

Alexander, M., Hu, R., Runtsch, M.C., Kagele, D.A., Mosbruger, T.L., Tolmachova, T., Seabra, M.C., Round, J.L., Ward, D.M., O'Connell, R.M., 2015. Exosome-delivered microRNAs modulate the inflammatory response to endotoxin. Nat. Commun. 6, 7321.

Arroyo, J.D., Chevillet, J.R., Kroh, E.M., Ruf, I.K., Pritchard, C.C., Gibson, D.F., Mitchell, P.S., Bennett, C.F., Pogosova-Agadjanyan, E.L., Stirewalt, D.L., Tait, J.F., Tewari, M., 2011. Argonaute2 complexes carry a population of circulating microRNAs independent of vesicles in human plasma. Proc. Natl. Acad. Sci. U.S.A. 108, 5003–5008.

Asangani, I.A., Rasheed, S.A., Nikolova, D.A., Leupold, J.H., Colburn, N.H., Post, S., Allgayer, H., 2008. MicroRNA-21 (miR-21) post-transcriptionally downregulates tumor suppressor Pdcd4 and stimulates invasion, intravasation and metastasis in colorectal cancer. Oncogene 27, 2128–2136.

Backes, C., Khaleeq, Q.T., Meese, E., Keller, A., 2016. miEAA: microRNA enrichment analysis and annotation. Nucleic Acids Res. 44, W110–W116.

Barabasi, A.L., Gulbahce, N., Loscalzo, J., 2011. Network medicine: a network-based approach to human disease. Nat. Rev. Genet. 12, 56–68.

Bartel, D.P., 2004. MicroRNAs: genomics, biogenesis, mechanism, and function. Cell 116, 281–297.

Berchem, G., Noman, M.Z., Bosseler, M., Paggetti, J., Baconnais, S., Le Cam, E., Nanbakhsh, A., Moussay, E., Mami-Chouaib, F., Janji, B., Chouaib, S., 2016. Hypoxic tumor-derived microvesicles negatively regulate NK cell function by a mechanism involving TGF-beta and miR23a transfer. Oncoimmunology 5, e1062968.

Bertero, T., Grosso, S., Robbe-Sermesant, K., Lebrigand, K., Henaoui, I.S., Puissegur, M.P., Fourre, S., Zaragosi, L.E., Mazure, N.M., Ponzio, G., Cardinaud, B., Barbry, P., Rezzonico, R., Mari, B., 2012. "Seed-Milarity" confers to hsa-miR-210 and hsa-miR-147b similar functional activity. PLoS One 7, e44919.

Bertero, T., Lu, Y., Annis, S., Hale, A., Bhat, B., Saggar, R., Wallace, W.D., Ross, D.J., Vargas, S.O., Graham, B.B., Kumar, R., Black, S.M., Fratz, S., Fineman, J.R., West, J.D., Haley, K.J., Waxman, A.B., Chau, B.N., Cottrill, K.A., Chan, S.Y., 2014a. Systems-level regulation of microRNA networks by miR-130/301 promotes pulmonary hypertension. J. Clin. Invest. 124, 3514–3528.

Bertero, T., Robbe-Sermesant, K., Le Brigand, K., Ponzio, G., Pottier, N., Rezzonico, R., Mazure, N.M., Barbry, P., Mari, B., 2014b. MicroRNA target identification: lessons from hypoxamiRs. Antioxid. Redox Signal. 21, 1249–1268.

Bertero, T., Cottrill, K., Krauszman, A., Lu, Y., Annis, S., Hale, A., Bhat, B., Waxman, A.B., Chau, B.N., Kuebler, W.M., Chan, S.Y., 2015a. The microRNA-130/301 family controls vasoconstriction in pulmonary hypertension. J. Biol. Chem. 290, 2069–2085.

Bertero, T., Cottrill, K.A., Annis, S., Bhat, B., Gochuico, B.R., Osorio, J.C., Rosas, I., Haley, K.J., Corey, K.E., Chung, R.T., Nelson Chau, B., Chan, S.Y., 2015b. A YAP/TAZ-miR-130/301 molecular circuit exerts systems-level control of fibrosis in a network of human diseases and physiologic conditions. Sci. Rep. 5, 18277.

Bertero, T., Cottrill, K.A., Lu, Y., Haeger, C.M., Dieffenbach, P., Annis, S., Hale, A., Bhat, B., Kaimal, V., Zhang, Y.Y., Graham, B.B., Kumar, R., Saggar, R., Wallace, W.D., Ross, D.J., Black, S.M., Fratz, S., Fineman, J.R., Vargas, S.O., Haley, K.J., Waxman, A.B., Chau, B.N., Fredenburgh, L.E., Chan, S.Y., 2015c. Matrix remodeling promotes pulmonary hypertension through feedback mechanoactivation of the YAP/TAZ-miR-130/301 circuit. Cell Rep. 13, 1016–1032.

Bertero, T., Oldham, W.M., Cottrill, K.A., Pisano, S., Vanderpool, R.R., Yu, Q., Zhao, J., Tai, Y., Tang, Y., Zhang, Y.Y., Rehman, S., Sugahara, M., Qi, Z., Gorcsan 3RD, J., Vargas, S.O., Saggar, R., Wallace, W.D., Ross, D.J., Haley, K.J., Waxman, A.B., Parikh, V.N., De Marco, T., Hsue, P.Y., Morris, A., Simon, M.A., Norris, K.A.,

Gaggioli, C., Loscalzo, J., Fessel, J., Chan, S.Y., 2016. Vascular stiffness mechanoactivates YAP/TAZ-dependent glutaminolysis to drive pulmonary hypertension. J. Clin. Invest. 126, 3313–3335.
Bodempudi, V., Hergert, P., Smith, K., Xia, H., Herrera, J., Peterson, M., Khalil, W., Kahm, J., Bitterman, P.B., Henke, C.A., 2014. miR-210 promotes IPF fibroblast proliferation in response to hypoxia. Am. J. Physiol. Lung Cell. Mol. Physiol. 307, L283–L294.
Boldin, M.P., Taganov, K.D., Rao, D.S., Yang, L., Zhao, J.L., Kalwani, M., Garcia-Flores, Y., Luong, M., Devrekanli, A., Xu, J., Sun, G., Tay, J., Linsley, P.S., Baltimore, D., 2011. miR-146a is a significant brake on autoimmunity, myeloproliferation, and cancer in mice. J. Exp. Med. 208, 1189–1201.
Bonci, D., 2010. MicroRNA-21 as therapeutic target in cancer and cardiovascular disease. Recent Pat. Cardiovasc. Drug Discov. 5, 156–161.
Bonnet, S., Archer, S.L., Allalunis-Turner, J., Haromy, A., Beaulieu, C., Thompson, R., Lee, C.T., Lopaschuk, G.D., Puttagunta, L., Harry, G., Hashimoto, K., Porter, C.J., Andrade, M.A., Thebaud, B., Michelakis, E.D., 2007. A mitochondria-K + channel axis is suppressed in cancer and its normalization promotes apoptosis and inhibits cancer growth. Cancer Cell 11, 37–51.
Brabletz, S., Brabletz, T., 2010. The ZEB/miR-200 feedback loop—a motor of cellular plasticity in development and cancer? EMBO Rep. 11, 670–677.
Brock, M., Haider, T.J., Vogel, J., Gassmann, M., Speich, R., Trenkmann, M., Ulrich, S., Kohler, M., Huber, L.C., 2015. The hypoxia-induced microRNA-130a controls pulmonary smooth muscle cell proliferation by directly targeting CDKN1A. Int. J. Biochem. Cell Biol. 61, 129–137.
Broderick, J.A., Zamore, P.D., 2011. MicroRNA therapeutics. Gene Ther. 18, 1104–1110.
Bruning, U., Cerone, L., Neufeld, Z., Fitzpatrick, S.F., Cheong, A., Scholz, C.C., Simpson, D.A., Leonard, M.O., Tambuwala, M.M., Cummins, E.P., Taylor, C.T., 2011. MicroRNA-155 promotes resolution of hypoxia-inducible factor 1alpha activity during prolonged hypoxia. Mol. Cell. Biol. 31, 4087–4096.
Bueno, M.J., Malumbres, M., 2011. MicroRNAs and the cell cycle. Biochim. Biophys. Acta 1812, 592–601.
Bueno, M.J., Perez De Castro, I., Gomez De Cedron, M., Santos, J., Calin, G.A., Cigudosa, J.C., Croce, C.M., Fernandez-Piqueras, J., Malumbres, M., 2008. Genetic and epigenetic silencing of microRNA-203 enhances ABL1 and BCR-ABL1 oncogene expression. Cancer Cell 13, 496–506.
Buffa, F.M., Camps, C., Winchester, L., Snell, C.E., Gee, H.E., Sheldon, H., Taylor, M., Harris, A.L., Ragoussis, J., 2011. microRNA-associated progression pathways and potential therapeutic targets identified by integrated mRNA and microRNA expression profiling in breast cancer. Cancer Res. 71, 5635–5645.
Calin, G.A., Dumitru, C.D., Shimizu, M., Bichi, R., Zupo, S., Noch, E., Aldler, H., Rattan, S., Keating, M., Rai, K., Rassenti, L., Kipps, T., Negrini, M., Bullrich, F., Croce, C.M., 2002. Frequent deletions and down-regulation of micro- RNA genes miR15 and miR16 at 13q14 in chronic lymphocytic leukemia. Proc. Natl. Acad. Sci. U.S.A. 99, 15524–15529.
Calin, G.A., Cimmino, A., Fabbri, M., Ferracin, M., Wojcik, S.E., Shimizu, M., Taccioli, C., Zanesi, N., Garzon, R., Aqeilan, R.I., Alder, H., Volinia, S., Rassenti, L., Liu, X., Liu, C.G., Kipps, T.J., Negrini, M., Croce, C.M., 2008. MiR-15a and miR-16-1 cluster functions in human leukemia. Proc. Natl. Acad. Sci. U.S.A. 105, 5166–5171.
Camps, C., Buffa, F.M., Colella, S., Moore, J., Sotiriou, C., Sheldon, H., Harris, A.L., Gleadle, J.M., Ragoussis, J., 2008. hsa-miR-210 is induced by hypoxia and is an independent prognostic factor in breast cancer. Clin. Cancer Res. 14, 1340–1348.

Carleton, M., Cleary, M.A., Linsley, P.S., 2007. MicroRNAs and cell cycle regulation. Cell Cycle 6, 2127–2132.
Caruso, P., Maclean, M.R., Khanin, R., McClure, J., Soon, E., Southgate, M., MacDonald, R.A., Greig, J.A., Robertson, K.E., Masson, R., Denby, L., Dempsie, Y., Long, L., Morrell, N.W., Baker, A.H., 2010. Dynamic changes in lung microRNA profiles during the development of pulmonary hypertension due to chronic hypoxia and monocrotaline. Arterioscler. Thromb. Vasc. Biol. 30, 716–723.
Cascio, S., D'Andrea, A., Ferla, R., Surmacz, E., Gulotta, E., Amodeo, V., Bazan, V., Gebbia, N., Russo, A., 2010. miR-20b modulates VEGF expression by targeting HIF-1 alpha and STAT3 in MCF-7 breast cancer cells. J. Cell. Physiol. 224, 242–249.
Chan, S.Y., Loscalzo, J., 2010. MicroRNA-210: a unique and pleiotropic hypoxamir. Cell Cycle 9, 1072–1083.
Chan, S.Y., Loscalzo, J., 2012. The emerging paradigm of network medicine in the study of human disease. Circ. Res. 111, 359–374.
Chan, S.Y., Zhang, Y.Y., Hemann, C., Mahoney, C.E., Zweier, J.L., Loscalzo, J., 2009. MicroRNA-210 controls mitochondrial metabolism during hypoxia by repressing the iron–sulfur cluster assembly proteins ISCU1/2. Cell Metab. 10, 273–284.
Chan, Y.C., Khanna, S., Roy, S., Sen, C.K., 2011. miR-200b targets Ets-1 and is down-regulated by hypoxia to induce angiogenic response of endothelial cells. J. Biol. Chem. 286, 2047–2056.
Chan, Y.C., Banerjee, J., Choi, S.Y., Sen, C.K., 2012. miR-210: the master hypoxamir. Microcirculation 19, 215–223.
Chang, C.J., Chao, C.H., Xia, W., Yang, J.Y., Xiong, Y., Li, C.W., Yu, W.H., Rehman, S.K., Hsu, J.L., Lee, H.H., Liu, M., Chen, C.T., Yu, D., Hung, M.C., 2011. p53 regulates epithelial–mesenchymal transition and stem cell properties through modulating miRNAs. Nat. Cell Biol. 13, 317–323.
Chau, B.N., Xin, C., Hartner, J., Ren, S., Castano, A.P., Linn, G., Li, J., Tran, P.T., Kaimal, V., Huang, X., Chang, A.N., Li, S., Kalra, A., Grafals, M., Portilla, D., Mackenna, D.A., Orkin, S.H., Duffield, J.S., 2012. MicroRNA-21 promotes fibrosis of the kidney by silencing metabolic pathways. Sci. Transl. Med. 4. 121ra18.
Chen, X., Ba, Y., Ma, L., Cai, X., Yin, Y., Wang, K., Guo, J., Zhang, Y., Chen, J., Guo, X., Li, Q., Li, X., Wang, W., Wang, J., Jiang, X., Xiang, Y., Xu, C., Zheng, P., Zhang, J., Li, R., Zhang, H., Shang, X., Gong, T., Ning, G., Zen, K., Zhang, C.Y., 2008. Characterization of microRNAs in serum: a novel class of biomarkers for diagnosis of cancer and other diseases. Cell Res. 18, 997–1006.
Chen, Z., Li, Y., Zhang, H., Huang, P., Luthra, R., 2010. Hypoxia-regulated microRNA-210 modulates mitochondrial function and decreases ISCU and COX10 expression. Oncogene 29, 4362–4368.
Chen, B., Li, H., Zeng, X., Yang, P., Liu, X., Zhao, X., Liang, S., 2012a. Roles of micro-RNA on cancer cell metabolism. J. Transl. Med. 10, 228.
Chen, X., Liang, H., Zhang, J., Zen, K., Zhang, C.Y., 2012b. Secreted microRNAs: a new form of intercellular communication. Trends Cell Biol. 22, 125–132.
Chen, Z., Li, C., Xu, Y., Li, Y., Yang, H., Rao, L., 2014. Circulating level of miR-378 predicts left ventricular hypertrophy in patients with aortic stenosis. PLoS One 9, e105702.
Cheng, Y., Zhang, C., 2010. MicroRNA-21 in cardiovascular disease. J. Cardiovasc. Transl. Res. 3, 251–255.
Cheng, C.J., Bahal, R., Babar, I.A., Pincus, Z., Barrera, F., Liu, C., Svoronos, A., Braddock, D.T., Glazer, P.M., Engelman, D.M., Saltzman, W.M., Slack, F.J., 2015. MicroRNA silencing for cancer therapy targeted to the tumour microenvironment. Nature 518, 107–110.

Chi, S.W., Zang, J.B., Mele, A., Darnell, R.B., 2009. Argonaute HITS-CLIP decodes microRNA-mRNA interaction maps. Nature 460, 479–486.
Chim, S.S., Shing, T.K., Hung, E.C., Leung, T.Y., Lau, T.K., Chiu, R.W., Lo, Y.M., 2008. Detection and characterization of placental microRNAs in maternal plasma. Clin. Chem. 54, 482–490.
Chin, L.J., Ratner, E., Leng, S., Zhai, R., Nallur, S., Babar, I., Muller, R.U., Straka, E., Su, L., Burki, E.A., Crowell, R.E., Patel, R., Kulkarni, T., Homer, R., Zelterman, D., Kidd, K.K., Zhu, Y., Christiani, D.C., Belinsky, S.A., Slack, F.J., Weidhaas, J.B., 2008. A SNP in a let-7 microRNA complementary site in the KRAS 3′ untranslated region increases non-small cell lung cancer risk. Cancer Res. 68, 8535–8540.
Chio, C.C., Lin, J.W., Cheng, H.A., Chiu, W.T., Wang, Y.H., Wang, J.J., Hsing, C.H., Chen, R.M., 2013. MicroRNA-210 targets antiapoptotic Bcl-2 expression and mediates hypoxia-induced apoptosis of neuroblastoma cells. Arch. Toxicol. 87, 459–468.
Choi, W.Y., Giraldez, A.J., Schier, A.F., 2007. Target protectors reveal dampening and balancing of Nodal agonist and antagonist by miR-430. Science 318, 271–274.
Chun, H.J., Bonnet, S., Chan, S.Y., 2017. Translating microRNA biology in pulmonary hypertension: it will take more than "miR" words. Am. J. Respir. Crit. Care Med. 195 (2), 167–178.
Cookson, V.J., Bentley, M.A., Hogan, B.V., Horgan, K., Hayward, B.E., Hazelwood, L.D., Hughes, T.A., 2012. Circulating microRNA profiles reflect the presence of breast tumours but not the profiles of microRNAs within the tumours. Cell. Oncol. (Dordr.) 35, 301–308.
Corapcioglu, M.E., Ogul, H., 2015. miSEA: microRNA set enrichment analysis. Biosystems 134, 37–42.
Cortez, M.A., Bueso-Ramos, C., Ferdin, J., Lopez-Berestein, G., Sood, A.K., Calin, G.A., 2011. MicroRNAs in body fluids—the mix of hormones and biomarkers. Nat. Rev. Clin. Oncol. 8, 467–477.
Cottrill, K.A., Chan, S.Y., Loscalzo, J., 2014. Hypoxamirs and mitochondrial metabolism. Antioxid. Redox Signal. 21, 1189–1201.
Courboulin, A., Paulin, R., Giguere, N.J., Saksouk, N., Perreault, T., Meloche, J., Paquet, E.R., Biardel, S., Provencher, S., Cote, J., Simard, M.J., Bonnet, S., 2011. Role for miR-204 in human pulmonary arterial hypertension. J. Exp. Med. 208, 535–548.
Covello, K.L., Kehler, J., Yu, H., Gordan, J.D., Arsham, A.M., Hu, C.J., Labosky, P.A., Simon, M.C., Keith, B., 2006. HIF-2alpha regulates Oct-4: effects of hypoxia on stem cell function, embryonic development, and tumor growth. Genes Dev. 20, 557–570.
Creemers, E.E., Tijsen, A.J., Pinto, Y.M., 2012. Circulating microRNAs: novel biomarkers and extracellular communicators in cardiovascular disease? Circ. Res. 110, 483–495.
Crosby, M.E., Kulshreshtha, R., Ivan, M., Glazer, P.M., 2009. MicroRNA regulation of DNA repair gene expression in hypoxic stress. Cancer Res. 69, 1221–1229.
Cui, H., Seubert, B., Stahl, E., Dietz, H., Reuning, U., Moreno-Leon, L., Ilie, M., Hofman, P., Nagase, H., Mari, B., Kruger, A., 2015. Tissue inhibitor of metalloproteinases-1 induces a pro-tumourigenic increase of miR-210 in lung adenocarcinoma cells and their exosomes. Oncogene 34, 3640–3650.
Cui, X., Kong, C., Zhu, Y., Zeng, Y., Zhang, Z., Liu, X., Zhan, B., Piao, C., Jiang, Z., 2016. miR-130b, an onco-miRNA in bladder cancer, is directly regulated by NF-kappaB and sustains NF-kappaB activation by decreasing Cylindromatosis expression. Oncotarget 7, 48547–48561.
Cummins, E.P., Taylor, C.T., 2005. Hypoxia-responsive transcription factors. Pflugers Arch. 450, 363–371.
D'Alessandra, Y., Devanna, P., Limana, F., Straino, S., Di Carlo, A., Brambilla, P.G., Rubino, M., Carena, M.C., Spazzafumo, L., De Simone, M., Micheli, B.,

Biglioli, P., Achilli, F., Martelli, F., Maggiolini, S., Marenzi, G., Pompilio, G., Capogrossi, M.C., 2010. Circulating microRNAs are new and sensitive biomarkers of myocardial infarction. Eur. Heart J. 31, 2765–2773.

Dang, K., Myers, K.A., 2015. The role of hypoxia-induced miR-210 in cancer progression. Int. J. Mol. Sci. 16, 6353–6372.

Dang, C.V., Kim, J.W., Gao, P., Yustein, J., 2008. The interplay between MYC and HIF in cancer. Nat. Rev. Cancer 8, 51–56.

Dantuma, N.P., Heinen, C., Hoogstraten, D., 2009. The ubiquitin receptor Rad23: at the crossroads of nucleotide excision repair and proteasomal degradation. DNA Repair (Amst) 8, 449–460.

Das, S., Ghosal, S., Sen, R., Chakrabarti, J., 2014. lnCeDB: database of human long noncoding RNA acting as competing endogenous RNA. PLoS One 9, e98965.

Dejean, E., Renalier, M.H., Foisseau, M., Agirre, X., Joseph, N., De Paiva, G.R., Al Saati, T., Soulier, J., Desjobert, C., Lamant, L., Prosper, F., Felsher, D.W., Cavaille, J., Prats, H., Delsol, G., Giuriato, S., Meggetto, F., 2011. Hypoxia-microRNA-16 downregulation induces VEGF expression in anaplastic lymphoma kinase (ALK)-positive anaplastic large-cell lymphomas. Leukemia 25, 1882–1890.

Dimova, D.K., Dyson, N.J., 2005. The E2F transcriptional network: old acquaintances with new faces. Oncogene 24, 2810–2826.

Doench, J.G., Sharp, P.A., 2004. Specificity of microRNA target selection in translational repression. Genes Dev. 18, 504–511.

Donker, R.B., Mouillet, J.F., Nelson, D.M., Sadovsky, Y., 2007. The expression of Argonaute2 and related microRNA biogenesis proteins in normal and hypoxic trophoblasts. Mol. Hum. Reprod. 13, 273–279.

Du, R., Sun, W., Xia, L., Zhao, A., Yu, Y., Zhao, L., Wang, H., Huang, C., Sun, S., 2012. Hypoxia-induced down-regulation of microRNA-34a promotes EMT by targeting the Notch signaling pathway in tubular epithelial cells. PLoS One 7, e30771.

Ebert, M.S., Sharp, P.A., 2012. Roles for microRNAs in conferring robustness to biological processes. Cell 149, 515–524.

Eliopoulos, A.G., Havaki, S., Gorgoulis, V.G., 2016. DNA damage response and autophagy: a meaningful partnership. Front. Genet. 7, 204.

Etheridge, A., Lee, I., Hood, L., Galas, D., Wang, K., 2011. Extracellular microRNA: a new source of biomarkers. Mutat. Res. 717, 85–90.

Fan, Y., Siklenka, K., Arora, S.K., Ribeiro, P., Kimmins, S., Xia, J., 2016. miRNet—dissecting miRNA-target interactions and functional associations through network-based visual analysis. Nucleic Acids Res. 44, W135–W141.

Fang, R., Xiao, T., Fang, Z., Sun, Y., Li, F., Gao, Y., Feng, Y., Li, L., Wang, Y., Liu, X., Chen, H., Liu, X.Y., Ji, H., 2012. MicroRNA-143 (miR-143) regulates cancer glycolysis via targeting hexokinase 2 gene. J. Biol. Chem. 287, 23227–23235.

Fasanaro, P., D'Alessandra, Y., Di Stefano, V., Melchionna, R., Romani, S., Pompilio, G., Capogrossi, M.C., Martelli, F., 2008. MicroRNA-210 modulates endothelial cell response to hypoxia and inhibits the receptor tyrosine kinase ligand Ephrin-A3. J. Biol. Chem. 283, 15878–15883.

Fasanaro, P., Greco, S., Lorenzi, M., Pescatori, M., Brioschi, M., Kulshreshtha, R., Banfi, C., Stubbs, A., Calin, G.A., Ivan, M., Capogrossi, M.C., Martelli, F., 2009. An integrated approach for experimental target identification of hypoxia-induced miR-210. J. Biol. Chem. 284, 35134–35143.

Fasanaro, P., Romani, S., Voellenkle, C., Maimone, B., Capogrossi, M.C., Martelli, F., 2012. ROD1 is a seedless target gene of hypoxia-induced miR-210. PLoS One 7, e44651.

Favaro, E., Ramachandran, A., McCormick, R., Gee, H., Blancher, C., Crosby, M., Devlin, C., Blick, C., Buffa, F., Li, J.L., Vojnovic, B., Pires Das Neves, R.,

Glazer, P., Iborra, F., Ivan, M., Ragoussis, J., Harris, A.L., 2010. MicroRNA-210 regulates mitochondrial free radical response to hypoxia and krebs cycle in cancer cells by targeting iron sulfur cluster protein ISCU. PLoS One 5, e10345.

Fiedler, J., Jazbutyte, V., Kirchmaier, B.C., Gupta, S.K., Lorenzen, J., Hartmann, D., Galuppo, P., Kneitz, S., Pena, J.T., Sohn-Lee, C., Loyer, X., Soutschek, J., Brand, T., Tuschl, T., Heineke, J., Martin, U., Schulte-Merker, S., Ertl, G., Engelhardt, S., Bauersachs, J., Thum, T., 2011. MicroRNA-24 regulates vascularity after myocardial infarction. Circulation 124, 720–730.

Fischer, K.R., Durrans, A., Lee, S., Sheng, J., Li, F., Wong, S.T., Choi, H., El Rayes, T., Ryu, S., Troeger, J., Schwabe, R.F., Vahdat, L.T., Altorki, N.K., Mittal, V., Gao, D., 2015. Epithelial-to-mesenchymal transition is not required for lung metastasis but contributes to chemoresistance. Nature 527, 472–476.

Foekens, J.A., Sieuwerts, A.M., Smid, M., Look, M.P., De Weerd, V., Boersma, A.W., Klijn, J.G., Wiemer, E.A., Martens, J.W., 2008. Four miRNAs associated with aggressiveness of lymph node-negative, estrogen receptor-positive human breast cancer. Proc. Natl. Acad. Sci. U.S.A. 105, 13021–13026.

Frankel, L.B., Christoffersen, N.R., Jacobsen, A., Lindow, M., Krogh, A., Lund, A.H., 2008. Programmed cell death 4 (PDCD4) is an important functional target of the microRNA miR-21 in breast cancer cells. J. Biol. Chem. 283, 1026–1033.

Gantier, M.P., McCoy, C.E., Rusinova, I., Saulep, D., Wang, D., Xu, D., Irving, A.T., Behlke, M.A., Hertzog, P.J., MacKay, F., Williams, B.R., 2011. Analysis of microRNA turnover in mammalian cells following Dicer1 ablation. Nucleic Acids Res. 39, 5692–5703.

Gao, P., Tchernyshyov, I., Chang, T.C., Lee, Y.S., Kita, K., Ochi, T., Zeller, K.I., De Marzo, A.M., Van Eyk, J.E., Mendell, J.T., Dang, C.V., 2009. c-Myc suppression of miR-23a/b enhances mitochondrial glutaminase expression and glutamine metabolism. Nature 458, 762–765.

Garzon, R., Liu, S., Fabbri, M., Liu, Z., Heaphy, C.E., Callegari, E., Schwind, S., Pang, J., Yu, J., Muthusamy, N., Havelange, V., Volinia, S., Blum, W., Rush, L.J., Perrotti, D., Andreeff, M., Bloomfield, C.D., Byrd, J.C., Chan, K., Wu, L.C., Croce, C.M., Marcucci, G., 2009. MicroRNA-29b induces global DNA hypomethylation and tumor suppressor gene reexpression in acute myeloid leukemia by targeting directly DNMT3A and 3B and indirectly DNMT1. Blood 113, 6411–6418.

Garzon, R., Marcucci, G., Croce, C.M., 2010. Targeting microRNAs in cancer: rationale, strategies and challenges. Nat. Rev. Drug Discov. 9, 775–789.

Gee, H.E., Camps, C., Buffa, F.M., Patiar, S., Winter, S.C., Betts, G., Homer, J., Corbridge, R., Cox, G., West, C.M., Ragoussis, J., Harris, A.L., 2010. hsa-mir-210 is a marker of tumor hypoxia and a prognostic factor in head and neck cancer. Cancer 116, 2148–2158.

Gee, H.E., Ivan, C., Calin, G.A., Ivan, M., 2014. HypoxamiRs and cancer: from biology to targeted therapy. Antioxid. Redox Signal. 21, 1220–1238.

Ghosh, G., Subramanian, I.V., Adhikari, N., Zhang, X., Joshi, H.P., Basi, D., Chandrashekhar, Y.S., Hall, J.L., Roy, S., Zeng, Y., Ramakrishnan, S., 2010. Hypoxia-induced microRNA-424 expression in human endothelial cells regulates HIF-alpha isoforms and promotes angiogenesis. J. Clin. Invest. 120, 4141–4154.

Giannakakis, A., Sandaltzopoulos, R., Greshock, J., Liang, S., Huang, J., Hasegawa, K., Li, C., O'Brien-Jenkins, A., Katsaros, D., Weber, B.L., Simon, C., Coukos, G., Zhang, L., 2008. miR-210 links hypoxia with cell cycle regulation and is deleted in human epithelial ovarian cancer. Cancer Biol. Ther. 7, 255–264.

Gill, S.E., Pape, M.C., Khokha, R., Watson, A.J., Leco, K.J., 2003. A null mutation for tissue inhibitor of metalloproteinases-3 (Timp-3) impairs murine bronchiole branching morphogenesis. Dev. Biol. 261, 313–323.

Godwin, J.G., Ge, X., Stephan, K., Jurisch, A., Tullius, S.G., Iacomini, J., 2010. Identification of a microRNA signature of renal ischemia reperfusion injury. Proc. Natl. Acad. Sci. U.S.A. 107, 14339–14344.
Gordan, J.D., Thompson, C.B., Simon, M.C., 2007. HIF and c-Myc: sibling rivals for control of cancer cell metabolism and proliferation. Cancer Cell 12, 108–113.
Gorospe, M., Tominaga, K., Wu, X., Fahling, M., Ivan, M., 2011. Post-transcriptional control of the hypoxic response by RNA-binding proteins and microRNAs. Front. Mol. Neurosci. 4, 7.
Greco, S., De Simone, M., Colussi, C., Zaccagnini, G., Fasanaro, P., Pescatori, M., Cardani, R., Perbellini, R., Isaia, E., Sale, P., Meola, G., Capogrossi, M.C., Gaetano, C., Martelli, F., 2009. Common micro-RNA signature in skeletal muscle damage and regeneration induced by Duchenne muscular dystrophy and acute ischemia. FASEB J. 23, 3335–3346.
Greco, S., Gaetano, C., Martelli, F., 2014. HypoxamiR regulation and function in ischemic cardiovascular diseases. Antioxid. Redox Signal. 21, 1202–1219.
Gregory, R.I., Yan, K.P., Amuthan, G., Chendrimada, T., Doratotaj, B., Cooch, N., Shiekhattar, R., 2004. The microprocessor complex mediates the genesis of microRNAs. Nature 432, 235–240. Epub 2004 Nov 7.
Grosso, S., Doyen, J., Parks, S.K., Bertero, T., Paye, A., Cardinaud, B., Gounon, P., Lacas-Gervais, S., Noel, A., Pouyssegur, J., Barbry, P., Mazure, N.M., Mari, B., 2013. MiR-210 promotes a hypoxic phenotype and increases radioresistance in human lung cancer cell lines. Cell Death Dis. 4. e544.
Guay, C., Regazzi, R., 2013. Circulating microRNAs as novel biomarkers for diabetes mellitus. Nat. Rev. Endocrinol. 9, 513–521.
Gunel, T., Zeybek, Y.G., Akcakaya, P., Kalelioglu, I., Benian, A., Ermis, H., Aydinli, K., 2011. Serum microRNA expression in pregnancies with preeclampsia. Genet. Mol. Res. 10, 4034–4040.
Gupta, S.K., Bang, C., Thum, T., 2010. Circulating microRNAs as biomarkers and potential paracrine mediators of cardiovascular disease. Circ. Cardiovasc. Genet. 3, 484–488.
Ha, M., Kim, V.N., 2014. Regulation of microRNA biogenesis. Nat. Rev. Mol. Cell Biol. 15, 509–524.
Hafner, M., Landthaler, M., Burger, L., Khorshid, M., Hausser, J., Berninger, P., Rothballer, A., Ascano, M., Jungkamp, A.C., Munschauer, M., Ulrich, A., Wardle, G.S., Dewell, S., Zavolan, M., Tuschl, T., 2010. PAR-CliP—a method to identify transcriptome-wide the binding sites of RNA binding proteins. J. Vis. Exp. (41), pii: 2034. http://dx.doi.org/10.3791/2034.
Hale, A., Lee, C., Annis, S., Min, P.K., Pande, R., Creager, M.A., Julian, C.G., Moore, L.G., Mitsialis, S.A., Hwang, S.J., Kourembanas, S., Chan, S.Y., 2014. An Argonaute 2 switch regulates circulating miR-210 to coordinate hypoxic adaptation across cells. Biochim. Biophys. Acta 1843, 2528–2542.
Hammer, S., To, K.K., Yoo, Y.G., Koshiji, M., Huang, L.E., 2007. Hypoxic suppression of the cell cycle gene CDC25A in tumor cells. Cell Cycle 6, 1919–1926.
Hansen, T.B., Jensen, T.I., Clausen, B.H., Bramsen, J.B., Finsen, B., Damgaard, C.K., Kjems, J., 2013. Natural RNA circles function as efficient microRNA sponges. Nature 495, 384–388.
Harper, A.R., Nayee, S., Topol, E.J., 2015. Protective alleles and modifier variants in human health and disease. Nat. Rev. Genet. 16, 689–701.
Harris, A.L., 2002. Hypoxia—a key regulatory factor in tumour growth. Nat. Rev. Cancer 2, 38–47.
Hayes, J., Peruzzi, P.P., Lawler, S., 2014. MicroRNAs in cancer: biomarkers, functions and therapy. Trends Mol. Med. 20, 460–469.
He, J., Wu, J., Xu, N., Xie, W., Li, M., Li, J., Jiang, Y., Yang, B.B., Zhang, Y., 2013. MiR-210 disturbs mitotic progression through regulating a group of mitosis-related genes. Nucleic Acids Res. 41, 498–508.

Helwak, A., Kudla, G., Dudnakova, T., Tollervey, D., 2013. Mapping the human miRNA interactome by CLASH reveals frequent noncanonical binding. Cell 153, 654–665.
Hensley, C.T., Wasti, A.T., Deberardinis, R.J., 2013. Glutamine and cancer: cell biology, physiology, and clinical opportunities. J. Clin. Invest. 123, 3678–3684.
Henze, A.T., Acker, T., 2010. Feedback regulators of hypoxia-inducible factors and their role in cancer biology. Cell Cycle 9, 2749–2763.
Hergenreider, E., Heydt, S., Treguer, K., Boettger, T., Horrevoets, A.J., Zeiher, A.M., Scheffer, M.P., Frangakis, A.S., Yin, X., Mayr, M., Braun, T., Urbich, C., Boon, R.A., Dimmeler, S., 2012. Atheroprotective communication between endothelial cells and smooth muscle cells through miRNAs. Nat. Cell Biol. 14, 249–256.
Hermeking, H., 2012. MicroRNAs in the p53 network: micromanagement of tumour suppression. Nat. Rev. Cancer 12, 613–626.
Ho, J.J., Metcalf, J.L., Yan, M.S., Turgeon, P.J., Wang, J.J., Chalsev, M., Petruzziello-Pellegrini, T.N., Tsui, A.K., He, J.Z., Dhamko, H., Man, H.S., Robb, G.B., Teh, B.T., Ohh, M., Marsden, P.A., 2012. Functional importance of Dicer protein in the adaptive cellular response to hypoxia. J. Biol. Chem. 287, 29003–29020.
Hong, Z., Chen, K.H., Dasgupta, A., Potus, F., Dunham-Snary, K., Bonnet, S., Tian, L., Fu, J., Breuils-Bonnet, S., Provencher, S., Wu, D., Mewburn, J., Ormiston, M.L., Archer, S.L., 2017. MicroRNA-138 and MicroRNA-25 down-regulate mitochondrial calcium uniporter, causing the pulmonary arterial hypertension cancer phenotype. Am. J. Respir. Crit. Care Med. 195, 515–529.
Hu, S., Huang, M., Li, Z., Jia, F., Ghosh, Z., Lijkwan, M.A., Fasanaro, P., Sun, N., Wang, X., Martelli, F., Robbins, R.C., Wu, J.C., 2010. MicroRNA-210 as a novel therapy for treatment of ischemic heart disease. Circulation 122, S124–S131.
Hu, J., Sun, T., Wang, H., Chen, Z., Wang, S., Yuan, L., Liu, T., Li, H.R., Wang, P., Feng, Y., Wang, Q., Mclendon, R.E., Friedman, A.H., Keir, S.T., Bigner, D.D., Rathmell, J., Fu, X.D., Li, Q.J., Wang, X.F., 2016. MiR-215 is induced post-transcriptionally via HIF-Drosha complex and mediates glioma-initiating cell adaptation to hypoxia by targeting KDM1B. Cancer Cell 29, 49–60.
Hua, Z., Lv, Q., Ye, W., Wong, C.K., Cai, G., Gu, D., Ji, Y., Zhao, C., Wang, J., Yang, B.B., Zhang, Y., 2006. MiRNA-directed regulation of VEGF and other angiogenic factors under hypoxia. PLoS One 1, e116.
Huang, J., Liang, Z., Yang, B., Tian, H., Ma, J., Zhang, H., 2007. Derepression of microRNA-mediated protein translation inhibition by apolipoprotein B mRNA-editing enzyme catalytic polypeptide-like 3G (APOBEC3G) and its family members. J. Biol. Chem. 282, 33632–33640.
Huang, X., Ding, L., Bennewith, K.L., Tong, R.T., Welford, S.M., Ang, K.K., Story, M., Le, Q.T., Giaccia, A.J., 2009. Hypoxia-inducible mir-210 regulates normoxic gene expression involved in tumor initiation. Mol. Cell 35, 856–867.
Humeau, M., Vignolle-Vidoni, A., Sicard, F., Martins, F., Bournet, B., Buscail, L., Torrisani, J., Cordelier, P., 2015. Salivary microRNA in pancreatic cancer patients. PLoS One 10, e0130996.
Iannone, L., Zhao, L., Dubois, O., Duluc, L., Rhodes, C.J., Wharton, J., Wilkins, M.R., Leiper, J., Wojciak-Stothard, B., 2014. miR-21/DDAH1 pathway regulates pulmonary vascular responses to hypoxia. Biochem. J. 462, 103–112.
Imtiyaz, H.Z., Simon, M.C., 2010. Hypoxia-inducible factors as essential regulators of inflammation. Curr. Top. Microbiol. Immunol. 345, 105–120.
Ingolia, N.T., Brar, G.A., Rouskin, S., Mcgeachy, A.M., Weissman, J.S., 2012. The ribosome profiling strategy for monitoring translation in vivo by deep sequencing of ribosome-protected mRNA fragments. Nat. Protoc. 7, 1534–1550.
Ishibashi, O., Ohkuchi, A., Ali, M.M., Kurashina, R., Luo, S.S., Ishikawa, T., Takizawa, T., Hirashima, C., Takahashi, K., Migita, M., Ishikawa, G., Yoneyama, K., Asakura, H.,

Izumi, A., Matsubara, S., Takeshita, T., 2012. Hydroxysteroid (17-beta) dehydrogenase 1 is dysregulated by miR-210 and miR-518c that are aberrantly expressed in preeclamptic placentas: a novel marker for predicting preeclampsia. Hypertension 59, 265–273.
Ivan, M., Kondo, K., Yang, H., Kim, W., Valiando, J., Ohh, M., Salic, A., Asara, J.M., Lane, W.S., Kaelin JR., W.G., 2001. HIFalpha targeted for VHL-mediated destruction by proline hydroxylation: implications for O2 sensing. Science 292, 464–468.
Iyer, N.V., Kotch, L.E., Agani, F., Leung, S.W., Laughner, E., Wenger, R.H., Gassmann, M., Gearhart, J.D., Lawler, A.M., Yu, A.Y., Semenza, G.L., 1998. Cellular and developmental control of O2 homeostasis by hypoxia-inducible factor 1 alpha. Genes Dev. 12, 149–162.
Izreig, S., Samborska, B., Johnson, R.M., Sergushichev, A., Ma, E.H., Lussier, C., Loginicheva, E., Donayo, A.O., Poffenberger, M.C., Sagan, S.M., Vincent, E.E., Artyomov, M.N., Duchaine, T.F., Jones, R.G., 2016. The miR-17 approximately 92 microRNA cluster is a global regulator of tumor metabolism. Cell Rep. 16, 1915–1928.
Jaakkola, P., Mole, D.R., Tian, Y.M., Wilson, M.I., Gielbert, J., Gaskell, S.J., Von Kriegsheim, A., Hebestreit, H.F., Mukherji, M., Schofield, C.J., Maxwell, P.H., Pugh, C.W., Ratcliffe, P.J., 2001. Targeting of HIF-alpha to the von Hippel–Lindau ubiquitylation complex by O2-regulated prolyl hydroxylation. Science 292, 468–472.
Jablonski, K.A., Gaudet, A.D., Amici, S.A., Popovich, P.G., Guerau-De-Arellano, M., 2016. Control of the inflammatory macrophage transcriptional signature by miR-155. PLoS One 11, e0159724.
Jackson, S.P., Bartek, J., 2009. The DNA-damage response in human biology and disease. Nature 461, 1071–1078.
Janssen, H.L., Reesink, H.W., Lawitz, E.J., Zeuzem, S., Rodriguez-Torres, M., Patel, K., Van Der Meer, A.J., Patick, A.K., Chen, A., Zhou, Y., Persson, R., King, B.D., Kauppinen, S., Levin, A.A., Hodges, M.R., 2013. Treatment of HCV infection by targeting microRNA. N. Engl. J. Med. 368, 1685–1694.
Jarry, J., Schadendorf, D., Greenwood, C., Spatz, A., Van Kempen, L.C., 2014. The validity of circulating microRNAs in oncology: five years of challenges and contradictions. Mol. Oncol. 8, 819–829.
Jiang, S., Zhang, L.F., Zhang, H.W., Hu, S., Lu, M.H., Liang, S., Li, B., Li, Y., Li, D., Wang, E.D., Liu, M.F., 2012. A novel miR-155/miR-143 cascade controls glycolysis by regulating hexokinase 2 in breast cancer cells. EMBO J. 31, 1985–1998.
Jiang, Y., Zhu, Y., Wang, X., Gong, J., Hu, C., Guo, B., Zhu, B., Li, Y., 2015. Temporal regulation of HIF-1 and NF-kappaB in hypoxic hepatocarcinoma cells. Oncotarget 6, 9409–9419.
Johnson, C.D., Esquela-Kerscher, A., Stefani, G., Byrom, M., Kelnar, K., Ovcharenko, D., Wilson, M., Wang, X., Shelton, J., Shingara, J., Chin, L., Brown, D., Slack, F.J., 2007. The let-7 microRNA represses cell proliferation pathways in human cells. Cancer Res. 67, 7713–7722.
Jung, E.J., Santarpia, L., Kim, J., Esteva, F.J., Moretti, E., Buzdar, A.U., Di Leo, A., Le, X.F., Bast JR., R.C., Park, S.T., Pusztai, L., Calin, G.A., 2012. Plasma microRNA 210 levels correlate with sensitivity to trastuzumab and tumor presence in breast cancer patients. Cancer 118, 2603–2614.
Jurkin, J., Schichl, Y.M., Koeffel, R., Bauer, T., Richter, S., Konradi, S., Gesslbauer, B., Strobl, H., 2010. miR-146a is differentially expressed by myeloid dendritic cell subsets and desensitizes cells to TLR2-dependent activation. J. Immunol. 184, 4955–4965.
Kai, A.K., Chan, L.K., Lo, R.C., Lee, J.M., Wong, C.C., Wong, J.C., Ng, I.O., 2016. Down-regulation of TIMP2 by HIF-1alpha/miR-210/HIF-3alpha regulatory feedback circuit enhances cancer metastasis in hepatocellular carcinoma. Hepatology 64, 473–487.

Karakas, M., Schulte, C., Appelbaum, S., Ojeda, F., Lackner, K.J., Munzel, T., Schnabel, R.B., Blankenberg, S., Zeller, T., 2017. Circulating microRNAs strongly predict cardiovascular death in patients with coronary artery disease-results from the large AtheroGene study. Eur. Heart J. 38 (7), 516–523.
Keane, F.K., Ratner, E.S., 2010. The KRAS-variant genetic test as a marker of increased risk of ovarian cancer. Rev. Obstet. Gynecol. 3, 118–121.
Kelly, T.J., Souza, A.L., Clish, C.B., Puigserver, P., 2011. A hypoxia-induced positive feedback loop promotes hypoxia-inducible factor 1alpha stability through miR-210 suppression of glycerol-3-phosphate dehydrogenase 1-like. Mol. Cell. Biol. 31, 2696–2706.
Kiga, K., Mimuro, H., Suzuki, M., Shinozaki-Ushiku, A., Kobayashi, T., Sanada, T., Kim, M., Ogawa, M., Iwasaki, Y.W., Kayo, H., Fukuda-Yuzawa, Y., Yashiro, M., Fukayama, M., Fukao, T., Sasakawa, C., 2014. Epigenetic silencing of miR-210 increases the proliferation of gastric epithelium during chronic Helicobacter pylori infection. Nat. Commun. 5, 4497.
Kim, H.W., Haider, H.K., Jiang, S., Ashraf, M., 2009. Ischemic preconditioning augments survival of stem cells via MIR-210 expression by targeting caspase-8 associated protein 2. J. Biol. Chem. 284, 33161–33168.
Kim, G.H., Ryan, J.J., Marsboom, G., Archer, S.L., 2011a. Epigenetic mechanisms of pulmonary hypertension. Pulm. Circ. 1, 347–356.
Kim, M., Kasinski, A.L., Slack, F.J., 2011b. MicroRNA therapeutics in preclinical cancer models. Lancet Oncol. 12, 319–321.
Kim, J., Kang, Y., Kojima, Y., Lighthouse, J.K., Hu, X., Aldred, M.A., Mclean, D.L., Park, H., Comhair, S.A., Greif, D.M., Erzurum, S.C., Chun, H.J., 2013a. An endothelial apelin-FGF link mediated by miR-424 and miR-503 is disrupted in pulmonary arterial hypertension. Nat. Med. 19, 74–82.
Kim, J.H., Park, S.G., Song, S.Y., Kim, J.K., Sung, J.H., 2013b. Reactive oxygen species-responsive miR-210 regulates proliferation and migration of adipose-derived stem cells via PTPN2. Cell Death Dis. 4. e588.
Kim, D., Sung, Y.M., Park, J., Kim, S., Kim, J., Ha, H., Bae, J.Y., Baek, D., 2016a. General rules for functional microRNA targeting. Nat. Genet. 48, 1517–1526.
Kim, Y.K., Kim, B., Kim, V.N., 2016b. Re-evaluation of the roles of DROSHA, export in 5, and DICER in microRNA biogenesis. Proc. Natl. Acad. Sci. U.S.A. 113, E1881–E1889.
Koppenol, W.H., Bounds, P.L., Dang, C.V., 2011. Otto Warburg's contributions to current concepts of cancer metabolism. Nat. Rev. Cancer 11, 325–337.
Kosaka, N., Iguchi, H., Yoshioka, Y., Takeshita, F., Matsuki, Y., Ochiya, T., 2010. Secretory mechanisms and intercellular transfer of microRNAs in living cells. J. Biol. Chem. 285, 17442–17452.
Kreeger, P.K., Lauffenburger, D.A., 2010. Cancer systems biology: a network modeling perspective. Carcinogenesis 31, 2–8.
Krock, B.L., Skuli, N., Simon, M.C., 2011. Hypoxia-induced angiogenesis: good and evil. Genes Cancer 2, 1117–1133.
Kroemer, G., Pouyssegur, J., 2008. Tumor cell metabolism: cancer's Achilles' heel. Cancer Cell 13, 472–482.
Kuijper, S., Turner, C.J., Adams, R.H., 2007. Regulation of angiogenesis by Eph-ephrin interactions. Trends Cardiovasc. Med. 17, 145–151.
Kulshreshtha, R., Ferracin, M., Negrini, M., Calin, G.A., Davuluri, R.V., Ivan, M., 2007a. Regulation of microRNA expression: the hypoxic component. Cell Cycle 6, 1426–1431.
Kulshreshtha, R., Ferracin, M., Wojcik, S.E., Garzon, R., Alder, H., Agosto-Perez, F.J., Davuluri, R., Liu, C.G., Croce, C.M., Negrini, M., Calin, G.A., Ivan, M., 2007b. A microRNA signature of hypoxia. Mol. Cell. Biol. 27, 1859–1867.

Kulshreshtha, R., Davuluri, R.V., Calin, G.A., Ivan, M., 2008. A microRNA component of the hypoxic response. Cell Death Differ. 15, 667–671.

Kurzynska-Kokorniak, A., Koralewska, N., Pokornowska, M., Urbanowicz, A., Tworak, A., Mickiewicz, A., Figlerowicz, M., 2015. The many faces of Dicer: the complexity of the mechanisms regulating Dicer gene expression and enzyme activities. Nucleic Acids Res. 43, 4365–4380.

Kusko, R.L., Brothers 2ND., J.F., Tedrow, J., Pandit, K., Huleihel, L., Perdomo, C., Liu, G., Juan-Guardela, B., Kass, D., Zhang, S., Lenburg, M., Martinez, F., Quackenbush, J., Sciurba, F., Limper, A., Geraci, M., Yang, I., Schwartz, D.A., Beane, J., Spira, A., Kaminski, N., 2016. Integrated genomics reveals convergent transcriptomic networks underlying chronic obstructive pulmonary disease and idiopathic pulmonary fibrosis. Am. J. Respir. Crit. Care Med. 194, 948–960.

Lal, A., Kim, H.H., Abdelmohsen, K., Kuwano, Y., Pullmann JR., R., Srikantan, S., Subrahmanyam, R., Martindale, J.L., Yang, X., Ahmed, F., Navarro, F., Dykxhoorn, D., Lieberman, J., Gorospe, M., 2008. p16(INK4a) translation suppressed by miR-24. PLoS One 3, e1864.

Lal, A., Navarro, F., Maher, C.A., Maliszewski, L.E., Yan, N., O'Day, E., Chowdhury, D., Dykxhoorn, D.M., Tsai, P., Hofmann, O., Becker, K.G., Gorospe, M., Hide, W., Lieberman, J., 2009. miR-24 Inhibits cell proliferation by targeting E2F2, MYC, and other cell-cycle genes via binding to "seedless" 3′'UTR microRNA recognition elements. Mol. Cell 35, 610–625.

Lawrie, C.H., Gal, S., Dunlop, H.M., Pushkaran, B., Liggins, A.P., Pulford, K., Banham, A.H., Pezzella, F., Boultwood, J., Wainscoat, J.S., Hatton, C.S., Harris, A.L., 2008. Detection of elevated levels of tumour-associated microRNAs in serum of patients with diffuse large B-cell lymphoma. Br. J. Haematol. 141, 672–675.

Le Brigand, K., Robbe-Sermesant, K., Mari, B., Barbry, P., 2010. MiRonTop: mining microRNAs targets across large scale gene expression studies. Bioinformatics 26, 3131–3132.

Lee, R.C., Feinbaum, R.L., Ambros, V., 1993. The C. elegans heterochronic gene lin-4 encodes small RNAs with antisense complementarity to lin-14. Cell 75, 843–854.

Lee, L.W., Zhang, S., Etheridge, A., Ma, L., Martin, D., Galas, D., Wang, K., 2010. Complexity of the microRNA repertoire revealed by next-generation sequencing. RNA 16, 2170–2180.

Lee, C., Mitsialis, S.A., Aslam, M., Vitali, S.H., Vergadi, E., Konstantinou, G., Sdrimas, K., Fernandez-Gonzalez, A., Kourembanas, S., 2012a. Exosomes mediate the cytoprotective action of mesenchymal stromal cells on hypoxia-induced pulmonary hypertension. Circulation 126, 2601–2611.

Lee, D.W., Futami, M., Carroll, M., Feng, Y., Wang, Z., Fernandez, M., Whichard, Z., Chen, Y., Kornblau, S., Shpall, E.J., Bueso-Ramos, C.E., Corey, S.J., 2012b. Loss of SHIP-1 protein expression in high-risk myelodysplastic syndromes is associated with miR-210 and miR-155. Oncogene 31, 4085–4094.

Lee, A., Mclean, D., Choi, J., Kang, H., Chang, W., Kim, J., 2014. Therapeutic implications of microRNAs in pulmonary arterial hypertension. BMB Rep. 47, 311–317.

Leone, G., Nuckolls, F., Ishida, S., Adams, M., Sears, R., Jakoi, L., Miron, A., Nevins, J.R., 2000. Identification of a novel E2F3 product suggests a mechanism for determining specificity of repression by Rb proteins. Mol. Cell. Biol. 20, 3626–3632.

Lewis, B.P., Burge, C.B., Bartel, D.P., 2005. Conserved seed pairing, often flanked by adenosines, indicates that thousands of human genes are microRNA targets. Cell 120, 15–20.

Li, H., Fan, J., Yin, Z., Wang, F., Chen, C., Wang, D.W., 2016. Identification of cardiac-related circulating microRNA profile in human chronic heart failure. Oncotarget 7, 33–45.

Liang, H., Studach, L., Hullinger, R.L., Xie, J., Andrisani, O.M., 2014. Down-regulation of RE-1 silencing transcription factor (REST) in advanced prostate cancer by hypoxia-induced miR-106b~25. Exp. Cell Res. 320, 188–199.
Lino Cardenas, C.L., Henaoui, I.S., Courcot, E., Roderburg, C., Cauffiez, C., Aubert, S., Copin, M.C., Wallaert, B., Glowacki, F., Dewaeles, E., Milosevic, J., Maurizio, J., Tedrow, J., Marcet, B., Lo-Guidice, J.M., Kaminski, N., Barbry, P., Luedde, T., Perrais, M., Mari, B., Pottier, N., 2013. miR-199a-5p Is upregulated during fibrogenic response to tissue injury and mediates TGFbeta-induced lung fibroblast activation by targeting caveolin-1. PLoS Genet. 9, e1003291.
Linsley, P.S., Schelter, J., Burchard, J., Kibukawa, M., Martin, M.M., Bartz, S.R., Johnson, J.M., Cummins, J.M., Raymond, C.K., Dai, H., Chau, N., Cleary, M., Jackson, A.L., Carleton, M., Lim, L., 2007. Transcripts targeted by the microRNA-16 family cooperatively regulate cell cycle progression. Mol. Cell. Biol. 27, 2240–2252.
Liu, N., Olson, E.N., 2010. MicroRNA regulatory networks in cardiovascular development. Dev. Cell 18, 510–525.
Liu, Q., Fu, H., Sun, F., Zhang, H., Tie, Y., Zhu, J., Xing, R., Sun, Z., Zheng, X., 2008. miR-16 family induces cell cycle arrest by regulating multiple cell cycle genes. Nucleic Acids Res. 36, 5391–5404.
Liu, G., Friggeri, A., Yang, Y., Milosevic, J., Ding, Q., Thannickal, V.J., Kaminski, N., Abraham, E., 2010. miR-21 mediates fibrogenic activation of pulmonary fibroblasts and lung fibrosis. J. Exp. Med. 207, 1589–1597.
Lorenzen, J.M., Volkmann, I., Fiedler, J., Schmidt, M., Scheffner, I., Haller, H., Gwinner, W., Thum, T., 2011. Urinary miR-210 as a mediator of acute T-cell mediated rejection in renal allograft recipients. Am. J. Transplant. 11, 2221–2227.
Lu, L.F., Gasteiger, G., Yu, I.S., Chaudhry, A., Hsin, J.P., Lu, Y., Bos, P.D., Lin, L.L., Zawislak, C.L., Cho, S., Sun, J.C., Leslie, C.S., Lin, S.W., Rudensky, A.Y., 2015. A single miRNA-mRNA interaction affects the immune response in a context- and cell-type-specific manner. Immunity 43, 52–64.
Ludwig, J.A., Weinstein, J.N., 2005. Biomarkers in cancer staging, prognosis and treatment selection. Nat. Rev. Cancer 5, 845–856.
Luo, R., Shao, X., Xu, P., Liu, Y., Wang, Y., Zhao, Y., Liu, M., Ji, L., Li, Y.X., Chang, C., Qiao, J., Peng, C., Wang, Y.L., 2014. MicroRNA-210 contributes to preeclampsia by downregulating potassium channel modulatory factor 1. Hypertension 64, 839–845.
Ma, X., Becker Buscaglia, L.E., Barker, J.R., Li, Y., 2011. MicroRNAs in NF-kappaB signaling. J. Mol. Cell Biol. 3, 159–166.
Mace, K.A., Hansen, S.L., Myers, C., Young, D.M., Boudreau, N., 2005. HOXA3 induces cell migration in endothelial and epithelial cells promoting angiogenesis and wound repair. J. Cell Sci. 118, 2567–2577.
Magenta, A., Cencioni, C., Fasanaro, P., Zaccagnini, G., Greco, S., Sarra-Ferraris, G., Antonini, A., Martelli, F., Capogrossi, M.C., 2011. miR-200c is upregulated by oxidative stress and induces endothelial cell apoptosis and senescence via ZEB1 inhibition. Cell Death Differ. 18, 1628–1639.
Maher, T.M., 2015. Aerobic glycolysis and the Warburg effect. An unexplored realm in the search for fibrosis therapies? Am. J. Respir. Crit. Care Med. 192, 1407–1409.
Marchi, S., Lupini, L., Patergnani, S., Rimessi, A., Missiroli, S., Bonora, M., Bononi, A., Corra, F., Giorgi, C., De Marchi, E., Poletti, F., Gafa, R., Lanza, G., Negrini, M., Rizzuto, R., Pinton, P., 2013. Downregulation of the mitochondrial calcium uniporter by cancer-related miR-25. Curr. Biol. 23, 58–63.
Mayeux, R., 2004. Biomarkers: potential uses and limitations. NeuroRx 1, 182–188.
Maynard, M.A., Evans, A.J., Hosomi, T., Hara, S., Jewett, M.A., Ohh, M., 2005. Human HIF-3alpha4 is a dominant-negative regulator of HIF-1 and is down-regulated in renal cell carcinoma. FASEB J. 19, 1396–1406.

McCormack, J.G., Denton, R.M., 1993. The role of intramitochondrial Ca2+ in the regulation of oxidative phosphorylation in mammalian tissues. Biochem. Soc. Trans. 21 (Pt. 3), 793–799.

McManus, D.D., Ambros, V., 2011. Circulating MicroRNAs in cardiovascular disease. Circulation 124, 1908–1910.

Medina, P.P., Nolde, M., Slack, F.J., 2010. OncomiR addiction in an in vivo model of microRNA-21-induced pre-B-cell lymphoma. Nature 467, 86–90.

Meloche, J., Pflieger, A., Vaillancourt, M., Paulin, R., Potus, F., Zervopoulos, S., Graydon, C., Courboulin, A., Breuils-Bonnet, S., Tremblay, E., Couture, C., Michelakis, E.D., Provencher, S., Bonnet, S., 2014. Role for DNA damage signaling in pulmonary arterial hypertension. Circulation 129, 786–797.

Meloche, J., Le Guen, M., Potus, F., Vinck, J., Ranchoux, B., Johnson, I., Antigny, F., Tremblay, E., Breuils-Bonnet, S., Perros, F., Provencher, S., Bonnet, S., 2015a. miR-223 reverses experimental pulmonary arterial hypertension. Am. J. Physiol. Cell Physiol. 309, C363–C372.

Meloche, J., Potus, F., Vaillancourt, M., Bourgeois, A., Johnson, I., Deschamps, L., Chabot, S., Ruffenach, G., Henry, S., Breuils-Bonnet, S., Tremblay, E., Nadeau, V., Lambert, C., Paradis, R., Provencher, S., Bonnet, S., 2015b. Bromodomain-containing protein 4: the epigenetic origin of pulmonary arterial hypertension. Circ. Res. 117, 525–535.

Memczak, S., Jens, M., Elefsinioti, A., Torti, F., Krueger, J., Rybak, A., Maier, L., Mackowiak, S.D., Gregersen, L.H., Munschauer, M., Loewer, A., Ziebold, U., Landthaler, M., Kocks, C., Le Noble, F., Rajewsky, N., 2013. Circular RNAs are a large class of animal RNAs with regulatory potency. Nature 495, 333–338.

Mercey, O., Popa, A., Cavard, A., Paquet, A., Chevalier, B., Pons, N., Magnone, V., Zangari, J., Brest, P., Zaragosi, L.E., Ponzio, G., Lebrigand, K., Barbry, P., Marcet, B., 2017. Characterizing isomiR variants within the microRNA-34/449 family. FEBS Lett. 591, 693–705.

Metallo, C.M., Vander Heiden, M.G., 2013. Understanding metabolic regulation and its influence on cell physiology. Mol. Cell 49, 388–398.

Michelakis, E.D., 2006. Spatio-temporal diversity of apoptosis within the vascular wall in pulmonary arterial hypertension: heterogeneous BMP signaling may have therapeutic implications. Circ. Res. 98, 172–175.

Michelakis, E.D., Hampl, V., Nsair, A., Wu, X., Harry, G., Haromy, A., Gurtu, R., Archer, S.L., 2002. Diversity in mitochondrial function explains differences in vascular oxygen sensing. Circ. Res. 90, 1307–1315.

Michelakis, E.D., Thebaud, B., Weir, E.K., Archer, S.L., 2004. Hypoxic pulmonary vasoconstriction: redox regulation of O2-sensitive K+ channels by a mitochondrial O2-sensor in resistance artery smooth muscle cells. J. Mol. Cell. Cardiol. 37, 1119–1136.

Michiels, C., 2004. Physiological and pathological responses to hypoxia. Am. J. Pathol. 164, 1875–1882.

Mitchell, P.S., Parkin, R.K., Kroh, E.M., Fritz, B.R., Wyman, S.K., Pogosova-Agadjanyan, E.L., Peterson, A., Noteboom, J., O'Briant, K.C., Allen, A., Lin, D.W., Urban, N., Drescher, C.W., Knudsen, B.S., Stirewalt, D.L., Gentleman, R., Vessella, R.L., Nelson, P.S., Martin, D.B., Tewari, M., 2008. Circulating microRNAs as stable blood-based markers for cancer detection. Proc. Natl. Acad. Sci. U.S.A. 105, 10513–10518.

Mittelbrunn, M., Gutierrez-Vazquez, C., Villarroya-Beltri, C., Gonzalez, S., Sanchez-Cabo, F., Gonzalez, M.A., Bernad, A., Sanchez-Madrid, F., 2011. Unidirectional transfer of microRNA-loaded exosomes from T cells to antigen-presenting cells. Nat. Commun. 2, 282.

Mizuno, Y., Tokuzawa, Y., Ninomiya, Y., Yagi, K., Yatsuka-Kanesaki, Y., Suda, T., Fukuda, T., Katagiri, T., Kondoh, Y., Amemiya, T., Tashiro, H., Okazaki, Y., 2009.

miR-210 promotes osteoblastic differentiation through inhibition of AcvR1b. FEBS Lett. 583, 2263–2268.
Mok, Y., Schwierzeck, V., Thomas, D.C., Vigorito, E., Rayner, T.F., Jarvis, L.B., Prosser, H.M., Bradley, A., Withers, D.R., Martensson, I.L., Corcoran, L.M., Blenkiron, C., Miska, E.A., Lyons, P.A., Smith, K.G., 2013. MiR-210 is induced by Oct-2, regulates B cells, and inhibits autoantibody production. J. Immunol. 191, 3037–3048.
Morin, R.D., O'Connor, M.D., Griffith, M., Kuchenbauer, F., Delaney, A., Prabhu, A.L., Zhao, Y., McDonald, H., Zeng, T., Hirst, M., Eaves, C.J., Marra, M.A., 2008. Application of massively parallel sequencing to microRNA profiling and discovery in human embryonic stem cells. Genome Res. 18, 610–621.
Moschos, S.A., Frick, M., Taylor, B., Turnpenny, P., Graves, H., Spink, K.G., Brady, K., Lamb, D., Collins, D., Rockel, T.D., Weber, M., Lazari, O., Perez-Tosar, L., Fancy, S.A., Lapthorn, C., Green, M.X., Evans, S., Selby, M., Jones, G., Jones, L., Kearney, S., Mechiche, H., Gikunju, D., Subramanian, R., Uhlmann, E., Jurk, M., Vollmer, J., Ciaramella, G., Yeadon, M., 2011. Uptake, efficacy, and systemic distribution of naked, inhaled short interfering RNA (siRNA) and locked nucleic acid (LNA) antisense. Mol. Ther. 19, 2163–2168.
Mutharasan, R.K., Nagpal, V., Ichikawa, Y., Ardehali, H., 2011. microRNA-210 is upregulated in hypoxic cardiomyocytes through Akt- and p53-dependent pathways and exerts cytoprotective effects. Am. J. Physiol. Heart Circ. Physiol. 301, H1519–H1530.
Nahid, M.A., Satoh, M., Chan, E.K., 2011. Mechanistic role of microRNA-146a in endotoxin-induced differential cross-regulation of TLR signaling. J. Immunol. 186, 1723–1734.
Nallamshetty, S., Chan, S.Y., Loscalzo, J., 2013. Hypoxia: a master regulator of microRNA biogenesis and activity. Free Radic. Biol. Med. 64, 20–30.
Nam, J.W., Rissland, O.S., Koppstein, D., Abreu-Goodger, C., Jan, C.H., Agarwal, V., Yildirim, M.A., Rodriguez, A., Bartel, D.P., 2014. Global analyses of the effect of different cellular contexts on microRNA targeting. Mol. Cell 53, 1031–1043.
Navickas, R., Gal, D., Laucevicius, A., Taparauskaite, A., Zdanyte, M., Holvoet, P., 2016. Identifying circulating microRNAs as biomarkers of cardiovascular disease: a systematic review. Cardiovasc. Res. 111, 322–337.
Noman, M.Z., Buart, S., Romero, P., Ketari, S., Janji, B., Mari, B., Mami-Chouaib, F., Chouaib, S., 2012. Hypoxia-inducible miR-210 regulates the susceptibility of tumor cells to lysis by cytotoxic T cells. Cancer Res. 72, 4629–4641.
Noman, M.Z., Janji, B., Hu, S., Wu, J.C., Martelli, F., Bronte, V., Chouaib, S., 2015. Tumor-promoting effects of myeloid-derived suppressor cells are potentiated by hypoxia-induced expression of miR-210. Cancer Res. 75, 3771–3787.
O'Connell, R.M., Chaudhuri, A.A., Rao, D.S., Baltimore, D., 2009. Inositol phosphatase SHIP1 is a primary target of miR-155. Proc. Natl. Acad. Sci. U.S.A. 106, 7113–7118.
O'Connell, R.M., Rao, D.S., Chaudhuri, A.A., Baltimore, D., 2010. Physiological and pathological roles for microRNAs in the immune system. Nat. Rev. Immunol. 10, 111–122.
Okada, N., Lin, C.P., Ribeiro, M.C., Biton, A., Lai, G., He, X., Bu, P., Vogel, H., Jablons, D.M., Keller, A.C., Wilkinson, J.E., He, B., Speed, T.P., He, L., 2014. A positive feedback between p53 and miR-34 miRNAs mediates tumor suppression. Genes Dev. 28, 438–450.
Olson, E.N., 2014. MicroRNAs as therapeutic targets and biomarkers of cardiovascular disease. Sci. Transl. Med. 6. 239ps3.
Orom, U.A., Lund, A.H., 2007. Isolation of microRNA targets using biotinylated synthetic microRNAs. Methods 43, 162–165.

Pandey, A., Shao, H., Marks, R.M., Polverini, P.J., Dixit, V.M., 1995. Role of B61, the ligand for the Eck receptor tyrosine kinase, in TNF-alpha-induced angiogenesis. Science 268, 567–569.

Paranjape, T., Heneghan, H., Lindner, R., Keane, F.K., Hoffman, A., Hollestelle, A., Dorairaj, J., Geyda, K., Pelletier, C., Nallur, S., Martens, J.W., Hooning, M.J., Kerin, M., Zelterman, D., Zhu, Y., Tuck, D., Harris, L., Miller, N., Slack, F., Weidhaas, J., 2011. A 3′-untranslated region KRAS variant and triple-negative breast cancer: a case–control and genetic analysis. Lancet Oncol. 12, 377–386.

Parikh, V.N., Jin, R.C., Rabello, S., Gulbahce, N., White, K., Hale, A., Cottrill, K.A., Shaik, R.S., Waxman, A.B., Zhang, Y.Y., Maron, B.A., Hartner, J.C., Fujiwara, Y., Orkin, S.H., Haley, K.J., Barabasi, A.L., Loscalzo, J., Chan, S.Y., 2012. MicroRNA-21 integrates pathogenic signaling to control pulmonary hypertension: results of a network bioinformatics approach. Circulation 125, 1520–1532.

Pegtel, D.M., Cosmopoulos, K., Thorley-Lawson, D.A., Van Eijndhoven, M.A., Hopmans, E.S., Lindenberg, J.L., De Gruijl, T.D., Wurdinger, T., Middeldorp, J.M., 2010. Functional delivery of viral miRNAs via exosomes. Proc. Natl. Acad. Sci. U.S.A. 107, 6328–6333.

Petersen, K.F., Dufour, S., Befroy, D., Garcia, R., Shulman, G.I., 2004. Impaired mitochondrial activity in the insulin-resistant offspring of patients with type 2 diabetes. N. Engl. J. Med. 350, 664–671.

Pharoah, P.D., Palmieri, R.T., Ramus, S.J., Gayther, S.A., Andrulis, I.L., Anton-Culver, H., Antonenkova, N., Antoniou, A.C., Goldgar, D., Beattie, M.S., Beckmann, M.W., Birrer, M.J., Bogdanova, N., Bolton, K.L., Brewster, W., Brooks-Wilson, A., Brown, R., Butzow, R., Caldes, T., Caligo, M.A., Campbell, I., Chang-Claude, J., Chen, Y.A., Cook, L.S., Couch, F.J., Cramer, D.W., Cunningham, J.M., Despierre, E., Doherty, J.A., Dork, T., Durst, M., Eccles, D.M., Ekici, A.B., Easton, D., Fasching, P.A., De Fazio, A., Fenstermacher, D.A., Flanagan, J.M., Fridley, B.L., Friedman, E., Gao, B., Sinilnikova, O., Gentry-Maharaj, A., Godwin, A.K., Goode, E.L., Goodman, M.T., Gross, J., Hansen, T.V., Harnett, P., Rookus, M., Heikkinen, T., Hein, R., Hogdall, C., Hogdall, E., Iversen, E.S., Jakubowska, A., Johnatty, S.E., Karlan, B.Y., Kauff, N.D., Kaye, S.B., Chenevix-Trench, G., Kelemen, L.E., Kiemeney, L.A., Kjaer, S.K., Lambrechts, D., Lapolla, J.P., Lazaro, C., Le, N.D., Leminen, A., Leunen, K., Levine, D.A., Lu, Y., Lundvall, L., Macgregor, S., Marees, T., Massuger, L.F., McLaughlin, J.R., Menon, U., Montagna, M., Moysich, K.B., Narod, S.A., Nathanson, K.L., Nedergaard, L., Ness, R.B., Nevanlinna, H., Nickels, S., Osorio, A., Paul, J., Pearce, C.L., Phelan, C.M., Pike, M.C., Radice, P., Rossing, M.A., Schildkraut, J.M., Sellers, T.A., Singer, C.F., Song, H., Stram, D.O., Sutphen, R., Lindblom, A., et al., 2011. The role of KRAS rs61764370 in invasive epithelial ovarian cancer: implications for clinical testing. Clin. Cancer Res. 17, 3742–3750.

Pineles, B.L., Romero, R., Montenegro, D., Tarca, A.L., Han, Y.M., Kim, Y.M., Draghici, S., Espinoza, J., Kusanovic, J.P., Mittal, P., Hassan, S.S., Kim, C.J., 2007. Distinct subsets of microRNAs are expressed differentially in the human placentas of patients with preeclampsia. Am. J. Obstet. Gynecol. 196 (261), e1–e6.

Poliseno, L., Salmena, L., Zhang, J., Carver, B., Haveman, W.J., Pandolfi, P.P., 2010. A coding-independent function of gene and pseudogene mRNAs regulates tumour biology. Nature 465, 1033–1038.

Polytarchou, C., Iliopoulos, D., Hatziapostolou, M., Kottakis, F., Maroulakou, I., Struhl, K., Tsichlis, P.N., 2011. Akt2 regulates all Akt isoforms and promotes resistance to hypoxia through induction of miR-21 upon oxygen deprivation. Cancer Res. 71, 4720–4731.

Popa, A., Lebrigand, K., Barbry, P., Waldmann, R., 2016. Pateamine A-sensitive ribosome profiling reveals the scope of translation in mouse embryonic stem cells. BMC Genomics 17, 52.

Pottier, N., Cauffiez, C., Perrais, M., Barbry, P., Mari, B., 2014. FibromiRs: translating molecular discoveries into new anti-fibrotic drugs. Trends Pharmacol. Sci. 35, 119–126.
Potus, F., Ruffenach, G., Dahou, A., Thebault, C., Breuils-Bonnet, S., Tremblay, E., Nadeau, V., Paradis, R., Graydon, C., Wong, R., Johnson, I., Paulin, R., Lajoie, A.C., Perron, J., Charbonneau, E., Joubert, P., Pibarot, P., Michelakis, E.D., Provencher, S., Bonnet, S., 2015. Downregulation of microRNA-126 contributes to the failing right ventricle in pulmonary arterial hypertension. Circulation 132, 932–943.
Puissegur, M.P., Mazure, N.M., Bertero, T., Pradelli, L., Grosso, S., Robbe-Sermesant, K., Maurin, T., Lebrigand, K., Cardinaud, B., Hofman, V., Fourre, S., Magnone, V., Ricci, J.E., Pouyssegur, J., Gounon, P., Hofman, P., Barbry, P., Mari, B., 2011. miR-210 is overexpressed in late stages of lung cancer and mediates mitochondrial alterations associated with modulation of HIF-1 activity. Cell Death Differ. 18, 465–478.
Pulkkinen, K., Malm, T., Turunen, M., Koistinaho, J., Yla-Herttuala, S., 2008. Hypoxia induces microRNA miR-210 in vitro and in vivo ephrin-A3 and neuronal pentraxin 1 are potentially regulated by miR-210. FEBS Lett. 582, 2397–2401.
Quiat, D., Olson, E.N., 2013. MicroRNAs in cardiovascular disease: from pathogenesis to prevention and treatment. J. Clin. Invest. 123, 11–18.
Raal, F.J., Santos, R.D., Blom, D.J., Marais, A.D., Charng, M.J., Cromwell, W.C., Lachmann, R.H., Gaudet, D., Tan, J.L., Chasan-Taber, S., Tribble, D.L., Flaim, J.D., Crooke, S.T., 2010. Mipomersen, an apolipoprotein B synthesis inhibitor, for lowering of LDL cholesterol concentrations in patients with homozygous familial hypercholesterolaemia: a randomised, double-blind, placebo-controlled trial. Lancet 375, 998–1006.
Rajasekaran, S., Rajaguru, P., Sudhakar Gandhi, P.S., 2015. MicroRNAs as potential targets for progressive pulmonary fibrosis. Front. Pharmacol. 6, 254.
Rane, S., He, M., Sayed, D., Vashistha, H., Malhotra, A., Sadoshima, J., Vatner, D.E., Vatner, S.F., Abdellatif, M., 2009. Downregulation of miR-199a derepresses hypoxia-inducible factor-1alpha and Sirtuin 1 and recapitulates hypoxia preconditioning in cardiac myocytes. Circ. Res. 104, 879–886.
Ratner, E., Lu, L., Boeke, M., Barnett, R., Nallur, S., Chin, L.J., Pelletier, C., Blitzblau, R., Tassi, R., Paranjape, T., Hui, P., Godwin, A.K., Yu, H., Risch, H., Rutherford, T., Schwartz, P., Santin, A., Matloff, E., Zelterman, D., Slack, F.J., Weidhaas, J.B., 2010. A KRAS-variant in ovarian cancer acts as a genetic marker of cancer risk. Cancer Res. 70, 6509–6515.
Raver-Shapira, N., Marciano, E., Meiri, E., Spector, Y., Rosenfeld, N., Moskovits, N., Bentwich, Z., Oren, M., 2007. Transcriptional activation of miR-34a contributes to p53-mediated apoptosis. Mol. Cell 26, 731–743.
Reid, G., Kirschner, M.B., Van Zandwijk, N., 2011. Circulating microRNAs: association with disease and potential use as biomarkers. Crit. Rev. Oncol. Hematol. 80, 193–208.
Rizzuto, R., De Stefani, D., Raffaello, A., Mammucari, C., 2012. Mitochondria as sensors and regulators of calcium signalling. Nat. Rev. Mol. Cell Biol. 13, 566–578.
Rodriguez, A., Vigorito, E., Clare, S., Warren, M.V., Couttet, P., Soond, D.R., Van Dongen, S., Grocock, R.J., Das, P.P., Miska, E.A., Vetrie, D., Okkenhaug, K., Enright, A.J., Dougan, G., Turner, M., Bradley, A., 2007. Requirement of bic/microRNA-155 for normal immune function. Science 316, 608–611.
Rothschild, S.I., 2014. microRNA therapies in cancer. Mol. Cell. Ther. 2, 7.
Rouault, T.A., Tong, W.H., 2008. Iron–sulfur cluster biogenesis and human disease. Trends Genet. 24, 398–407.
Ruffenach, G., Chabot, S., Tanguay, V.F., Courboulin, A., Boucherat, O., Potus, F., Meloche, J., Pflieger, A., Breuils-Bonnet, S., Nadeau, V., Paradis, R., Tremblay, E., Girerd, B., Hautefort, A., Montani, D., Fadel, E., Dorfmuller, P., Humbert, M., Perros, F., Paulin, R., Provencher, S., Bonnet, S., 2016. Role for runt-related

transcription factor 2 in proliferative and calcified vascular lesions in pulmonary arterial hypertension. Am. J. Respir. Crit. Care Med. 194, 1273–1285.

Rupaimoole, R., Han, H.D., Lopez-Berestein, G., Sood, A.K., 2011. MicroRNA therapeutics: principles, expectations, and challenges. Chin. J. Cancer 30, 368–370.

Rupaimoole, R., Wu, S.Y., Pradeep, S., Ivan, C., Pecot, C.V., Gharpure, K.M., Nagaraja, A.S., Armaiz-Pena, G.N., Mcguire, M., Zand, B., Dalton, H.J., Filant, J., Miller, J.B., Lu, C., Sadaoui, N.C., Mangala, L.S., Taylor, M., Van Den Beucken, T., Koch, E., Rodriguez-Aguayo, C., Huang, L., Bar-Eli, M., Wouters, B.G., Radovich, M., Ivan, M., Calin, G.A., Zhang, W., Lopez-Berestein, G., Sood, A.K., 2014. Hypoxia-mediated downregulation of miRNA biogenesis promotes tumour progression. Nat. Commun. 5, 5202.

Ryan, J.J., Archer, S.L., 2015. Emerging concepts in the molecular basis of pulmonary arterial hypertension: part I: metabolic plasticity and mitochondrial dynamics in the pulmonary circulation and right ventricle in pulmonary arterial hypertension. Circulation 131, 1691–1702.

Sakao, S., Taraseviciene-Stewart, L., Lee, J.D., Wood, K., Cool, C.D., Voelkel, N.F., 2005. Initial apoptosis is followed by increased proliferation of apoptosis-resistant endothelial cells. FASEB J. 19, 1178–1180.

Salceda, S., Caro, J., 1997. Hypoxia-inducible factor 1alpha (HIF-1alpha) protein is rapidly degraded by the ubiquitin-proteasome system under normoxic conditions. Its stabilization by hypoxia depends on redox-induced changes. J. Biol. Chem. 272, 22642–22647.

Sayed, D., He, M., Hong, C., Gao, S., Rane, S., Yang, Z., Abdellatif, M., 2010. MicroRNA-21 is a downstream effector of AKT that mediates its antiapoptotic effects via suppression of Fas ligand. J. Biol. Chem. 285, 20281–20290.

Schmiedel, J.M., Klemm, S.L., Zheng, Y., Sahay, A., Bluthgen, N., Marks, D.S., Van Oudenaarden, A., 2015. Gene expression. MicroRNA control of protein expression noise. Science 348, 128–132.

Schultz, J., Lorenz, P., Gross, G., Ibrahim, S., Kunz, M., 2008. MicroRNA let-7b targets important cell cycle molecules in malignant melanoma cells and interferes with anchorage-independent growth. Cell Res. 18, 549–557.

Seitz, H., 2009. Redefining microRNA targets. Curr. Biol. 19, 870–873.

Semenza, G.L., 2000. HIF-1: mediator of physiological and pathophysiological responses to hypoxia. J. Appl. Physiol. 88 (1985), 1474–1480.

Semenza, G.L., 2007. Oxygen-dependent regulation of mitochondrial respiration by hypoxia-inducible factor 1. Biochem. J. 405, 1–9.

Semenza, G.L., 2011. Oxygen sensing, homeostasis, and disease. N. Engl. J. Med. 365, 537–547.

Semenza, G.L., 2012. Hypoxia-inducible factors in physiology and medicine. Cell 148, 399–408.

Semenza, G.L., 2013. HIF-1 mediates metabolic responses to intratumoral hypoxia and oncogenic mutations. J. Clin. Invest. 123, 3664–3671.

Semenza, G.L., 2014. Hypoxia-inducible factor 1 and cardiovascular disease. Annu. Rev. Physiol. 76, 39–56.

Seok, J.K., Lee, S.H., Kim, M.J., Lee, Y.M., 2014. MicroRNA-382 induced by HIF-1alpha is an angiogenic miR targeting the tumor suppressor phosphatase and tensin homolog. Nucleic Acids Res. 42, 8062–8072.

Shang, C., Hong, Y., Guo, Y., Liu, Y.H., Xue, Y.X., 2014. MiR-210 up-regulation inhibits proliferation and induces apoptosis in glioma cells by targeting SIN3A. Med. Sci. Monit. 20, 2571–2577.

Sheedy, F.J., Palsson-McDermott, E., Hennessy, E.J., Martin, C., O'Leary, J.J., Ruan, Q., Johnson, D.S., Chen, Y., O'Neill, L.A., 2010. Negative regulation of TLR4 via

targeting of the proinflammatory tumor suppressor PDCD4 by the microRNA miR-21. Nat. Immunol. 11, 141–147.
Shen, G., Li, X., Jia, Y.F., Piazza, G.A., Xi, Y., 2013a. Hypoxia-regulated microRNAs in human cancer. Acta Pharmacol. Sin. 34, 336–341.
Shen, J., Xia, W., Khotskaya, Y.B., Huo, L., Nakanishi, K., Lim, S.O., Du, Y., Wang, Y., Chang, W.C., Chen, C.H., Hsu, J.L., Wu, Y., Lam, Y.C., James, B.P., Liu, X., Liu, C.G., Patel, D.J., Hung, M.C., 2013b. EGFR modulates microRNA maturation in response to hypoxia through phosphorylation of AGO2. Nature 497, 383–387.
Shi, L., Kojonazarov, B., Elgheznawy, A., Popp, R., Dahal, B.K., Bohm, M., Pullamsetti, S.S., Ghofrani, H.A., Godecke, A., Jungmann, A., Katus, H.A., Muller, O.J., Schermuly, R.T., Fisslthaler, B., Seeger, W., Fleming, I., 2016. miR-223-IGF-IR signalling in hypoxia- and load-induced right-ventricular failure: a novel therapeutic approach. Cardiovasc. Res. 111, 184–193.
Shin, C., Nam, J.W., Farh, K.K., Chiang, H.R., Shkumatava, A., Bartel, D.P., 2010. Expanding the microRNA targeting code: functional sites with centered pairing. Mol. Cell 38, 789–802.
Singh, P.K., Mehla, K., Hollingsworth, M.A., Johnson, K.R., 2011. Regulation of aerobic glycolysis by microRNAs in cancer. Mol. Cell. Pharmacol. 3, 125–134.
Small, E.M., Frost, R.J., Olson, E.N., 2010. MicroRNAs add a new dimension to cardiovascular disease. Circulation 121, 1022–1032.
Solaini, G., Baracca, A., Lenaz, G., Sgarbi, G., 2010. Hypoxia and mitochondrial oxidative metabolism. Biochim. Biophys. Acta 1797, 1171–1177.
Sounni, N.E., Cimino, J., Blacher, S., Primac, I., Truong, A., Mazzucchelli, G., Paye, A., Calligaris, D., Debois, D., De Tullio, P., Mari, B., De Pauw, E., Noel, A., 2014. Blocking lipid synthesis overcomes tumor regrowth and metastasis after antiangiogenic therapy withdrawal. Cell Metab. 20, 280–294.
Steinberg, F., Zhuang, L., Beyeler, M., Kalin, R.E., Mullis, P.E., Brandli, A.W., Trueb, B., 2010. The FGFRL1 receptor is shed from cell membranes, binds fibroblast growth factors (FGFs), and antagonizes FGF signaling in Xenopus embryos. J. Biol. Chem. 285, 2193–2202.
Steinfeld, I., Navon, R., Ach, R., Yakhini, Z., 2013. miRNA target enrichment analysis reveals directly active miRNAs in health and disease. Nucleic Acids Res. 41, e45.
Stenvang, J., Petri, A., Lindow, M., Obad, S., Kauppinen, S., 2012. Inhibition of microRNA function by antimiR oligonucleotides. Silence 3, 1.
Sun, F., Fu, H., Liu, Q., Tie, Y., Zhu, J., Xing, R., Sun, Z., Zheng, X., 2008. Downregulation of CCND1 and CDK6 by miR-34a induces cell cycle arrest. FEBS Lett. 582, 1564–1568.
Sutendra, G., Michelakis, E.D., 2014. The metabolic basis of pulmonary arterial hypertension. Cell Metab. 19, 558–573.
Suzuki, H.I., Yamagata, K., Sugimoto, K., Iwamoto, T., Kato, S., Miyazono, K., 2009. Modulation of microRNA processing by p53. Nature 460, 529–533.
Tadokoro, H., Umezu, T., Ohyashiki, K., Hirano, T., Ohyashiki, J.H., 2013. Exosomes derived from hypoxic leukemia cells enhance tube formation in endothelial cells. J. Biol. Chem. 288, 34343–34351.
Taganov, K.D., Boldin, M.P., Chang, K.J., Baltimore, D., 2006. NF-kappaB-dependent induction of microRNA miR-146, an inhibitor targeted to signaling proteins of innate immune responses. Proc. Natl. Acad. Sci. U.S.A. 103, 12481–12486.
Taguchi, A., Yanagisawa, K., Tanaka, M., Cao, K., Matsuyama, Y., Goto, H., Takahashi, T., 2008. Identification of hypoxia-inducible factor-1 alpha as a novel target for miR-17-92 microRNA cluster. Cancer Res. 68, 5540–5545.
Takeshita, F., Patrawala, L., Osaki, M., Takahashi, R.U., Yamamoto, Y., Kosaka, N., Kawamata, M., Kelnar, K., Bader, A.G., Brown, D., Ochiya, T., 2010. Systemic delivery

of synthetic microRNA-16 inhibits the growth of metastatic prostate tumors via downregulation of multiple cell-cycle genes. Mol. Ther. 18, 181–187.

Tarasov, V., Jung, P., Verdoodt, B., Lodygin, D., Epanchintsev, A., Menssen, A., Meister, G., Hermeking, H., 2007. Differential regulation of microRNAs by p53 revealed by massively parallel sequencing: miR-34a is a p53 target that induces apoptosis and G1-arrest. Cell Cycle 6, 1586–1593.

Taylor, C.T., 2008. Interdependent roles for hypoxia inducible factor and nuclear factor-kappaB in hypoxic inflammation. J. Physiol. 586, 4055–4059.

Taylor, C.T., Doherty, G., Fallon, P.G., Cummins, E.P., 2016. Hypoxia-dependent regulation of inflammatory pathways in immune cells. J. Clin. Invest. 126, 3716–3724.

Thienpont, B., Steinbacher, J., Zhao, H., D'Anna, F., Kuchnio, A., Ploumakis, A., Ghesquiere, B., Van Dyck, L., Boeckx, B., Schoonjans, L., Hermans, E., Amant, F., Kristensen, V.N., Koh, K.P., Mazzone, M., Coleman, M.L., Carell, T., Carmeliet, P., Lambrechts, D., 2016. Tumour hypoxia causes DNA hypermethylation by reducing TET activity. Nature 537, 63–68.

Thum, T., Chau, N., Bhat, B., Gupta, S.K., Linsley, P.S., Bauersachs, J., Engelhardt, S., 2011. Comparison of different miR-21 inhibitor chemistries in a cardiac disease model. J. Clin. Invest. 121, 461–462. author reply 462–3.

Tiberio, P., Callari, M., Angeloni, V., Daidone, M.G., Appierto, V., 2015. Challenges in using circulating miRNAs as cancer biomarkers. Biomed. Res. Int. 2015, 731479.

Tsuchiya, S., Fujiwara, T., Sato, F., Shimada, Y., Tanaka, E., Sakai, Y., Shimizu, K., Tsujimoto, G., 2011. MicroRNA-210 regulates cancer cell proliferation through targeting fibroblast growth factor receptor-like 1 (FGFRL1). J. Biol. Chem. 286, 420–428.

Tsujimoto, Y., Shimizu, S., 2007. Role of the mitochondrial membrane permeability transition in cell death. Apoptosis 12, 835–840.

Tuder, R.M., Davis, L.A., Graham, B.B., 2012. Targeting energetic metabolism: a new frontier in the pathogenesis and treatment of pulmonary hypertension. Am. J. Respir. Crit. Care Med. 185, 260–266.

Ulitsky, I., Laurent, L.C., Shamir, R., 2010. Towards computational prediction of microRNA function and activity. Nucleic Acids Res. 38, e160.

Valadi, H., Ekstrom, K., Bossios, A., Sjostrand, M., Lee, J.J., Lotvall, J.O., 2007. Exosome-mediated transfer of mRNAs and microRNAs is a novel mechanism of genetic exchange between cells. Nat. Cell Biol. 9, 654–659.

Van Den Beucken, T., Koch, E., Chu, K., Rupaimoole, R., Prickaerts, P., Adriaens, M., Voncken, J.W., Harris, A.L., Buffa, F.M., Haider, S., Starmans, M.H., Yao, C.Q., Ivan, M., Ivan, C., Pecot, C.V., Boutros, P.C., Sood, A.K., Koritzinsky, M., Wouters, B.G., 2014. Hypoxia promotes stem cell phenotypes and poor prognosis through epigenetic regulation of DICER. Nat. Commun. 5, 5203.

Van Dongen, S., Abreu-Goodger, C., Enright, A.J., 2008. Detecting microRNA binding and siRNA off-target effects from expression data. Nat. Methods 5, 1023–1025.

Van Dyck, E., Stasiak, A.Z., Stasiak, A., West, S.C., 1999. Binding of double-strand breaks in DNA by human Rad52 protein. Nature 398, 728–731.

Vander Heiden, M.G., Cantley, L.C., Thompson, C.B., 2009. Understanding the Warburg effect: the metabolic requirements of cell proliferation. Science 324, 1029–1033.

Vickers, K.C., Palmisano, B.T., Shoucri, B.M., Shamburek, R.D., Remaley, A.T., 2011. MicroRNAs are transported in plasma and delivered to recipient cells by high-density lipoproteins. Nat. Cell Biol. 13, 423–433.

Virtue, A., Wang, H., Yang, X.F., 2012. MicroRNAs and toll-like receptor/interleukin-1 receptor signaling. J. Hematol. Oncol. 5, 66.

Vlachos, I.S., Paraskevopoulou, M.D., Karagkouni, D., Georgakilas, G., Vergoulis, T., Kanellos, I., Anastasopoulos, I.L., Maniou, S., Karathanou, K., Kalfakakou, D., Fevgas, A., Dalamagas, T., Hatzigeorgiou, A.G., 2015. DIANA-TarBase v7.0: indexing more than half a million experimentally supported miRNA:mRNA interactions. Nucleic Acids Res. 43, D153–D159.

Voellenkle, C., Rooij, J., Guffanti, A., Brini, E., Fasanaro, P., Isaia, E., Croft, L., David, M., Capogrossi, M.C., Moles, A., Felsani, A., Martelli, F., 2012. Deep-sequencing of endothelial cells exposed to hypoxia reveals the complexity of known and novel microRNAs. RNA 18, 472–484.

Voellenkle, C., Garcia-Manteiga, J.M., Pedrotti, S., Perfetti, A., De Toma, I., Da Silva, D., Maimone, B., Greco, S., Fasanaro, P., Creo, P., Zaccagnini, G., Gaetano, C., Martelli, F., 2016. Implication of long noncoding RNAs in the endothelial cell response to hypoxia revealed by RNA-sequencing. Sci. Rep. 6, 24141.

Wan, G., Mathur, R., Hu, X., Zhang, X., Lu, X., 2011. miRNA response to DNA damage. Trends Biochem. Sci. 36, 478–484.

Wang, G.L., Semenza, G.L., 1995. Purification and characterization of hypoxia-inducible factor 1. J. Biol. Chem. 270, 1230–1237.

Wang, F., Fu, X.D., Zhou, Y., Zhang, Y., 2009. Down-regulation of the cyclin E1 oncogene expression by microRNA-16-1 induces cell cycle arrest in human cancer cells. BMB Rep. 42, 725–730.

Wang, F., Xiong, L., Huang, X., Zhao, T., Wu, L.Y., Liu, Z.H., Ding, X., Liu, S., Wu, Y., Zhao, Y., Wu, K., Zhu, L.L., Fan, M., 2013. miR-210 suppresses BNIP3 to protect against the apoptosis of neural progenitor cells. Stem Cell Res. 11, 657–667.

Wang, H., Flach, H., Onizawa, M., Wei, L., McManus, M.T., Weiss, A., 2014a. Negative regulation of Hif1a expression and TH17 differentiation by the hypoxia-regulated microRNA miR-210. Nat. Immunol. 15, 393–401.

Wang, J., Zhao, J., Shi, M., Ding, Y., Sun, H., Yuan, F., Zou, Z., 2014b. Elevated expression of miR-210 predicts poor survival of cancer patients: a systematic review and meta-analysis. PLoS One 9, e89223.

Wang, J., Zhang, X., Shi, J., Cao, P., Wan, M., Zhang, Q., Wang, Y., Kridel, S.J., Liu, W., Xuaa, J., Sui, G., 2016. Fatty acid synthase is a primary target of MiR-15a and MiR-16-1 in breast cancer. Oncotarget 7, 78566–78576.

Wang, S., Wu, W., Claret, F.X., 2017. Mutual regulation of microRNAs and DNA methylation in human cancers. Epigenetics 1–11. http://dx.doi.org/10.1080/15592294.2016.1273308 [Epub ahead of print].

Warburg, O., 1956. On respiratory impairment in cancer cells. Science 124, 269–270.

Wei, C., Henderson, H., Spradley, C., Li, L., Kim, I.K., Kumar, S., Hong, N., Arroliga, A.C., Gupta, S., 2013. Circulating miRNAs as potential marker for pulmonary hypertension. PLoS One 8, e64396.

White, K., Loscalzo, J., Chan, S.Y., 2012. Holding our breath: the emerging and anticipated roles of microRNA in pulmonary hypertension. Pulm. Circ. 2, 278–290.

White, K., Lu, Y., Annis, S., Hale, A.E., Chau, B.N., Dahlman, J.E., Hemann, C., Opotowsky, A.R., Vargas, S.O., Rosas, I., Perrella, M.A., Osorio, J.C., Haley, K.J., Graham, B.B., Kumar, R., Saggar, R., Wallace, W.D., Ross, D.J., Khan, O.F., Bader, A., Gochuico, B.R., Matar, M., Polach, K., Johannessen, N.M., Prosser, H.M., Anderson, D.G., Langer, R., Zweier, J.L., Bindoff, L.A., Systrom, D., Waxman, A.B., Jin, R.C., Chan, S.Y., 2015. Genetic and hypoxic alterations of the microRNA-210-ISCU1/2 axis promote iron–sulfur deficiency and pulmonary hypertension. EMBO Mol. Med. 7, 695–713.

Wiedemann, M., Trueb, B., 2000. Characterization of a novel protein (FGFRL1) from human cartilage related to FGF receptors. Genomics 69, 275–279.

Wightman, B., Ha, I., Ruvkun, G., 1993. Posttranscriptional regulation of the heterochronic gene lin-14 by lin-4 mediates temporal pattern formation in C. elegans. Cell 75, 855–862.

Wigle, D.A., Jurisica, I., Radulovich, N., Pintilie, M., Rossant, J., Liu, N., Lu, C., Woodgett, J., Seiden, I., Johnston, M., Keshavjee, S., Darling, G., Winton, T., Breitkreutz, B.J., Jorgenson, P., Tyers, M., Shepherd, F.A., Tsao, M.S., 2002. Molecular profiling of non-small cell lung cancer and correlation with disease-free survival. Cancer Res. 62, 3005–3008.

Witwer, K.W., 2015. Circulating microRNA biomarker studies: pitfalls and potential solutions. Clin. Chem. 61, 56–63.

Wood, S.M., Wiesener, M.S., Yeates, K.M., Okada, N., Pugh, C.W., Maxwell, P.H., Ratcliffe, P.J., 1998. Selection and analysis of a mutant cell line defective in the hypoxia-inducible factor-1 alpha-subunit (HIF-1alpha). Characterization of hif-1alpha-dependent and -independent hypoxia-inducible gene expression. J. Biol. Chem. 273, 8360–8368.

Wulfken, L.M., Moritz, R., Ohlmann, C., Holdenrieder, S., Jung, V., Becker, F., Herrmann, E., Walgenbach-Brunagel, G., Von Ruecker, A., Muller, S.C., Ellinger, J., 2011. MicroRNAs in renal cell carcinoma: diagnostic implications of serum miR-1233 levels. PLoS One 6, e25787.

Xie, N., Tan, Z., Banerjee, S., Cui, H., Ge, J., Liu, R.M., Bernard, K., Thannickal, V.J., Liu, G., 2015. Glycolytic reprogramming in myofibroblast differentiation and lung fibrosis. Am. J. Respir. Crit. Care Med. 192, 1462–1474.

Xiong, L., Wang, F., Huang, X., Liu, Z.H., Zhao, T., Wu, L.Y., Wu, K., Ding, X., Liu, S., Wu, Y., Zhao, Y., Zhu, L.L., Fan, M., 2012. DNA demethylation regulates the expression of miR-210 in neural progenitor cells subjected to hypoxia. FEBS J. 279, 4318–4326.

Xu, T.X., Zhao, S.Z., Dong, M., Yu, X.R., 2016. Hypoxia responsive miR-210 promotes cell survival and autophagy of endometriotic cells in hypoxia. Eur. Rev. Med. Pharmacol. Sci. 20, 399–406.

Yang, J.S., Lai, E.C., 2011. Alternative miRNA biogenesis pathways and the interpretation of core miRNA pathway mutants. Mol. Cell 43, 892–903.

Yang, Y., Wang, J.K., 2016. The functional analysis of MicroRNAs involved in NF-kappaB signaling. Eur. Rev. Med. Pharmacol. Sci. 20, 1764–1774.

Yang, W., Sun, T., Cao, J., Liu, F., Tian, Y., Zhu, W., 2012. Downregulation of miR-210 expression inhibits proliferation, induces apoptosis and enhances radiosensitivity in hypoxic human hepatoma cells in vitro. Exp. Cell Res. 318, 944–954.

Yang, D., Sun, Y., Hu, L., Zheng, H., Ji, P., Pecot, C.V., Zhao, Y., Reynolds, S., Cheng, H., Rupaimoole, R., Cogdell, D., Nykter, M., Broaddus, R., Rodriguez-Aguayo, C., Lopez-Berestein, G., Liu, J., Shmulevich, I., Sood, A.K., Chen, K., Zhang, W., 2013a. Integrated analyses identify a master microRNA regulatory network for the mesenchymal subtype in serous ovarian cancer. Cancer Cell 23, 186–199.

Yang, W., Wei, J., Sun, T., Liu, F., 2013b. Effects of knockdown of miR-210 in combination with ionizing radiation on human hepatoma xenograft in nude mice. Radiat. Oncol. 8, 102.

Ying, Q., Liang, L., Guo, W., Zha, R., Tian, Q., Huang, S., Yao, J., Ding, J., Bao, M., Ge, C., Yao, M., Li, J., He, X., 2011. Hypoxia-inducible microRNA-210 augments the metastatic potential of tumor cells by targeting vacuole membrane protein 1 in hepatocellular carcinoma. Hepatology 54, 2064–2075.

Yoshino, H., Yonemori, M., Miyamoto, K., Tatarano, S., Kofuji, S., Nohata, N., Nakagawa, M., Enokida, H., 2017. microRNA-210-3p depletion by CRISPR/Cas9 promoted tumorigenesis through revival of TWIST1 in renal cell carcinoma. Oncotarget. http://dx.doi.org/10.18632/oncotarget.14930 [Epub ahead of print].

Yoshioka, Y., Kosaka, N., Ochiya, T., Kato, T., 2012. Micromanaging Iron Homeostasis: hypoxia-inducible micro-RNA-210 suppresses iron homeostasis-related proteins. J. Biol. Chem. 287, 34110–34119.
Zeng, L., Liu, J., Wang, Y., Wang, L., Weng, S., Tang, Y., Zheng, C., Cheng, Q., Chen, S., Yang, G.Y., 2011. MicroRNA-210 as a novel blood biomarker in acute cerebral ischemia. Front. Biosci. (Elite Ed.) 3, 1265–1272.
Zeng, Y., Zhang, X., Kang, K., Chen, J., Wu, Z., Huang, J., Lu, W., Chen, Y., Zhang, J., Wang, Z., Zhai, Y., Qu, J., Ramchandran, R., Raj, J.U., Wang, J., Gou, D., 2016. MicroRNA-223 attenuates hypoxia-induced vascular remodeling by targeting RhoB/MLC2 in pulmonary arterial smooth muscle cells. Sci. Rep. 6, 24900.
Zhang, Z., Sun, H., Dai, H., Walsh, R.M., Imakura, M., Schelter, J., Burchard, J., Dai, X., Chang, A.N., Diaz, R.L., Marszalek, J.R., Bartz, S.R., Carleton, M., Cleary, M.A., Linsley, P.S., Grandori, C., 2009. MicroRNA miR-210 modulates cellular response to hypoxia through the MYC antagonist MNT. Cell Cycle 8, 2756–2768.
Zhang, Y., Liu, D., Chen, X., Li, J., Li, L., Bian, Z., Sun, F., Lu, J., Yin, Y., Cai, X., Sun, Q., Wang, K., Ba, Y., Wang, Q., Wang, D., Yang, J., Liu, P., Xu, T., Yan, Q., Zhang, J., Zen, K., Zhang, C.Y., 2010. Secreted monocytic miR-150 enhances targeted endothelial cell migration. Mol. Cell 39, 133–144.
Zhang, X., Ng, W.L., Wang, P., Tian, L., Werner, E., Wang, H., Doetsch, P., Wang, Y., 2012a. MicroRNA-21 modulates the levels of reactive oxygen species by targeting SOD3 and TNFalpha. Cancer Res. 72, 4707–4713.
Zhang, Y., Fei, M., Xue, G., Zhou, Q., Jia, Y., Li, L., Xin, H., Sun, S., 2012b. Elevated levels of hypoxia-inducible microRNA-210 in pre-eclampsia: new insights into molecular mechanisms for the disease. J. Cell. Mol. Med. 16, 249–259.
Zhang, Q., Kandic, I., Faughnan, M.E., Kutryk, M.J., 2013. Elevated circulating microRNA-210 levels in patients with hereditary hemorrhagic telangiectasia and pulmonary arteriovenous malformations: a potential new biomarker. Biomarkers 18, 23–29.
Zhang, D., Cao, X., Li, J., Zhao, G., 2015. MiR-210 inhibits NF-kappaB signaling pathway by targeting DR6 in osteoarthritis. Sci. Rep. 5, 12775.
Zhou, R., Hu, G., Gong, A.Y., Chen, X.M., 2010. Binding of NF-kappaB p65 subunit to the promoter elements is involved in LPS-induced transactivation of miRNA genes in human biliary epithelial cells. Nucleic Acids Res. 38, 3222–3232.
Zhu, X.M., Han, T., Sargent, I.L., Yin, G.W., Yao, Y.Q., 2009. Differential expression profile of microRNAs in human placentas from preeclamptic pregnancies vs normal pregnancies. Am. J. Obstet. Gynecol. 200 (661), e1–e7.

CHAPTER FOUR

Understanding the Genomic Ultraconservations: T-UCRs and Cancer

Linda Fabris*, George A. Calin*,†,1
*The University of Texas MD Anderson Cancer Center, Houston, TX, United States
†Center for RNA Interference and Non-Coding RNAs, The University of Texas MD Anderson Cancer Center, Houston, TX, United States
[1]Corresponding author: e-mail address: gcalin@mdanderson.org

Contents

Abstract

Transcribed ultraconserved regions (T-UCRs) are genomic regions conserved across large evolutionary distances, which encode for noncoding RNAs that serve as regulators of gene expression. Although T-UCRs have been linked to multiple aspects of mammalian gene regulation, the roles of their extreme evolutionary conservation remain largely unexplained. Growing body of literature is now focusing on T-UCRs as potential cancer biomarkers or as new drug targets. Here we present an overview of the discoveries so far published about the role of T-UCR in cancer and disease.

1. INTRODUCTION

The exact number of protein-coding genes presents in the genome has been shrinking since the human draft genome sequence was published more than 15 years ago. While initial predictions from the pre-sequencing era were at 100,000 protein-coding genes, the first publications on human genome sequences in 2001 predicted 26,000–30,000 genes (Lander et al., 2001;

International Review of Cell and Molecular Biology, Volume 333
ISSN 1937-6448
http://dx.doi.org/10.1016/bs.ircmb.2017.04.004

Venter et al., 2001), and further through an evolutionary comparison the number was reduced to 20,500 (Pertea and Salzberg, 2010), while now it has been accepted as closer to 19,000 (Ezkurdia et al., 2014).

The fact that protein-coding genes represent less than 2% of the genome has been even more surprising with the advent of tiling resolution genomic microarrays and whole genome and transcriptome sequencing technologies that determined that at least 90% of the genome is actively transcribed (Consortium et al., 2007).

The human transcriptome has been found to be more than a collection of protein-coding genes and their splice variants, and the presence of extensive antisense, overlapping, and noncoding RNA (ncRNA) highlight the major role of the "Dark Matter" of the human genome in cell proliferation and metabolism and in each single of the 10 hallmark of cancer (Hanahan and Weinberg, 2011).

ncRNAs fall into several classes based on their length, biogenesis, polarity (sense or antisense), and putative functions, but are divided in two groups based only on their transcript size for the sake of simplicity: small and long noncoding RNA. Small ncRNAs are shorter than 200 nucleotides and represent a broad range of RNA species processed from longer precursors and mostly associated with 5′ or 3′ regions of protein-coding genes. This class includes MicroRNAs (MiRNAs), Piwi-interacting RNAs (piRNAs), endogenous small interfering RNAs (siRNAs), and some bacterial regulatory RNAs. Transfer RNAs, ribosomal RNAs, and small nuclear (sno) RNAs are also classified as noncoding and included in this group (Spizzo et al., 2012). In contrast, long ncRNAs (lncRNAs) are mRNA-like transcripts at least 200 nt long, mostly transcribed by RNA polymerase II that do not undergo major mRNA processing steps and lack significant open reading frames (Spizzo et al., 2012).

2. LONG NONCODING RNA: HETEROGENEOUS AND STILL LESS UNDERSTOOD TRANSCRIPTS

Currently, a universal classification scheme to define lncRNA does not exist and therefore there are numerous synonyms describing the same type of transcripts, creating confusion, and misinterpretation, since most of them can easily fall on more than one category. For all the purposes of this review, we will consider the transcript longer than 200 nt, with a particular focus on transcribed ultraconserved regions (T-UCRs).

Many lncRNA were discovered between 1980s and 1990s by chance, screening cDNA library for clones of interest expressing new genes important for a specific biological process, as happened for H19 and Xist genes (Bartolomei et al., 1991; Brown et al., 1991). The catalog of noncoding genes has grown tremendously over the past decade, in large part due to the FANTOM and ENCODE transcript mapping projects (Djebali et al., 2012; Harrow et al., 2012) and to new computation methods for transcriptome assembly and lncRNA annotation (Guttman et al., 2010; Mattick and Rinn, 2015).

Among the spectrum of lncRNAs, an intriguing class of lncRNA has been brought to light, T-UCR. Ultraconserved regions (UCRs) were initially described as stretches of sequence longer than 200 bp with 100% conservation between the orthologous regions of the human, rat, and mouse genomes and in large part identified also in chicken and dogs, but without any protein-coding ability (Bejerano et al., 2004). UCRs include 481 transcriptionally active regions, supposed to have biological indispensable regulatory functions, and are defined as a specific class of lncRNA. The transcription products of UCRs, the T-UCR are 200–779 nucleotides long and were originally classified into three categories, nonexonic, exonic, and possibly exonic, according to their overlap with known protein-coding genes (Bejerano et al., 2004; Calin et al., 2007). Since then, they have been reannotated into a more descriptive set of five categories: intergenic (38.7%), intronic (42.6%), exonic (4.2%), partly exonic (5%), or exonic containing (5.6%) (Mestdagh et al., 2010) (Fig. 1A). Furthermore, UCRs are frequently located in fragile sites and cancer-associated genomic regions (CAGRs), thus it has been suggested that these regions could be candidate genes for cancer susceptibility (Rossi et al., 2008).

3. WHY ULTRACONSERVATION?

The role of UCR is still one of the most fascinating mysteries of biology, more than a decade after their identification (Bejerano et al., 2004). Why these fully conserved large genomic regions never changed even a single nucleotide during more than 75 million of years of evolution is an unsolved puzzle. It has been suggested that ultraconserved sequences are a subset of extremely evolutionarily constrained developmental enhancer rather than a functionally distinct enhancer class (Visel et al., 2008). UCRs are transcribed mostly as lncRNAs, and recently has been proposed that some of these transcripts act as enhancer RNAs (eRNAs) that increase

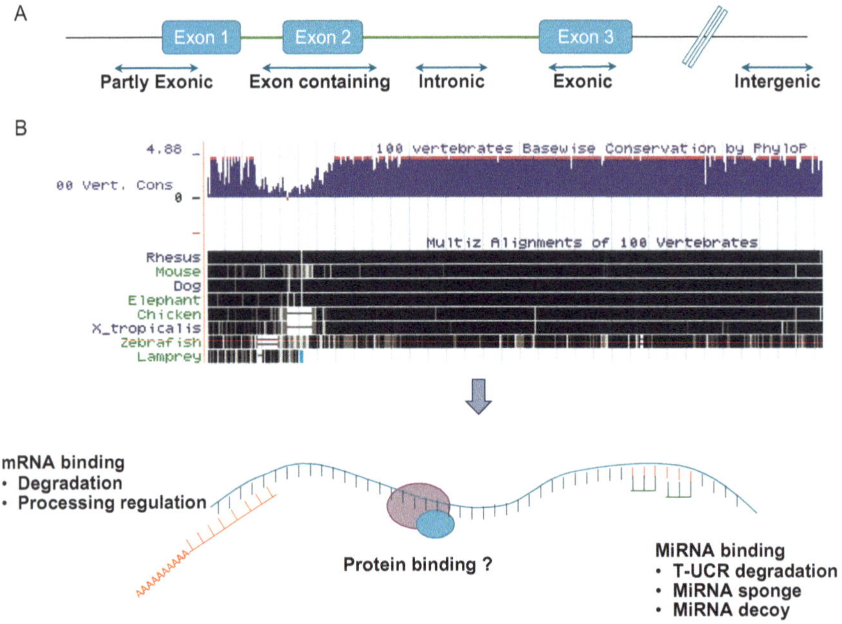

Fig. 1 Transcription of T-UCRs. T-UCRs, as lncRNAs, are classified basing on their genomic location respect to protein-coding genes (A). Genome browser schematic representation of T-UCRs and their potential mechanism of action (B).

the expression of protein-coding genes located in *cis* (on the same chromosome as the UCR) or in *trans* (on other chromosomes) (Orom et al., 2010). Further, the discovery of cancer-specific mutations in UCRs (Wojcik et al., 2010) opened up their potential role as oncogenes or tumor suppressors. Despite these advances, our understanding remains rudimentary as to why large pieces of genomic sequences were "frozen" during evolution.

4. REGULATION MECHANISM OF T-UCR

Due to their high percentage of conservation through evolution, alteration of T-UCRs in human tumors is an unexpected finding, assuming that their evolutional conservation is based on biologically fundamental regulation in the cell environment. Single base pair mutation or deletion have been found in the genomic sequence of T-UCRs (Wojcik et al., 2010). T-UCR are also differentially expressed in both solid and hematological malignancies compared to the corresponding normal tissue (Calin et al., 2007). T-UCR involvement in cancers has been found to occur through three potential different mechanisms: altered interactions with MicroRNA, direct binding to target mRNA, or hypermethylation of CpG island promoters (Calin et al., 2007; Lujambio et al., 2010).

Many T-UCRs show significant complementarity with MicroRNA and experimental results demonstrate that they are effectively downstream target or interactors of MicroRNAs. In vitro in MEG01 leukemia cells, uc.160 was observed to be a direct target of MiR-155, and using luciferase reporter assays in the same model, MiR-24-1 and MiR-29b were able to directly interact and downregulate uc.346A and uc.348, respectively (Calin et al., 2007). In neuroblastoma, nine T-UCRs were proposed to be regulated by five complementary MicroRNAs (Scaruffi et al., 2009), while more recently the interregulation between uc.173 and MiR-291a-3p has emerged to have a role in neuronal apoptosis (Nan et al., 2016). In gastric cancer, MiR-153 directly binds and suppresses uc.416, acting as a tumor suppressor (Goto et al., 2016), while in bladder cancer uc.8 seems to act as a decoy on MiR-596 through their direct binding, thus introducing a possible regulatory loop between T-UCRs and MiRNAs (Olivieri et al., 2016).

T-UCR expression was also found to be correlated with methylation of CpG island promoters. Indeed, three T-UCRs (uc.283A, uc.346, uc.160) had one CpG island within a 2000-bp distance upstream and their expressions have been restored in HCT-116 colon cancer or LNCaP prostate cancer cells by treatment with 5-aza-2-deoxycytidine (5-Aza-dC) (Hudson et al., 2013; Lujambio et al., 2010). To note, these CpG islands were always unmethylated in normal tissue. Another confirmation comes from both gastric and prostate cancer, where another T-UCR, uc.158, is downregulated due to a methylation-associated mechanism (Goto et al., 2016). Additionally, hypoxia is able to induce T-UCRs, such as uc.63, uc.73, uc.106, uc.134, and uc.475 (later named HINCUT, hypoxia-induced noncoding ultraconserved transcripts), at least partially though HIF-1 control (Ferdin et al., 2013). HIF-1α-binding sites have been found upstream of uc.475, while for the other T-UCRs, the mechanism of induction by hypoxia is still unknown.

On the other hand, recent evidence highlight that T-UCRs, in turn, can directly bind and regulate mRNAs, such as for uc.338 which directly interacts with the 3′UTR of TIMP-1 mRNA and acts as its negative regulator (Wang et al., 2017) (Fig. 1B). Another mechanism involves a negative regulation of primary MiRNA (the longest precursor of a MicroRNA, the pri-miR) processing by a T-UCR (Liz et al., 2014); uc.283 controls pri-miRNA processing through RNA:RNA complementarity with the lower stem region of the pri-miR-195, preventing pri-miRNA cleavage by Drosha.

The complex T-UCRs activities reflect their potential importance in cell homeostasis; the mutual regulation between ultraconserved transcripts and MicroRNAs is only the tip of the whole new noncoding regulation of cell homeostasis iceberg that has been discovered so far.

5. T-UCR IN DISEASE AND CANCER

Given the aberrant regulation of T-UCRs in cancer, great efforts have been made to characterize the roles of T-UCRs in cancer. Several evidences demonstrate their potential role as biomarker in different types of cancer. However, little is known about their biogenesis or function in normal tissue. To note, most of the aberrant T-UCR expression in cancer are directed toward an upregulation of these transcript, and thus their low levels in normal tissue are so far difficult to study. A correlation between copy number variation (CNV) and genomic UCRs has recently been described (McCole et al., 2014). The low detectable expression levels of most T-UCR can be explained with their aberrant dosage of CNV in cancer. Indeed, CNV inherited by healthy individuals are significantly depleted in UCR, while cancer CNV, on the contrary, are enriched. This observation suggests a potential dosage sensitivity of cells to UCR, which can be a prerequisite for oncogenic development.

Recently, several studies have underlined the widespread role of T-UCRs in multiple types of cancers, and distinctive signature have been published for chronic lymphocytic leukemia (CLL) (Calin et al., 2007), hepatocellular carcinoma (HCC) (Braconi et al., 2011; Calin et al., 2007; Carotenuto et al., 2016), prostate cancer (Bao et al., 2016; Hudson et al., 2013), neuroblastoma (Mestdagh et al., 2010; Watters et al., 2013), colorectal carcinoma (CRC) (Hanahan and Weinberg, 2011; Sana et al., 2012; Wang et al., 2017), cervical cancer (Li et al., 2016), bladder cancer (Olivieri et al., 2016; Terreri et al., 2016), pancreatic adenocarcinoma (Jiang et al., 2016b), and breast cancer (Marini et al., 2016) (Table 1).

First studies were published by Calin and colleagues, who identified a unique expression profile of T-UCRs in leukemia and carcinomas when comparing cancer with normal cells (Calin et al., 2007; Wojcik et al., 2010). In CLL, uc.349 and uc.352, which map to chromosome 13q21, known to be linked to occurrence of familial CLL, are deregulated in malignant B-CLL CD5-positive cells. This observation not only confirms the

Table 1 T-UCR as Potential Biomarker in Disease and Cancer

Disease	T-UCR	Alteration	Mechanism	References
CLL	uc.349	Down	?	Calin et al. (2007)
	uc.352	Up	?	Calin et al. (2007)
	uc.348	Up	Complementary to mir-29 family	Calin et al. (2007)
Prostate cancer	uc.368	Mut	SNP causes different secondary structure	Bao et al. (2016)
	uc.287	Up	Target of androgen receptor signaling	Bao et al. (2016)
Colorectal cancer	uc.338	Up	Inhibition of TIMP-1	Wang et al. (2017)
	uc.338	Down	?	Sana et al. (2012)
Neuroblastoma	uc.460, uc.350, uc.379	Up	upregulated upon MYCN activation in amplified tumors	Mestdagh et al. (2010)
	uc.209, uc.330, uc.412	Up	Complementary to MiR-877	Mestdagh et al. (2010)
HCC	uc.158	Up	ceRNA for MiR-193b	Carotenuto et al. (2016)
	uc.338	Up	Inhibition $p16^{INK4}$	Braconi et al. (2011)
Pancreatic adenocarcinoma	uc.190, uc.233, uc.270	Up	?	Jiang et al. (2016a)
Bladder cancer	uc.8	Up	Binds to MiR-596, inhibits YY1, increases MMP9	Olivieri et al. (2016) Terreri et al. (2016)
Cervical cancer	uc.206	Up	Binding to TP53	Li et al. (2016)

importance of T-UCRs location in CAGRs but also for the first time underlines the importance of T-UCRs in cancer susceptibility.

In HCC, uc.158 has been described as a downstream mediator of the Wnt/β-catenin pathway in human, mouse, and rats and its inhibition in vitro in HepG2 and SW1 cells is sufficient to reduce anchorage-independent cell growth, 3D-spheroid formation and spheroid-based cell migration, and increase apoptosis by acting as a competitive endogenous RNA for MiR-193b (Carotenuto et al., 2016). While no potential mediator downstream target has been analyzed, the levels of MiR-193b are inversely correlated to uc.158 also in patients, confirming their interregulation. Moreover, uc.158 can potentially be a successful biomarker, being able to differentiate between HCC and normal liver.

In HepG2 cells, expression of uc.338 is significantly increased compared to nonmalignant human hepatocytes and its inhibition results in increase expression of genes involved in G1/S transition, such as p16INK4a and concomitant reduction of CDK4 and CDK6 activity (Braconi et al., 2011).

In prostate cancer, multistage genetic association studies demonstrated that a single-nucleotide polymorphism (SNP) in an UCR is significantly associated with biochemical recurrences in localized prostate cancer patients and with prostate cancer-specific mortality in advanced cancer patients (Bao et al., 2016), probably by impairing uc.368 RNA secondary structure. Disease-specific expression pattern has been also observed for uc.106, uc.477, uc.363, uc.454 in association with Gleason score and extraprostatic extension (Hudson et al., 2013). Using LnCap cells the authors also demonstrated a significant increase of uc.287 expression after treatment with androgens, suggesting that this transcript could be a target of the androgen receptor signaling pathway, and thus potentially relevant in both hormone refractory and responsive cancer patients (Hudson et al., 2013).

Deregulation of T-UCR was also identified in neuroblastoma: a signature of seven T-UCRs (uc.347, uc.350, uc.279, uc.460, uc.379, uc.446, and uc.364) is significantly upregulated in highly aggressive MYCN-amplified tumors compared with nonamplified tumors (Mestdagh et al., 2010) and correlates with overall and event-free survival. uc.460, uc.350, and uc.379 are also MYCN responsive, being significantly upregulated upon MYCN activation (Mestdagh et al., 2010). In the same model, T-UCRs are significantly deregulated after treatment with all-*trans*-retinoic acid (ATRA), which lead to neuroblastoma cells differentiation and decreased MYC transcription (Watters et al., 2013). None of the significant T-UCRs were in common with the previous study (Mestdagh et al.,

2010), but uc.300 was the most downregulated by the treatment. Indeed, uc.300 is overexpressed in neuroblastoma and seems to play a role in cell proliferation and invasion. The beneficial role of ATRA treatment in neuroblastoma patients' seems to be based on uc.300 downregulation, thus inhibiting cell proliferation and invasion. T-UCRs are also potential targets for treatment of neuropathic pain: four different T-UCRs are differentially expressed in spinal cord under spinal nerve ligation (Jiang et al., 2016a); uc.305 and uc.189 were significantly upregulated, while on the contrary, uc.46 and uc.217 were downregulated. The gene ontology and signaling pathway analyses for the T-UCRs host genes indicated these T-UCRs are involved in several intracellular activities such as Ephrin receptor activity, soluble NSF attachment protein receptor (SNARE) interactions in vesicular transport pathway, and WNT signaling pathway, thus confirming the potential of T-UCR as biomarkers or drug targets.

In colorectal cancer, uc.338 appears to be a key transcript, even though it has been found to be both significantly downregulated (Sana et al., 2012) and upregulated (Wang et al., 2017) in CRC tissue, leading to some doubts about the true mechanism through which this transcript is dysregulated in CRC. Furthermore, uc.338 has been found upregulated in different types of cancer cell lines, including biliary, pancreatic, colorectal, prostatic, and breast cancers (Braconi et al., 2011). uc.338 also has a significant role in regulating migration and invasion in lung cancer (Gao et al., 2016), thus supporting the oncogenic role of this lncRNA.

Other T-UCRs have been found to be important in colon tumorigenesis: overexpression of uc.261 contribute to intestinal mucosa barrier damage, by inhibiting the expression and assembly of tight junction proteins such as occludin, claudin-1, and zonula occluden-1, and thus reducing transepithelial electrical resistance (Qian et al., 2016).

In cervical cancer, uc.206 is significantly upregulated and acts as a possible oncogene regulating TP53 expression through the direct binding with its 3′UTR. The inverse correlation between uc.206 and TP53 has also been described in patients (Li et al., 2016), and the same transcript has also been described to have a significant role in Barrett's esophagus carcinogenesis (Fassan et al., 2014).

In bladder cancer, after a genome-wide profiling to evaluate the expression of T-UCRs, uc.8 was identified as the most highly upregulated ultraconserved transcript compared to normal tissue. uc.8 through its interaction with MiR-596, inhibited by YY1 (Terreri et al., 2016) and increased expression of MMP9, contributing to invasive potential in bladder cancer cells (Olivieri et al., 2016). The importance of the relation between T-UCR and MiR-596 in bladder cancer has been elucidated lately: uc.8 silencing increases MiR-596 expression, which in turn increases uc.201 and uc.283 levels, demonstrating the importance of understanding a complex network more than the role of a single T-UCR transcript in cancer (Terreri et al., 2016).

In pancreatic ductal adenocarcinoma (PDAC), using a qPCR array to profile T-UCRs in pancreatic cancer specimens, cell lines, and KrasLSL-G12D mice, uc.190, uc.233, and uc.270 were consistently overexpressed in all three models, and their knockdown significantly reduced the proliferation of MIA PaCa-2 and PANC-1 cells suggesting their possible role as regulators of PDAC oncogenesis (Jiang et al., 2016b).

In breast cancer, uc.63 acts as an oncogene and seems to be responsible for promoting cell survival and tumor growth (Marini et al., 2016). uc.63 has already been described as an hypoxia-induced noncoding transcript (Ferdin et al., 2013), and its expression has been proposed as a potential useful tool for breast cancer prognosis.

The results collected so far stimulate the research toward T-UCRs, in particular to understand the fine mechanisms of regulation in human carcinogenesis. T-UCRs have been described to control mRNA expression, MiRNA stability, and most probably also protein activity (Fig. 1B). So far, it seems clear that T-UCRs are promising cancer biomarkers or potential predictors of response to therapy. Targeting directly T-UCRs in cancer seems on the other hand, more difficult at the moment; this is primary due to the large extent of unknown functions they have that still need to be discovered.

6. CONCLUSIONS AND FUTURE PERSPECTIVES

The interest in the contribution of ncRNAs to the genesis and progression of cancer exponentially increased in the last decade. Given the limited number of protein-coding genes in the human genome, growing evidence suggest that the vast majority of the genome is transcribed into RNA with no or low protein-coding potential. From this interest emerges a completely new area of research aimed to the development of new drugs or new molecular marker identification. One important challenge is to identify the roles of ncRNA, and in particular the roles of T-UCR in tissue homeostasis. Understanding their role and mechanism of action will help to design better approaches to their regulation in carcinogenesis and to design therapeutic strategies or drug combination with the aim to not only kill cancer cells but also to overcome resistance.

Despite T-UCRs being discovered more than 10 years ago, few advances have been done in this front and therapeutic options targeting T-UCRs are still not available. This family of noncoding RNAs includes a wide range of transcripts, associated by conservation and not through function. Discerning the mechanistic role of these transcripts in normal cells will help in identifying the role of their aberrations in cancer. Also, understanding how T-UCR work in normal cells and malignant cells will help us to understand why and how genomic ultraconservation was maintained through millions of years of evolution.

ACKNOWLEDGMENTS

We thank to Maitri Shah for the comments on the manuscript.

Work in Dr. Calin's laboratory is supported by National Institutes of Health (NIH/NCATS) grant UH3TR00943-01 through the NIH Common Fund, Office of Strategic Coordination (OSC), the NIH/NCI grant 1 R01 CA182905-01, a U54 grant—UPR/MDACC Partnership for Excellence in Cancer Research 2016 Pilot Project, a CLL Moonshot Flagship project and the Estate of C.G. Johnson, Jr.

REFERENCES

Bao, B.Y., Lin, V.C., Yu, C.C., Yin, H.L., Chang, T.Y., Lu, T.L., Lee, H.Z., Pao, J.B., Huang, C.Y., Huang, S.P., 2016. Genetic variants in ultraconserved regions associate with prostate cancer recurrence and survival. Sci. Rep. 6, 22124.

Bartolomei, M.S., Zemel, S., Tilghman, S.M., 1991. Parental imprinting of the mouse H19 gene. Nature 351, 153–155.

Bejerano, G., Pheasant, M., Makunin, I., Stephen, S., Kent, W.J., Mattick, J.S., Haussler, D., 2004. Ultraconserved elements in the human genome. Science 304, 1321–1325.

Braconi, C., Valeri, N., Kogure, T., Gasparini, P., Huang, N., Nuovo, G.J., Terracciano, L., Croce, C.M., Patel, T., 2011. Expression and functional role of a transcribed noncoding RNA with an ultraconserved element in hepatocellular carcinoma. Proc. Natl. Acad. Sci. U.S.A. 108, 786–791.

Brown, C.J., Ballabio, A., Rupert, J.L., Lafreniere, R.G., Grompe, M., Tonlorenzi, R., Willard, H.F., 1991. A gene from the region of the human X inactivation centre is expressed exclusively from the inactive X chromosome. Nature 349, 38–44.

Calin, G.A., Liu, C.G., Ferracin, M., Hyslop, T., Spizzo, R., Sevignani, C., Fabbri, M., Cimmino, A., Lee, E.J., Wojcik, S.E., et al., 2007. Ultraconserved regions encoding ncRNAs are altered in human leukemias and carcinomas. Cancer Cell 12, 215–229.

Carotenuto, P., Fassan, M., Pandolfo, R., Lampis, A., Vicentini, C., Cascione, L., Paulus-Hock, V., Boulter, L., Guest, R., Quagliata, L., et al., 2016. Wnt signalling modulates transcribed-ultraconserved regions in hepatobiliary cancers. Gut. published online: 12 September 2016. http://dx.doi.org/10.1136/gutjnl-2016-312278.

ENCODE Project Consortium, Birney, E., Stamatoyannopoulos, J.A., Dutta, A., Guigo, R., Gingeras, T.R., Margulies, E.H., Weng, Z., Snyder, M., Dermitzakis, E.T., et al., 2007. Identification and analysis of functional elements in 1% of the human genome by the ENCODE pilot project. Nature 447, 799–816.

Djebali, S., Davis, C.A., Merkel, A., Dobin, A., Lassmann, T., Mortazavi, A., Tanzer, A., Lagarde, J., Lin, W., Schlesinger, F., et al., 2012. Landscape of transcription in human cells. Nature 489, 101–108.

Ezkurdia, I., Juan, D., Rodriguez, J.M., Frankish, A., Diekhans, M., Harrow, J., Vazquez, J., Valencia, A., Tress, M.L., 2014. Multiple evidence strands suggest that there may be as few as 19,000 human protein-coding genes. Hum. Mol. Genet. 23, 5866–5878.

Fassan, M., Dall'Olmo, L., Galasso, M., Braconi, C., Pizzi, M., Realdon, S., Volinia, S., Valeri, N., Gasparini, P., Baffa, R., et al., 2014. Transcribed ultraconserved noncoding RNAs (T-UCR) are involved in Barrett's esophagus carcinogenesis. Oncotarget 5, 7162–7171.

Ferdin, J., Nishida, N., Wu, X., Nicoloso, M.S., Shah, M.Y., Devlin, C., Ling, H., Shimizu, M., Kumar, K., Cortez, M.A., et al., 2013. HINCUTs in cancer: hypoxia-induced noncoding ultraconserved transcripts. Cell Death Differ. 20, 1675–1687.

Gao, X., Gao, X., Li, C., Zhang, Y., Gao, L., 2016. Knockdown of long noncoding RNA uc.338 by siRNA inhibits cellular migration and invasion in human lung cancer cells. Oncol. Res. 24, 337–343.

Goto, K., Ishikawa, S., Honma, R., Tanimoto, K., Sakamoto, N., Sentani, K., Oue, N., Teishima, J., Matsubara, A., Yasui, W., 2016. The transcribed-ultraconserved regions in prostate and gastric cancer: DNA hypermethylation and microRNA-associated regulation. Oncogene 35, 3598–3606.

Guttman, M., Garber, M., Levin, J.Z., Donaghey, J., Robinson, J., Adiconis, X., Fan, L., Koziol, M.J., Gnirke, A., Nusbaum, C., et al., 2010. Ab initio reconstruction of cell type-specific transcriptomes in mouse reveals the conserved multi-exonic structure of lincRNAs. Nat. Biotechnol. 28, 503–510.

Hanahan, D., Weinberg, R.A., 2011. Hallmarks of cancer: the next generation. Cell 144, 646–674.

Harrow, J., Frankish, A., Gonzalez, J.M., Tapanari, E., Diekhans, M., Kokocinski, F., Aken, B.L., Barrell, D., Zadissa, A., Searle, S., et al., 2012. GENCODE: the reference human genome annotation for The ENCODE Project. Genome Res. 22, 1760–1774.

Hudson, R.S., Yi, M., Volfovsky, N., Prueitt, R.L., Esposito, D., Volinia, S., Liu, C.G., Schetter, A.J., Van Roosbroeck, K., Stephens, R.M., et al., 2013. Transcription signatures encoded by ultraconserved genomic regions in human prostate cancer. Mol. Cancer 12, 13.

Jiang, B.C., Yang, T., He, L.N., Tao, Y.X., Gao, Y.J., 2016a. Altered T-UCRs expression profile in the spinal cord of mice with neuropathic pain. Transl. Perioper. Pain Med. 1, 1–10.

Jiang, J., Azevedo-Pouly, A.C., Redis, R.S., Lee, E.J., Gusev, Y., Allard, D., Sutaria, D.S., Badawi, M., Elgamal, O.A., Lerner, M.R., et al., 2016b. Globally increased ultraconserved noncoding RNA expression in pancreatic adenocarcinoma. Oncotarget 7, 53165–53177.

Lander, E.S., Linton, L.M., Birren, B., Nusbaum, C., Zody, M.C., Baldwin, J., Devon, K., Dewar, K., Doyle, M., FitzHugh, W., et al., 2001. Initial sequencing and analysis of the human genome. Nature 409, 860–921.

Li, Q., Li, X., Wang, C., 2016. Uc.206 regulates cell proliferation and apoptosis by targeting P53 in cervical cancer cells. Neoplasma 63, 411–418.

Liz, J., Portela, A., Soler, M., Gomez, A., Ling, H., Michlewski, G., Calin, G.A., Guil, S., Esteller, M., 2014. Regulation of pri-miRNA processing by a long noncoding RNA transcribed from an ultraconserved region. Mol. Cell 55, 138–147.

Lujambio, A., Portela, A., Liz, J., Melo, S.A., Rossi, S., Spizzo, R., Croce, C.M., Calin, G.A., Esteller, M., 2010. CpG island hypermethylation-associated silencing of non-coding RNAs transcribed from ultraconserved regions in human cancer. Oncogene 29, 6390–6401.

Marini, A., Lena, A.M., Panatta, E., Ivan, C., Han, L., Liang, H., Annicchiarico-Petruzzelli, M., Di Daniele, N., Calin, G.A., Candi, E., et al., 2016. Ultraconserved long non-coding RNA uc.63 in breast cancer. Oncotarget. http://dx.doi.org/10.18632/oncotarget.10572.

Mattick, J.S., Rinn, J.L., 2015. Discovery and annotation of long noncoding RNAs. Nat. Struct. Mol. Biol. 22, 5–7.

McCole, R.B., Fonseka, C.Y., Koren, A., Wu, C.T., 2014. Abnormal dosage of ultraconserved elements is highly disfavored in healthy cells but not cancer cells. PLoS Genet. 10, e1004646.

Mestdagh, P., Fredlund, E., Pattyn, F., Rihani, A., Van Maerken, T., Vermeulen, J., Kumps, C., Menten, B., De Preter, K., Schramm, A., et al., 2010. An integrative genomics screen uncovers ncRNA T-UCR functions in neuroblastoma tumours. Oncogene 29, 3583–3592.

Nan, A., Zhou, X., Chen, L., Liu, M., Zhang, N., Zhang, L., Luo, Y., Liu, Z., Dai, L., Jiang, Y., 2016. A transcribed ultraconserved noncoding RNA, Uc.173, is a key molecule for the inhibition of lead-induced neuronal apoptosis. Oncotarget 7, 112–124.

Olivieri, M., Ferro, M., Terreri, S., Durso, M., Romanelli, A., Avitabile, C., De Cobelli, O., Messere, A., Bruzzese, D., Vannini, I., et al., 2016. Long non-coding RNA containing ultraconserved genomic region 8 promotes bladder cancer tumorigenesis. Oncotarget 7, 20636–20654.

Orom, U.A., Derrien, T., Beringer, M., Gumireddy, K., Gardini, A., Bussotti, G., Lai, F., Zytnicki, M., Notredame, C., Huang, Q., et al., 2010. Long noncoding RNAs with enhancer-like function in human cells. Cell 143, 46–58.

Pertea, M., Salzberg, S.L., 2010. Between a chicken and a grape: estimating the number of human genes. Genome Biol. 11, 206.

Qian, X.X., Peng, J.C., Xu, A.T., Zhao, D., Qiao, Y.Q., Wang, T.R., Shen, J., Ran, Z.H., 2016. Noncoding transcribed ultraconserved region (T-UCR) uc.261 participates in intestinal mucosa barrier damage in Crohn's disease. Inflamm. Bowel Dis. 22, 2840–2852.

Rossi, S., Sevignani, C., Nnadi, S.C., Siracusa, L.D., Calin, G.A., 2008. Cancer-associated genomic regions (CAGRs) and noncoding RNAs: bioinformatics and therapeutic implications. Mamm. Genome 19, 526–540.

Sana, J., Hankeova, S., Svoboda, M., Kiss, I., Vyzula, R., Slaby, O., 2012. Expression levels of transcribed ultraconserved regions uc.73 and uc.388 are altered in colorectal cancer. Oncology 82, 114–118.

Scaruffi, P., Stigliani, S., Moretti, S., Coco, S., De Vecchi, C., Valdora, F., Garaventa, A., Bonassi, S., Tonini, G.P., 2009. Transcribed-ultra conserved region expression is associated with outcome in high-risk neuroblastoma. BMC Cancer 9, 441.

Spizzo, R., Almeida, M.I., Colombatti, A., Calin, G.A., 2012. Long non-coding RNAs and cancer: a new frontier of translational research? Oncogene 31, 4577–4587.

Terreri, S., Durso, M., Colonna, V., Romanelli, A., Terracciano, D., Ferro, M., Perdona, S., Castaldo, L., Febbraio, F., de Nigris, F., et al., 2016. New cross-talk layer between Ultraconserved non-coding RNAs, MicroRNAs and polycomb protein YY1 in bladder Cancer. Genes 7, 127.

Venter, J.C., Adams, M.D., Myers, E.W., Li, P.W., Mural, R.J., Sutton, G.G., Smith, H.O., Yandell, M., Evans, C.A., Holt, R.A., et al., 2001. The sequence of the human genome. Science 291, 1304–1351.

Visel, A., Prabhakar, S., Akiyama, J.A., Shoukry, M., Lewis, K.D., Holt, A., Plajzer-Frick, I., Afzal, V., Rubin, E.M., Pennacchio, L.A., 2008. Ultraconservation identifies a small subset of extremely constrained developmental enhancers. Nat. Genet. 40, 158–160.

Wang, C., Wang, Z., Zhou, J., Liu, S., Wu, C., Huang, C., Ding, Y., 2017. TUC.338 promotes invasion and metastasis in colorectal cancer. International Journal of Cancer 140, 1457–1464.

Watters, K.M., Bryan, K., Foley, N.H., Meehan, M., Stallings, R.L., 2013. Expressional alterations in functional ultra-conserved non-coding RNAs in response to all-trans retinoic acid-induced differentiation in neuroblastoma cells. BMC Cancer 13, 184.
Wojcik, S.E., Rossi, S., Shimizu, M., Nicoloso, M.S., Cimmino, A., Alder, H., Herlea, V., Rassenti, L.Z., Rai, K.R., Kipps, T.J., et al., 2010. Non-codingRNA sequence variations in human chronic lymphocytic leukemia and colorectal cancer. Carcinogenesis 31, 208–215.

CHAPTER FIVE

Following MicroRNAs Through the Cancer Metastatic Cascade

Eleonora Grisard, Milena S. Nicoloso[1]
CRO-Aviano, National Cancer Institute, Aviano, Italy
[1]Corresponding author: e-mail address: mnicoloso@cro.it

Contents

Abstract

Approximately a decade ago the first MicroRNAs (MiRNAs) participating in cancer metastasis were identified and metastmiRs were initially only a handful. Since those first reports, MiRNA research has explosively thrived, mainly due to their revolutionary mechanism of action and the hope of having at hand a novel tool to control cancer aggressiveness. This has ultimately led to delineate an *almost* impenetrable regulatory network: hundreds of MiRNAs transversally dominating every aspect of normal and cancer biology, each MiRNA having hundreds of targets and context-dependent activity.

Providing a comprehensive description of MiRNA roles in cancer metastasis is a daunting task; nevertheless, we still believe that grasping the big picture of MiRNAs in cancer metastasis can give a different perspective on the potential insights and approaches that MiRNAs can offer to understand cancer complexity (e.g., as predictive and prognostic markers) and to tackle cancer metastasis (e.g., as therapeutic targets or tools).

This chapter presents a schematic overview of the role of MiRNAs in governing cancer metastasis, describing step by step the cellular and molecular processes whereby

International Review of Cell and Molecular Biology, Volume 333
ISSN 1937-6448
http://dx.doi.org/10.1016/bs.ircmb.2017.04.005

cancer cells conquer distant organs and can grow as secondary tumors at different distant sites, and for each step, we will introduce how MiRNAs impinge on each one of them.

We deeply apologize with our colleagues for any of their research work that, for clarity, for our effort to streamline and due to space limitations, we did not cite.

1. INTRODUCTION

Cancer is one of the leading causes of death across the world, with 14 million new cases in 2012 and 8.2 million cancer-related deaths worldwide (American Cancer Society, 2015). Late-stage cancer presentation and diagnosis are common, rendering metastasis the main cause of cancer-related deaths (Cancer Facts Sheet, 2017).

Metastasis is the process whereby a tumor diffuses to other anatomical sites that are not contiguous to its site of origin. This can occur early during the clinical history of cancer patients or even decades after the primary tumor has been removed, mainly depending on the cancer type. Most breast cancer metastases are detected 10–15 years after diagnosis or surgical resection except HER2-positive or triple-negative breast cancers, which relapse earlier (<5 years), whereas median time for prostate cancer metastatic relapse after resection is 16 years (Freedland and Moul, 2007; Giancotti, 2013; Karrison et al., 1999). This wide range of differences in tumor behaviors has challenged scientists who developed different theories for metastatic progression, such as the linear theory-in which cells within the primary tumor accumulate progressively the traits that lead to aggressiveness, and the parallel theory that postulates the independent progression of metastases, which arise from early disseminated tumor cells following different molecular and genetic pathways (Klein, 2009; Koscielny and Tubiana, 2010).

In the late 1970s, cancer "metastatic potential" was considered as the statistical probability of developing metastasis based on the size of the primary tumor (Sugarbaker, 1979). Yet, clinical oncologists have always been aware of the difficulty of translating these data (gathered from series of patients) to individual patients, and thus cancer research has ever since been keen in grasping the complexity of each cancer type and of each individual's metastatic potential. This concerted struggle of clinical, basic, and translational cancer researchers to tackle the riddle of cancer metastasis has led to a surge of molecular discoveries with potential clinical relevance but also revealed a daunting molecular intricacy of the cancer metastatic process, for which there is a strong need of analysis, interpretation, and validation.

A major breakthrough in cancer research occurred with the discovery of MicroRNA (MiRNA) alterations in human cancer (Calin et al., 2002). MiRNAs are master gene regulators that are involved in every aspect of cell biology, including proliferation, differentiation, and survival and that contribute to normal development, cell homeostasis, and disease (Vidigal and Ventura, 2015). Moreover, MiRNA expression is tightly controlled and displays a high degree of developmental and tissue specificity (Almeida et al., 2011). At the molecular level, MiRNAs are a class of small noncoding RNAs 22 nucleotide long that function as guide molecules in RNA silencing (Ha and Kim, 2014). The biogenesis of MiRNAs starts with the transcription by RNA polymerase II of long primary transcripts, termed pri-miRNAs (primary MiRNAs) in the nucleus; then, pri-miRNAs are cleaved by the "microprocessor complex" formed by the RNAse III Drosha and Dgcr8 (DiGeorge Critical Region 8)/Pasha in a 70 nucleotides precursor, known as pre-miRNA. Finally, in the cytoplasm, pre-miRNAs are cleaved by the RNAse III Dicer producing the mature 22 nucleotides MiRNA form (Bartel, 2004).

By means of complete or partial complementarity of their 5′-end seven-nucleotide "seed sequence," MiRNAs bind to specific response elements carried by mRNAs (but also by other RNAs species; see below) primarily within their 3′-UTRs but also within their 5′-UTR and their coding sequence (Filipowicz et al., 2008). Binding of MiRNAs to target mRNAs, via Ago (Argonaute) proteins, results in either mRNA direct cleavage by nucleases like Dicer of the RISC (RNA-Induced Silencing Complex) or inhibition of their translation, ultimately affecting protein expression levels. Being the seed sequence of MiRNAs so small, bioinformatics analyses and experimental evidence support the notion that one single MiRNA can act on hundreds of target mRNAs (Lim et al., 2005; Xie et al., 2005). However, MiRNA activity goes far beyond sequence complementarity, being also tightly controlled by regulation of MiRNA-binding proteins, such as the phosphorylation cycles of Ago2 (Argonaute 2) (Golden et al., 2017). The above-mentioned ability of one MiRNA to target multiple proteins and the fact that one protein can be controlled by several MiRNAs make MiRNAs the rheostat switches that directly or indirectly fine-tune protein levels, inducing whole proteome changes to rapidly address promptly every cell requirement (Baek et al., 2008; Selbach et al., 2008). Consequently, derangements of a small number of MiRNAs may severely affect cellular circuitries and drive cells toward transformation (Sotiropoulou et al., 2009), and it is not surprising that pathological conditions, including cancer, show specific MiRNA expression patterns.

As shown by the MiR-155 transgenic mouse model, MiRNAs can drive tumor initiation (Costinean et al., 2006) effectively; at least some MiRNAs are driving forces and act as oncogenes and tumor suppressors (Kent and Mendell, 2006) by controlling the expression of cancer-related protein-coding genes. Yet, understanding whether each MiRNA found deregulated in cancer and consequently during cancer metastasis has a driving role or is just an innocent bystander is not trivial.

Overall MiRNA deregulation in cancer follows the mechanisms of protein-coding oncogenes and tumor-suppressor gene deregulation and can occur by virtue of gene amplification and deletions, as shown by the fact that MiRNAs are enriched in cancer-associated genomic regions (Calin et al., 2004). Alternative mechanisms responsible for MiRNA deregulation are epigenetic regulation (Kunej et al., 2011), which is particularly strong and occurs at higher frequency than for protein-coding genes (Weber et al., 2007), transcriptional regulation by oncogenes such as c-MYC (Chang et al., 2008; O'Donnell et al., 2005), and tumor suppressors like p53 (He et al., 2007a,b) or posttranscriptional regulation. This latter mechanism is mediated by the enzymes of the MiRNA maturation machinery, such as the RNAses III Drosha and Dicer, which process MiRNA precursors in the nucleus and in the cytoplasm, respectively (Thomson et al., 2006), and which can also be found deregulated in cancer.

Given their pervasive roles in all aspects of human cell biology, it is not surprising that, soon after their connection with cancer, MiRNAs were also found to act as the "microsteering wheel" of cancer metastasis (Nicoloso et al., 2009) and MiRNAs with an activator or a suppressive role in metastasis establishment are generally termed *metastamiRs* (Hurst et al., 2009).

For sake of clarity, notwithstanding which theory of the metastatic cascade is true, in this review we will attempt to describe linearly the steps that a cancer cell undertakes to become invasive and aggressive and that ultimately concur to the outburst of life-threatening metastases, focusing for each step on the switches tuned by MiRNAs. We would like to advise the reader that each MiRNA–target interaction reported in this review is truly the result of several experimental approaches and must always be contextualized by referring to the original research articles.

At the cellular level, metastasis takes place when cancer cells succeed in leaving the primary tumor site, traveling via the circulation to a distant tissue site, and therein are able to thrive and establish a secondary tumor (Gupta and Massagué, 2006). This process can be arbitrarily divided in several steps, which are collectively termed the *metastatic cascade* and include: (i) detachment and local invasion, (ii) intravasation, (iii) survival in the circulation, (iv) arrest at distant sites and extravasation, (v) micrometastasis formation, and finally (vi) metastatic colonization (Fidler, 2003). Neoangiogenesis, the ability of cancer cells to recruit neoformed vessels, although presented at the end of the cascade, is essential both in the first and in the last steps. An overview and summary of the MiRNAs involved in these steps is also provided by the different panels of Fig. 1 and Table 1.

2. DETACHMENT AND LOCAL INVASION (FIG. 1A)

Tissue architecture integrity, and in particular the integrity of epithelial basement membranes, guarantees that a noninvasive tumor will remain confined at the primary tumor site, and thus it is a first impediment preventing cell invasiveness (Hu et al., 2008). To become invasive, tumor cells must detach from the basement membrane and invade the surrounding stroma, two processes that simultaneously require *cell plasticity*, *increased motility*, and *ability to remodel the extracellular matrix* (ECM).

To detach from their epithelium of origin, tumor cells must first lose cell–cell adhesive contacts and cell–ECM contacts, which depend, respectively, on adherent- and tight-junction proteins (e.g., E-cadherin and Zonulin-1, Occludin, Claudin) and on integrin repertoire. At the same time, the loss of cell–ECM contacts or the adhesion to an inappropriate location in a normal cell triggers a form of programmed cell death, termed *anoikis* (Gilmore, 2005). Anoikis prevents the survival of aberrant cells that have detached from their native location, representing an important barrier to cancer metastasis. Thus, the plasticity of cancer cells to modulate their repertoire of membrane proteins can favor anoikis resistance and the subsequent steps of metastasis (Paoli et al., 2013). Among the first metastamiRs discovered, MiR-373 and MiR-520c control breast, prostate, colon, and glioma cancer cell detachment and invasion primarily by modulating cell–ECM interactions through the repression of *CD44* (Huang et al., 2008; Wei et al., 2016a). *CD44* is a cell surface receptor for Hyaluronan, a major component of the ECM, which acts as a metastasis suppressor in several cancer types. MiR-124, which is downregulated in oral cell squamous carcinoma, directly targets *ITGB1* (Integrin-β1) and affects cell adhesion and motility (Hunt et al., 2011). Loss of cell–cell junctions is instead achieved by MiR-9 direct targeting of *CDH1* (E-cadherin) in breast cancer cells (Khew-Goodall and Goodall, 2010). This, in turn, leads to activation of β-catenin signaling, Vegf (Vascular endothelial growth factor) upregulation, and increased angiogenesis (Khew-Goodall and Goodall, 2010). Another MiRNA targeting *CDH1* is MiR-224 (Ling et al., 2016). MiR-224 controls

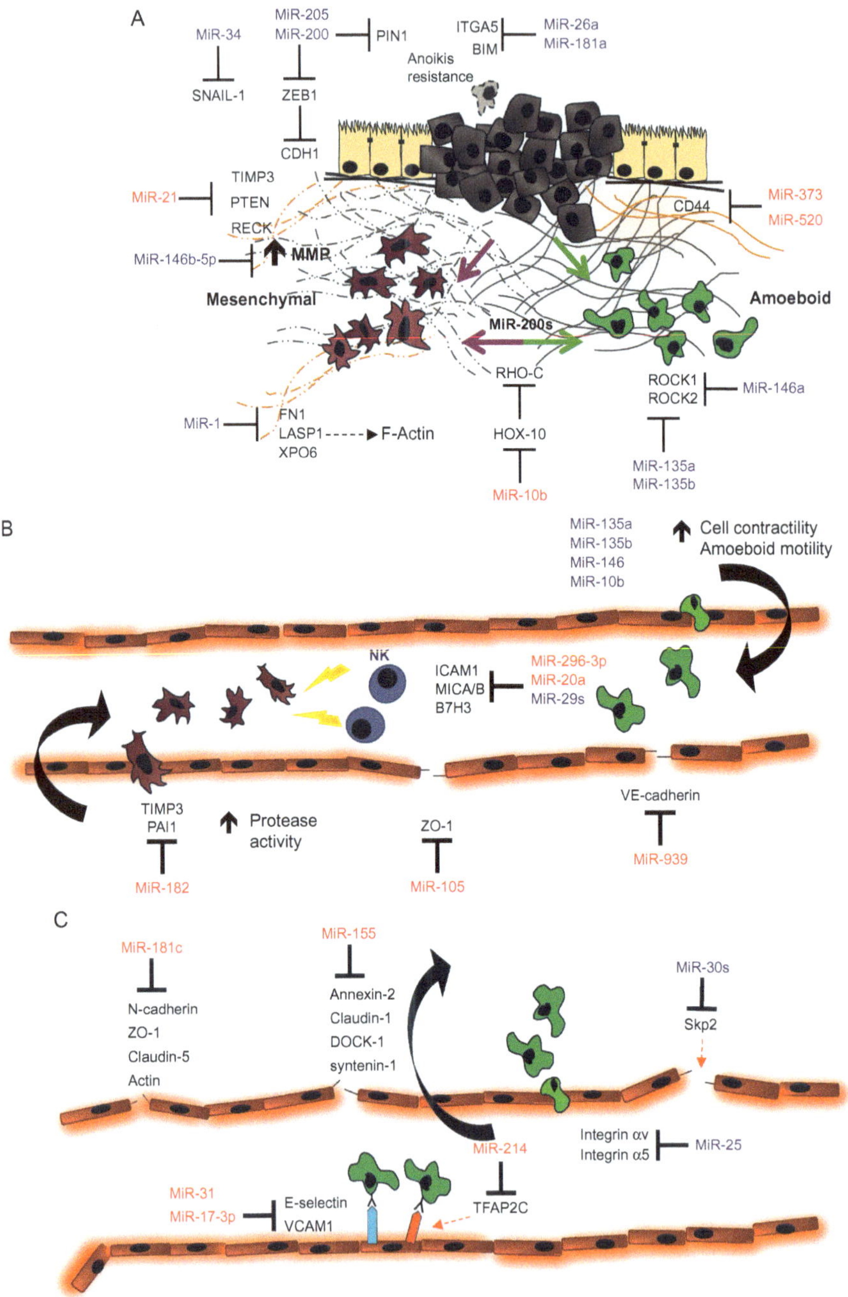

Fig. 1 See legend on opposite page.

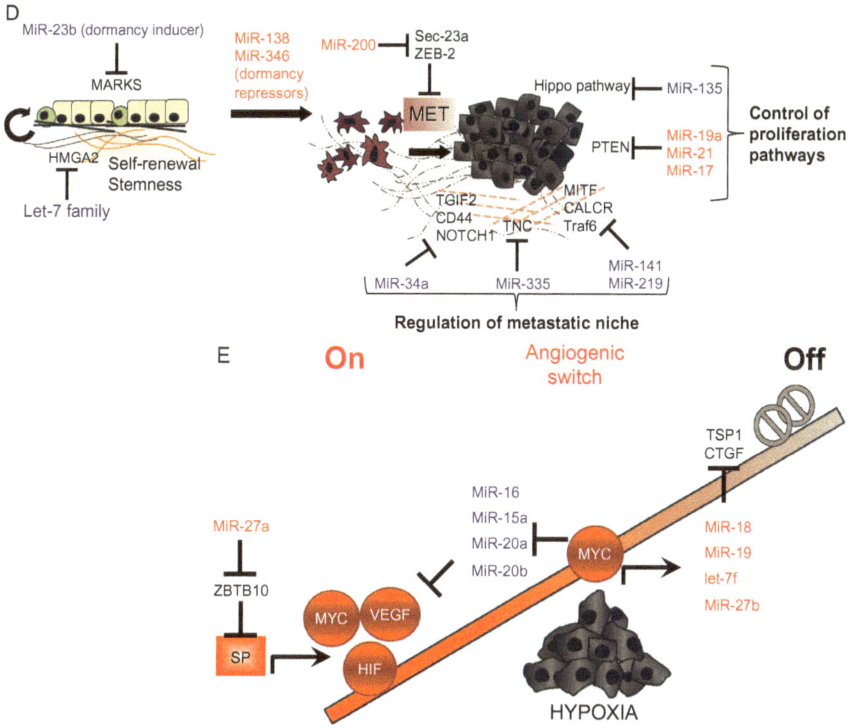

Fig. 1 *Overview of the MiRNAs in the different steps of the metastatic cascade*. The figure shows the network of MiRNAs and protein-coding genes regulating the sequential steps of the metastatic cascade grouped in different panels. Upregulated MiRs, *red*; downregulated MiRs, *blue*. *AMT*, amoeboid-to-mesenchymal transition; *CTCs*, circulating tumor cells; *DTCs*, disseminated tumor cells; *ECM*, extracellular matrix; *EMT*, epithelial-to-mesenchymal transition; *MAT*, mesenchymal-to-amoeboid transition; *MET*, mesenchymal-to-epithelial transition; *MMP*, matrix metalloproteinases; *NK*, natural killer cells. (A) *Migration and invasion*. Migration of tumor cells occurs by two different mechanisms. Mesenchymal migration depends on integrin-based adhesions, cell protrusions like filopodia, and ECM degradation by proteases. MiR-1 is involved the regulation of filopodia formation, MiR-21 and MiR-146-5p regulate MMP function, and MiR-373 and MiR-520c regulate cell–ECM interactions. Mesenchymal migration is related to EMT, which is regulated by MiR-200/205 and MiR-34 via a reciprocal feedback loop with Zeb-1 and Snail-1 transcription factors. Amoeboid migration involves Rho–ROCK-regulated actomyosin contractility and high cell deformability that are controlled by MiR-10b, MiR-146a, and MiR-135a/b. MiR-200 family members modulate MAT/AMT. (B) *Intravasation and survival in the circulation*. Intravasation requires the alteration of the endothelial barrier and increased cell deformability. MiR-105, MiR-939, and MiR-182 directly alter the permeability of endothelial barrier; MiR-10, MiR-146a, and MiR-135a/b, which modulate Rho–ROCK-dependent cell contractility, may regulate also tumor cell intravasation. During circulation in bloodstream, circulating tumor cells (CTCs) must withstand immune attack. MiR-296-3p, MiR-20a, and MiR-29s are involved

(Continued)

in vitro cell migration, invasion, and in vivo metastasis formation of colorectal cancer cells and correlates with colorectal cancer stage (Ling et al., 2016). By concomitantly targeting *CDH1* and *SMAD4* (Mothers Against Decapentaplegic Homolog 4) (Ling et al., 2016), which promotes the prometastatic activity of TGF-β (Transforming Growth Factor-β) (Zhang et al., 2010), MiR-224 likely acts in the detachment of cells from their primary site.

Several MiRNAs that are lost during cancer progression control anoikis of particular control anoikis. Of particular relevance is the role played in this step by MiR-200b. This MiRNA belongs to the MiR-200 family that will be later described as primarily involved in the regulation of the epithelial–mesenchymal transition (EMT) and its reversal, mesenchymal to epithelial transition (MET). In particular, MiR-200b promotes anoikis of breast cancer cells by downregulating the levels of *PIN1* (Zhang et al., 2013b), a peptidyl-prolylisomerase known to affect anchorage-independent growth (Tan et al., 2010). Also the tumor suppressor MiR-26a can promote anoikis; in fact, its loss in hepatocellular carcinoma favors cancer invasiveness by inducing *ITGA5* (Integrin α5) (Zhang et al., 2015b). In breast cancer, MiR-181a-5p, which is induced by TGF-β, inhibits anoikis by targeting the proapoptotic protein Bim (Taylor et al., 2013), ultimately promoting metastasis. Instead, this same MiRNA in normal breast cells induces anoikis by acting on autophagy (Wei et al., 2016b).

Fig. 1—Cont'd in immune tolerance to NK cells. (C) *Arrest at distant organ site and extravasation.* To arrest at a distant site, CTCs adhere to endothelial cells via the specific binding to endothelial adhesion molecules. The binding of CTCs to E-selectin, VCAM1, and other adhesion molecules is regulated by MiR-31, MiR-17-3p, and MiR-214. As for intravasation, tumor cell extravasation is facilitated by the alteration of cell permeability, which is directly regulated by MiR-30s, MiR-155, and MiR-181c. (D) *Micrometastasis formation and metastatic colonization.* After extravasation in the target organ, disseminated tumor cells (DTCs) must adapt to a foreign microenvironment. If this adaptation delays, DTCs encounter a period of dormancy, during which they reside in the target organ as single cells or as micrometastases, without giving metastatic outgrowth. Tumor dormancy is regulated by MiR-23b, MiR-138, and MiR-346 (*left part*). Metastatic colonization requires active proliferation of DTCs, which can be regulated by MiRNAs in different manners: MiR-135, MiR-19a, MiR-21, and MiR-17 directly regulate proliferation pathways; MiR-34a, MiR-335, MiR-141, and MiR-219 regulate cell–ECM and cell–cell interactions in the metastatic niche, and MiR-200 regulates MET, the reverse process of EMT, which involves cell differentiation and exit from dormancy. (E) *Angiogenesis.* Tumors subvert the balance between pro- and antiangiogenic factors and turn on the angiogenic switch by producing high amounts of proangiogenic factors both at the primary tumor site and in distant organs during colonization. MiR-17–MiR-92 cluster, MiR-27a/b, and Let-7 act as proangiogenic MiRNAs, and MiR-16, MiR-15, MiR-20a, and MiR-20b act as antiangiogenic MiRNAs.

Table 1 Examples of MiRNA Functions in the Different Steps of the Metastatic Cascade

Metastatic Step	Function	MiRNA	Targets	Cancer Type	References
Invasion	ECM remodeling	MiR-182	*FOXO3* *MITF*	Melanoma	Segura et al. (2009)
		MiR-21	*RECK* *TIMP3*	Glioma Breast cancer Hepatocellular carcinoma Squamous carcinoma	Gabriely et al. (2008), Song et al. (2010), Meng et al. (2007)
	ECM remodeling, cell contractility	MiR-1	*FN1* *LASP1* *XPO6*	Prostate cancer	Hudson et al. (2011)
		MiR-146b-5p MiR-146	*MMP-16* *F-actin*	Glioma T-Cell acute Lymphoblastic leukemia	Li et al. (2013c), Correia et al. (2016)
	Cell–ECM interactions	MiR-373 MiR-520c	*CD44* *TGFBR2*	Prostate cancer Colon cancer Glioma	Voorhoeve et al. (2006), Huang et al. (2008), Wei et al. (2016a)
	Cell–cell interactions	MiR-188-3p	*MLLT4* (Afadin)	Colorectal cancer	Pichler et al. (2017)
		MiR-224	*SMAD4* *CDH1*	Colorectal cancer	Ling et al. (2016)
	EMT Cell–cell interactions Hippo pathway	MiR-9	*CDH1* *LIFR*	Breast cancer	Ma et al. (2010a,b), Chen et al. (2012a)

Continued

Table 1 Examples of MiRNA Functions in the Different Steps of the Metastatic Cascade—cont'd

Metastatic Step	Function	MiRNA	Targets	Cancer Type	References
	EMT	MiR-34 MiR-30a MiR-30b MiR-30c MiR-30d	*SNAIL1*	Colorectal cancer Non-small cell lung cancer Ovarian cancer	Siemens et al. (2011), Kim et al. (2011), Kumarswamy et al. (2012), Ye et al. (2015), Ouzounova et al. (2013)
		MiR-214	*FGFR1*	Colorectal cancer	Chen et al. (2014)
		MiR-210 MiR-218	*FOXN3* *SETD2* *ZEB2* *CDH2*	Colorectal cancer Brain tumors	Mudduluru et al. (2015)
	EMT Cell contractility Anoikis	MiR-200 MiR-205 MiR-200a MiR-200c	*ZEB1*, *ZEB2* *MARCKS* *VEGFR1* *FHOD1* *PPM1F* *PIN1*	Breast cancer Melanoma Adenocarcinoma Pancreatic cancer	Gregory et al. (2008), Huang et al. (2015), Roybal et al. (2011), Zhang et al. (2013b)
	EMT Cell plasticity Contractility	MiR-155	*RHOA*	Breast cancer	Kong et al. (2008)
		MiR-10b	*HOX-10*	Breast cancer	Ma et al. (2007)
	Cell contractility	MiR-146a	*ROCK1*	Hormone refractory prostate cancer	Lin et al. (2008)

	EMT Cell contractility Hippo pathway	MiR-135a MiR-135b	*ROCK1* *ROCK2* *LZTS1* *LATS2* *NDR2* *MOB1b* *β-TRCP* *SIAH1* *FOXN3*	Prostate cancer Non-small cell lung cancer Colorectal cancer	Kroiss et al. (2015), Mudduluru et al. (2015), Lin et al. (2013)
	Cell–ECM interactions Cell contractility	MiR-340	*CTNNB1* *ROCK1* *MYC*	Breast cancer	Mohammadi-Yeganeh et al. (2016)
		MiR-124	*ITGB1* *ROCK2* *EXH2*	Squamous carcinoma Hepatocellular carcinoma	Kato et al. (2014), Zheng et al. (2012)
	Cell migration MAPK pathway	MiR-23a/ 24–2/27a cluster	*SPRY2*	Breast cancer Liver metastasis	Li et al. (2013b)
	EMT miscellaneous	MiR-215 MiR-192 MiR-194 MiR-215	*ZEB2* *MDM2* *RUNX1*	Clear cell renal cell carcinoma Gastric cancer Non-small cell lung cancer	Khella et al. (2013), Li et al. (2015b)

Continued

Table 1 Examples of MiRNA Functions in the Different Steps of the Metastatic Cascade—cont'd

Metastatic Step	Function	MiRNA	Targets	Cancer Type	References
Intravasation	Disruption of endothelial barrier	MiR-105 (in exosomes secreted by cancer cells)	*ZO-1*	Breast cancer	Zhou et al. (2014)
		MiR-939	*CDH5* (VE-cadherin)	Breast cancer	Di Modica et al. (2017)
	Pleiotropic	Let-7 Lin-28	*HMGA*	Breast cancer	Dangi-Garimella et al. (2009)
		MiR-373 MiR-520	*TGFBR2*	ER—breast cancer	Keklikoglou et al. (2012)
		MiR-21	*PDCD4*	Sarcoma	Asangani et al. (2008)
	Vessel permeability	MiR-182	*PAI1* *TIMP3* *RSU1* *MTSS1*	Colorectal cancer	Sachdeva et al. (2014)
Survival in the circulation	Immune tolerance	MiR-296-3p	*ICAM1*	Prostate cancer	Liu et al. (2013)
		MiR-20a	*MICA/B*	Ovarian cancer	Xie et al. (2014)
		MiR-29a MiR-29b MiR-29c	*B7H3*	Neuroblastoma Sarcoma Brain tumor	Xu et al. (2009)
	Immune response	MiR-155 MiR-223 MiR-146a	*MEF2C* *IRAK1* *TRAF6*	Immune cells	Lu and Liston (2009), Johnnidis et al. (2008), Bhaumik et al. (2008)

Arrest at distant organ site	Cell–cell adhesion Endothelial barrier integrity	MiR-214 MiR-148b	*ITGA3* *TFAP2C ITGA5* *ALCAM*	Melanoma	Penna et al. (2011), Orso et al. (2016)
	Cell–cell adhesion	MiR-31 MiR-17-3p	*SELE* (E-Selectin) *ICAM1*		Suárez et al. (2010)
Extravasation	Endothelial cell recruitment	MiR-199a-3p MiR-199q-5p MiR-1908	*DNAJA4* *APOE*	Melanoma	Pencheva et al. (2012)
	Endothelial cell recruitment Disruption of endothelial barrier Disruption of blood–brain barrier	MiR-155	*ANXA2* *CLDN1* (Claudin-1) *DOCK-1* *SDCBP* (Syntenin-1) *ALDH1A1* *PIR* *PDCD4*	Breast cancer Neuroinflammation	Thomsen et al. (2015), Lopez-Ramirez et al. (2014)
	Disruption of endothelial barrier	MiR-30a MiR-30b MiR-30c MiR-30d MiR-30e	*SKP2*	Melanoma	Qi et al. (2015)
	Disruption of blood–brain barrier	MiR-181c (in extravesicles secreted by cancer cells)	*CDH2* *ZO-1* *CLDN5* *ACTB* *PDPK1*	Breast cancer	Tominaga et al. (2015)
	Cell–ECM interactions Cytoskeletal dynamics	MiR-25	*ITGA5* *ITGA5*	Prostate cancer	Zoni et al. (2015)

Continued

Table 1 Examples of MiRNA Functions in the Different Steps of the Metastatic Cascade—cont'd

Metastatic Step	Function	MiRNA	Targets	Cancer Type	References
Micrometastasis formation	Tumor dormancy	MiR-23b (in exosomes secreted by BM-MSCs)	*MARCKS*	Breast cancer	Ono et al. (2014)
		MiR-138 MiR-346	Not identified	Breast cancer	Gao et al. (2014)
Metastatic colonization	Cell metabolism	MiR-122 (in extra vesicles secreted by cancer cells)	*PKM2*	Breast cancer	Fong et al. (2015)
	Proliferation	MiR-19a (in exosomes secreted by astrocytes) MiR-19a MiR-21 MiR-17	*PTEN*	Breast cancer Brain metastasis CRC liver metastasis	Alečković and Kang (2015), Zhang et al. (2015a), Zhang et al. (2012)
		MiR-335 MiR-126	*TNC* *SOX4*	Breast cancer	Tavazoie et al. (2008)
	MET	MiR-200	*SEC23A* *ZEB2*	Breast cancer metastasis in lung and liver	Dykxhoorn et al. (2009), Korpal et al. (2011)
	Osteoclastogenesis self-renewal	MiR-34a	*TGIF2* *CD44* *NOTCH1*	Prostate cancer Breast cancer Breast cancer bone metastasis	Krzeszinski et al. (2014), Li et al. (2015a), Liu et al. (2011), Kang et al. (2015)
		MiR-141 MiR-219	*MITF* *CALCR* *TRAF6*	Breast cancer Bladder cancer	Ell et al. (2013)

	Stroma remodeling by cancer: cancer-associated fibroblasts	MiR-31 MiR-214 MiR-155	*CCL5*	Ovarian cancer	Mitra et al. (2012)
	Hippo pathway	MiR-135b	*LZTS1* *LATS2* *NDR2* *MOB1* *β-TRCP*	Lung and colorectal cancer	Lin et al. (2013), Mudduluru et al. (2015)
	Self-renewal and differentiation	Let-7 Lin-28	*HMGA2* *H-RAS* *MMP1* *OPN* *BACH1*	Breast cancer stem cells Non-small cell lung cancer	Yu et al. (2007), Yun et al. (2011), Alam et al. (2015)
Angiogenesis	Induction of tumor angiogenesis	MiR-194	*TPS1*	Colorectal cancer	Sundaram et al. (2011)
		Let-7	*TPS1*		Kuehbacher et al. (2007)
		MiR-17-92 cluster	*TPS1* *CTGF* *TGF-β*	Myc-induced cancers Adenocarcinoma	Dews et al. (2006, 2010)
	Endothelial cell recruitment and activation	MiR-9 (secreted in microvesicles by tumor cells)	*IGFBP2* *PITPNC1* *MERTK* *SOCS5*	Colorectal cancer	Zhuang et al. (2012)

Continued

Table 1 Examples of MiRNA Functions in the Different Steps of the Metastatic Cascade—cont'd

Metastatic Step	Function	MiRNA	Targets	Cancer Type	References
		MiR-214	*HDGF*	Hepatocellular carcinoma	Shih et al. (2012)
		MiR-150 (monocyte-derived microvesicles)	*MYB*	Sarcoma	Li et al. (2013a)
		MiR-130a	*GAX* *HOXA5*	Endothelial cells	Chen and Gorski (2008)
		MiR-221 MiR-222	*KIT*	Endothelial cells	Poliseno et al. (2006)
	Endothelial cell recruitment and vascular integrity	MiR-126	*VCAM-1* *SPRED* *PIK3R2*		Fish et al. (2008), Harris et al. (2008), Png et al. (2012)
	HIF pathway	MiR-382	*PTEN*		Seok et al. (2014)
		MiR-107 MiR-26 MiR-210	(Predicted) *CDC25A* *Cyclin-E1* *CDK6* *NPTX1* (Pentraxin)		Kulshreshtha et al. (2006)
		MiR-107	*HIF1B*	Colorectal cancer	Yamakuchi et al. (2010)
	VEGF expression	MiR-15a MiR-16 MiR-20a MiR-20b	*HIF1A* *STAT3*	Breast cancer	Cascio et al. (2010), Yamakuchi et al. (2010), Hua et al. (2006)
		MiR-27a MiR-27b	*SP1* *SP3* *SP4*	Breast cancer	Mertens-Talcott et al. (2007)

Cancer cells that have survived the detachment from their original site of origin can move through the surrounding stroma adopting migration mechanisms similar to those that occur in normal cells during physiological processes, such as wound healing (Friedl and Bröcker, 2000). However, different from normal cells, tumor cells are unable to terminate these processes. The two main patterns of motility used by tumor cells are individual cell migration, when cell–cell junctions are lost, and collective cell migration, when cell–cell adhesions are retained (Friedl and Alexander, 2011).

Individual cell migration can occur in two different morphological variants: mesenchymal and amoeboid (Friedl and Wolf, 2003). Mesenchymal migration is typical of connective tissue tumors, such as fibrosarcomas, and of epithelial tumors following dedifferentiation (Polette et al., 1998). Cells that undergo mesenchymal migration have a fibroblast-like spindle-shaped morphology and move via a migration cycle that is dependent on integrin-based adhesion and traction forces with the formation of actin-based cell protrusions, like lamellipodia or filopodia (Friedl and Alexander, 2011). Besides affecting integrin cell repertoire, several MiRNAs deregulated in cancer affect directly or indirectly *actin remodeling*. In particular, MiR-1 inhibits prostate cancer invasion by targeting actin-binding partners *LASP1* (LIM and SH3 domain Protein 1) and *XPO6* (Exportin 6), which mediate F-actin (filamentous actin) organization, thus impairing filopodia formation (Hudson et al., 2011); instead, MiR-21 increases breast cancer invasion and metastasis regulating cell contractility by targeting *TPM1* (Tropomyosin 1) (Zhu et al., 2008). Cell–cell junctions can be also regulated by MLLT4/Afadin, a protein that connects the actin cytoskeleton to the transmembrane protein Nectin (Takai and Nakanishi, 2003). MiR-188-3p overexpression in colorectal cancer cells can target *MLLT4* and promotes colorectal cancer cell in vitro migratory properties and in vivo metastatic potential. Consistently, MiR-188-3p is significantly associated with colorectal cancer prognosis (Pichler et al., 2017). Mesenchymal migration involves changes in composition and remodeling of the surrounding ECM via de novo synthesis of ECM components and the recruitment of *proteases*, such as MMPs (matrix metalloproteinases) (Friedl and Alexander, 2011). Many metastamiRs participating in this initial step of the metastatic cascade influence ECM composition and the activity of MMPs both directly and indirectly; for example, MiR-1 loss increases the expression of *FN1* (Fibronectin 1) (Hudson et al., 2011), an ECM component typical of mesenchymal state. MiR-21 promotes invasiveness of tumor cells by downregulating two MMP inhibitors, *RECK* (REversion-inducing-Cysteine-rich protein with Kazal motifs) and *TIMP3* (Tissue Inhibitor of

MMPs 3), as shown in gliomas (Gabriely et al., 2008) and breast cancer (Song et al., 2010), but also by downregulating *PTEN*. In fact, loss of *PTEN*, among other pathways, unleashes FAK1 (Focal Adhesion Kinase 1) signaling, which ultimately reaches the nucleus to induce *MMP-2* and *MMP-9* transcription (Meng et al., 2007). MiR-373 and MiR-520c, mentioned earlier in cell detachment, act on ECM remodeling in fibrosarcomas cells by indirectly activating MMP-9 via suppression of *mTOR* and *SIRT1* (Sirtuin 1). Their function though, as it occurs for almost all MiRNAs is context dependent. In fact, MiR-373 and MiR-520c in estrogen receptor negative breast cancer cells have a metastasis-inhibiting role by inhibiting both NFkB (Nuclear Factor kappa light-chain enhancer of activated B cells) and TGF-βR2 (TGF-β Receptor 2) signaling, blocking both cell invasion and intravasation (see below) (Keklikoglou et al., 2012). Instead, MiR-146b-5p inhibits glioma cell migration and invasion by directly downregulating *MMP-16* mRNA (Li et al., 2013c).

Mesenchymal migration is associated with EMT, a dedifferentiation process occurring in epithelial cells during embryonic development, tissue repair, and tumorigenic transformation (Thiery et al., 2009). Consistently, EMT traits in various cancer types have been linked to increased invasiveness and metastasis (Byles et al., 2012; Deckers et al., 2006; Gulhati et al., 2011; Kurrey et al., 2005; Yang and Weinberg, 2008). EMT causes epithelial cells to lose their epithelial features, such as presence of adherent- and tight-junction proteins, cuboidal shape, and almost total immobility, to gain the traits of mesenchymal cells, such as fibroblast-like shape, loss of junctional molecules, and increased motility (Thiery et al., 2009). Transcription factors, such as Snail-1, Slug, Twist, and Zeb-1 and -2 (Zinc Finger E-box-binding homeobox 1 and 2), promote the EMT program in part by directly downregulating the transcription of CDH1, a typical marker of epithelial state (Thiery et al., 2009). Moreover, Snail-1 and Zeb-1 transcription factors through a double-negative feedback loop sustain EMT by repressing two MiRNA families' exponents of the epithelial differentiation state, i.e., MiR-34 and MiR-200, respectively (Gregory et al., 2008; Siemens et al., 2011). In fact, MiR-34 and MiR-200 are lost in the initial steps of the metastatic cascade but are later regeared by cancer cells to promote metastatic colonization via MET (see below). In this first step, loss of MiR-200 also contributes to actin remodeling and therefore to cell motility and invasiveness by suppressing *WAVE3*, a member of the WASP (Wiskott–Aldrich syndrome protein)/WAVE actin cytoskeleton-remodeling family of proteins (Sossey-Alaoui et al., 2009). Also the loss of several members of the MiR-30 family contributes to cell invasiveness by unleashing the EMT program (Ouzounova et al., 2013). For example, MiR-30a is lost in most non-small cell lung cancers where it would otherwise posttranscriptionally repress *SNAIL1*, inhibiting EMT and finally cell invasiveness (Kumarswamy et al., 2012). MiR-30 family's role in EMT is further underscored by the fact that MiR-30b, MiR30c, and MiR-30d are downregulated in ovarian cancer cell lines upon TGF-β-induced EMT, whereas MiR-30d restoration reverses the effects of TGF-β reducing ovarian cancer cell invasive and migratory properties (Ye et al., 2015).

Core components of the Hippo pathway regulate at different levels the metastatic cascade, starting from the induction of EMT to stem cell renewal and finally to MET during micrometastasis (Janse van Rensburg and Yang, 2016). MiR-9, already described to act on E-cadherin, increases breast cancer migration, invasion, and metastatic colonization also by downregulating *LIFR* (Leukemia Inhibitory Factor Receptor), a metastasis suppressor acting upstream of the Hippo signaling cascade (Chen et al., 2012a). MiR-135b promotes non-small lung cancer in vitro migration and invasion, and in vivo metastasis by downregulating multiple genes involved in the of the Hippo pathway, such as *LATS2* (Large Tumor Suppressor kinase 2), *BTRCP* (β-Transducin Repeat-Containing E3 ubiquitin protein ligase), *NDR2* (Nuclear Dbf2-related kinase 2), as well as *LZTS1* (Leucine Zipper putative Tumor Suppressor 1) (Lin et al., 2013).

By comparing colorectal cancer metastatic samples with their relative primary tumors and normal tissues, Mudduluru and colleagues found a MiRNA signature comprising MiR-210, MiR-135a, and MiR-218 inducing the EMT proteome. In particular, MiR-210 and MiR-135 induced, whereas MiR-218 repressed, invasion and metastasis of colorectal cancer cells by targeting *SETD2* (SET-containing domain 2), *SIAH1*, and *FOX3N* (Forkhead box protein N3), or *CDH2* and *ZEB2*, respectively (Mudduluru et al., 2015), therefore acting at different levels on cell–cell junctions and EMT.

The second type of individual migration adopted by invasive tumor cells, the *amoeboid* migration, is similar to the movement of the single-celled organism amoeba *Dictyostelium discoideum* (Friedl and Alexander, 2011). Different from mesenchymal migration, amoeboid migration is not dependent on integrin-adhesion dynamics and relies mainly on actomyosin contractility that is regulated by the Rho family of small GTPases and its kinase effectors of the ROCK (Rho-associated protein kinase) family (Friedl and Alexander, 2011). Amoeboid migrating tumor cells, rather than destroying the surrounding ECM, penetrates it in a compressed form, with a rounded or elliptical shape and high degree of deformability (Friedl and Alexander, 2011), which has long been associated with increased metastatic potential in melanoma cells (Ochalek et al., 1988). Amoeboid migration is mainly found in lymphomas, myeloid leukemia cells (Verschueren et al., 1991), and neuroendocrine tumors (Rintoul and Sethi, 2001), yet, it is likely to occur in other solid tumors. In addition, tumor cells that can easily switch between the mesenchymal and amoeboid migration strategies in response to changing microenvironment are thus intrinsically more aggressive because they can adapt to different microenvironments.

In the presence of a loose and soft ECM, tumor cells prefer the amoeboid type of migration; in contrast, when ECM is dense and stiff, the prevailing type of migration is the mesenchymal one (Friedl and Alexander, 2011). Although it is difficult to detect the specific type of motility adopted by cancer cells in vivo, several MiRNAs contribute to the balance between mesenchymal and amoeboid migration/invasion by targeting either Rho GTPases or ROCK kinases. For example, MiR-155 targets *RHOA* and regulates TGF-β-induced EMT, migration, and invasion in breast cancer (Kong et al., 2008). MiR-10b, downstream of Twist, promotes tumor cell invasion and metastasis by indirectly regulating *RHOC* levels (Ma et al., 2007). The dynamics of Rho–ROCK kinases governing actin cytoskeleton are regulated at the posttranslational level by E3 ubiquitin-ligase Smurf1 that degrades RhoA (Tian et al., 2011). In breast cancer, loss of *SMURF1* induces a rounded cell morphology that is attributable to mesenchymal amoeboid transition (MAT), increasing cell invasion and intravasation (Sahai et al., 2007). By targeting *SMURF1*, MiR-497 inhibits ovarian cancer invasion and migration (Wang et al., 2014), while, in colorectal cancer, MiR-497 inhibits invasion by targeting *IGF1R* (Insulin Growth Factor 1 Receptor) (Guo et al., 2012).

Also members of the MiR-146 family fine-tune MAT in a tissue-specific manner. In fact, while MiR-146b-5p loss in glioma cells favors mesenchymal motility by unleashing its target *MMP-16* (Li et al., 2013c), in T-cell acute lymphoblastic leukemia, MiR-146b likely favors cell propelled amoeboid motility by increasing the levels of F-actin (Correia et al., 2016). Similarly, in hormone refractory prostate cancer, MiR-146a loss regulates actin remodeling, MAT, and its reverse processes, the amoeboid–mesenchymal transition (AMT), by directly targeting *ROCK1* (Lin et al., 2008). Instead, in hormone-sensitive prostate cancer, the androgen-regulated MiR-135a can directly target both *ROCK1* and *ROCK2* (Kroiss et al., 2015). In breast cancer, *ROCK1* is regulated by the tumor suppressor MiR-340, which acts also on *CTNNB1* and *MYC*, affecting at different levels the Wnt pathway and ultimately inhibiting migration and metastasis (Mohammadi-Yeganeh et al., 2016). Interestingly, in melanoma cell lines, distinct MiR-200 family members promote the switch between mesenchymal and amoeboid migration (Elson-Schwab et al., 2010), ultimately inducing cell invasiveness rather than suppressing it, as reported for other cancers (Jurmeister et al., 2012; Roybal et al., 2011; Yu et al., 2010).

As anticipated, invasive tumors can also adopt a collective migration pattern in which multicellular groups (clusters, strands, or files) originate from the primary mass preserving cell–cell adhesions. Moving multicellular groups are characterized by one or more "leader cells" that undergo a partial EMT, i.e., the acquisition of a mesenchymal morphology without losing cell–cell adhesions. These cells form the tip of the multicellular strand and invade the surrounding stroma by using integrins and proteases (Friedl and Alexander, 2011; Krakhmal et al., 2015). Collective migration has been observed in colorectal carcinoma (Nabeshima, 1999), rhabdomyosarcoma, and other solid tumors (Yilmaz and Christofori, 2010). Recent findings indicate that MiR-21 regulates breast cancer collective migration via well-defined spatiotemporal dynamics (Dean et al., 2015). In addition, during collective migration of squamous carcinomas, MiR-124 modulates, by targeting Integrin-β1, the interactions between leader cells and the ECM (Kato et al., 2014). Interestingly, this MiRNA is lost in hepatocellular carcinomas with poor prognosis, and in this context, MiR-124 inhibits EMT together with stress fibers, filopodia, and lamellipodia formation by targeting *ROCK2* and *EZH2* (Enhancer of Zeste Homolog 2) (Zheng et al., 2012). Also in breast cancer, MiR-124 loss correlates with worse histological grade, but in this case, MiR-124 impinges on EMT by targeting of *SLUG* (Liang et al., 2013).

Environmental cues and cell–cell contacts during migration and invasion in unison lead to abnormal activation of MAPKs (Mitogen-Activated Protein Kinases) (Dhillon et al., 2007). Many MiRNAs act on this signaling cascade at different levels. In particular, MiR-23a/MiR-24-2/MiR-27a cluster can activate p44/42 MAPK and promote breast cancer migration, invasion, and hepatic metastasis by downregulating *SPRY2* (Li et al., 2013b), which is a known negative regulator of MAPK (Cabrita and Christofori, 2008). Consistently, this MiRNA cluster is increased in breast cancer lymph node metastasis compared to normal tissue (Li et al., 2013b). In a different setting, *SPRY2* is also targeted by the metastamiR activator MiR-21, which stimulates colon cancer cell migration by acting on cell protrusion formation (Sayed et al., 2008).

Among the transmembrane receptors triggering internal signaling cascades and motility, there is FGFR1 (Fibroblast Growth Factor Receptor 1) (Turner and Grose, 2010). In colorectal cancer, MiR-214 is significantly downregulated and can directly target *FGFR1*, abolishing in vitro cell proliferation, migration, and invasion as well as in vivo tumor growth and liver metastasis (Chen et al., 2014) in prostate cancer; instead, *FGFR1* is targeted by MiR-573 inhibiting in vitro migration, invasion, TGF-β1-induced EMT, and in vivo lung metastasis (Wang et al., 2015).

3. INTRAVASATION (FIG. 1B)

To reach distant organs, tumor cells that have detached from the primary mass and locally invaded the surrounding stroma must penetrate endothelial barriers to enter in blood or lymphatic vessels, thus becoming circulating tumor cells (CTCs). To this end, metastatic cancer cells commonly use transendothelial migration, the same process used by leukocytes during inflammation and immune response (Muller, 2011; Reymond et al., 2013). This process requires tumor cells to reach the vessels and to transiently alter endothelial integrity and permeability. The vessels generated by tumor cells through local neoangiogenesis present weak cell–cell junctions and are more permeable on macromolecular diffusion than normal tissue vessels, and this facilitates intravasation and also passive shedding in the circulation of cancer cells (Utoguchi et al., 1996). However, active disruption of endothelial integrity greatly increases the number of cancer cells entering in blood or lymphatic vessels and increases metastasis (Anderberg et al., 2013). MiRNAs can contribute in various manners to the alteration of endothelial barrier; for example, MiR-373 and MiR-520c, in the context of estrogen receptor negative breast cancers, abrogate intravasation by affecting TGFβR2 levels and their loss sustains cancer cell TGF-β signaling and induces the transcription and secretion of PAI1 (Plasminogen Activator Inhibitor 1), ANGPTL4 (Angiopoietin-like protein 4), and PTHRP (Parathyroid hormone-related protein), which are secreted factors that ultimately affect vascular integrity (Keklikoglou et al., 2012).

In addition to protein factors (i.e., VEGF and TGF-β) that disrupt endothelial junctions during intravasation (Drabsch and Ten Dijke, 2011;

Hoeben et al., 2004; Muller, 2011), cancer cells can also secrete MiRNAs with similar function. For example, MiR-105 is secreted in exosomes by breast cancer cells and increases vascular permeability and metastasis by directly targeting, inside endothelial cells, the tight-junction protein ZO-1 (Zonulin-1) (Zhou et al., 2014). Similarly, exosome-secreted MiR-939 when delivered to endothelial cells can increase breast cancer intravasation and metastasis by targeting *CDH5* (VE-cadherin) (Di Modica et al., 2017).

Cancer cells can undermine the integrity of the endothelium layer also by secreting proteases, such as MMPs, collagenases, and uPA (urokinase Plasminogen Activator) (Deryugina and Quigley, 2015; Kim et al., 1998; Muller, 2011). In sarcoma cells, deletion of MiR-182 decreases the number of CTCs that travel in blood circulation, whereas its overexpression increases CTCs, suggesting a role for this MiRNA in the regulation of metastatic cell intravasation (Sachdeva et al., 2014). Interestingly, among the targets of MiR-182, there are *PAI1* and *TIMP1*, which can repress metastasis by affecting the activity of MMP-2, MMP-9, and uPA (Sachdeva et al., 2014). Other MiRNAs, such as MiR-21 and MiR-146b-5p regulating at different levels of MMP activity (Gabriely et al., 2008; Li et al., 2013c; Meng et al., 2007; Song et al., 2010), likely intervene in tumor cells intravasation. Additionally, in colorectal cancer, MiR-21 stimulates invasion, intravasation, and metastasis by indirectly acting on uPA via PDCD4 (Programmed Cell Death Protein 4) and SP transcription factors (Asangani et al., 2008; Leupold et al., 2007).

Finally, to successfully intravasate, cancer cells must be able to squeeze themselves and move through the spaces between endothelial cells (Muller, 2011), involving processes of actomyosin contractility similar to those of amoeboid motility. Thus, MiRNAs that regulate amoeboid migration by targeting *RHO* or *ROCK*, such as MiR-155 (Kong et al., 2008), MiR-10b (Ma et al., 2007), and MiR-497 via Smurf1 (Wang et al., 2014), might also drive the intravasation of invasive cancer cells.

4. SURVIVAL IN THE CIRCULATION (FIG. 1B)

Metastatic tumor cells that succeed in intravasation, the above-mentioned CTCs, travel via the blood or lymphatic circulation. During their transit, CTCs encounter several obstacles, such as sheer forces of the circulation, collision with host cells, and attack of the immune system, which can ultimately affect their survival and their further ability to establish metastatic foci (Leber and Efferth, 2009; Wirtz et al., 2011). As a consequence, the major part of CTCs die, indicating that metastatic spread is a very inefficient process and that a selection for particularly resistant and aggressive tumor cells takes place (Valastyan and Weinberg, 2011). The primary role in immune surveillance against CTCs is played by NK (natural killer) cells, which can remove CTCs before they extravasate (Hanna, 1982; Wirtz et al., 2011). Several transmembrane proteins expressed by CTCs can induce early activation of NK cytotoxicity, such as Icam1 (Intercellular adhesion molecule 1) (Barber et al., 2004). CTCs can hide from NK cells, which become tolerant to CTCs' presence in the circulation, by MiRNA direct modulation of cell membrane molecules and NK triggers. For example, MiR-296-3p promotes survival in the circulation and prostate cancer metastasis by inhibiting the expression of *ICAM1* (Liu et al., 2013). MiR-20a downregulates MICA/B proteins, which are other ligands of Nkgd2 (Natural killer group member d) receptors present in NK cells (Xie et al., 2014); instead, MiR-29a, MiR-29b, and MiR-29c loss can upregulate the levels of B7-H3, a surface glycoprotein that inhibits NK cells and T-cell activation (Xu et al., 2009). A number of other MiRNAs regulate not only T- and B-cell activation but also innate and adaptive response, thus being possibly involved in the regulation of CTCs survival in the circulation. For example, MiR-155 is necessary for T-lymphocyte differentiation and activation, and for B-lymphocyte development (Lu and Liston, 2009), whereas MiR-223 is required for granulocyte proliferation and activation (Johnnidis et al., 2008). Interestingly, MiR-146 family members, initially described in the first step of the metastatic cascade (Correia et al., 2016; Li et al., 2013c), can control TLR (Toll-Like Receptor) and cytokine signaling via a negative feedback loop involving downregulation of Irak1 (IL-1 receptor-associated kinase 1) and Traf6 (TNF receptor-associated factor 6) protein levels (Taganov et al., 2006).

5. ARREST AT DISTANT ORGAN SITE AND EXTRAVASATION (FIG. 1C)

To exit the circulation, CTCs must arrest and attempt the penetration of the endothelium. In large vessels, the arrest of CTCs occurs actively via the engagement of adhesion molecules expressed by endothelial cells, in particular, P-L and E-selectin (Borsig et al., 2001). Consistently, metastatic properties of colorectal cancer cells are greatly enhanced if they express CD44 surface variant isoforms that bind to E-selectins on endothelial cells

(Napier et al., 2007). The arrest of cancer cells also depends on the interaction between integrin family ligands expressed on tumor cell surface and their counterreceptors, ICAM1 and VCAM1 (Vascular Adhesion Molecule 1) on endothelial cells (Konstantopoulos and Thomas, 2009). Interestingly, E-selectin and ICAM1 expressed on endothelial cells are direct targets of MiR-31 and MiR-17-3p, respectively (Suárez et al., 2010), which may play a role in the interaction between CTCs and endothelial cells. The active interactions between CTCs and endothelial cells are necessary not only for CTC arrest at distant organ sites but also for their extravasation. Overexpression of MiR-214 in melanoma cells increases their in vitro and in vivo extravasation capability; interestingly, MiR-214-overexpressing melanoma cells can induce the in vitro formation of fenestrations in an endothelial monolayer compared to control cells (Penna et al., 2011). Among the targets of MiR-214, there are *ITGA3* (Integrin $\alpha 3$) and transcription factor *TFAP2C* (Transcription Factor AP-2 γ) that induces the transcription of the *ALCAM* adhesion molecule and the loss of MiR-148b, which is also a negative regulator of *ALCAM*. Thus, MiR-214 and MiR-148b antagonize each other in enabling or blocking, respectively, melanoma CTC adhesion to the endothelium and extravasation (Penna et al., 2013).

If somehow intravasation of tumor cells is facilitated by the presence of a leaky and tortuous neovasculature at the primary tumor site (Utoguchi et al., 1996), CTCs that attempt extravasation at a secondary site must face a highly functional and almost impermeable barrier. To solve this problem, circulating breast cancer cells can increase vasculature permeability by secreting soluble factors, such as TGF-β-induced Angiopoietin-like 4 (Padua et al., 2008), Angiopoietin-2, and MMP-3 and MMP-10 (Huang et al., 2009). Another soluble factor promoting destabilization of endothelial barrier during CTC extravasation is VEGF, which exerts its function in an Src-dependent manner (Weis et al., 2004). Also MiRNAs play a role in the regulation of vascular permeability at the secondary tumor site. For example, overexpression of MiR-30a, MiR-30b, MiR-30c, MiR-30d, and MiR-30e in the lungs of melanoma bearing mice postpones metastasis occurrence by preserving the integrity of pulmonary vessels; in fact, lung fibroblast-derived MiR-30 family members target SKP2 that is directly involved in the disruption of vessel integrity (Qi et al., 2015). To produce brain metastases, breast cancer cells must penetrate blood–brain barrier, a highly specialized barrier that separates blood from brain extracellular fluid. Metastatic breast cancer cells can induce the disruption of blood–brain barrier by secreting extravesicles containing MiR-181c. Interestingly, these extravesicles containing MiR-181c alter the localization of junctional molecules of the endothelial blood–brain barrier, such as N-cadherin, Claudin-5, and Zonulin-1, as well as the organization of actin filaments (Tominaga et al., 2015).

Once again, also during extravasation, CTCs need to display a high degree of intrinsic deformability that is guaranteed, at least in melanoma-extravasating cells, by low CdC42 and high Sox2 transcription factor expression levels (Chen et al., 2016). Similarly, in breast cancer cells, extravasation and metastasis are promoted by the cation channel TRPV4 (Transient Receptor Potential Cation Channel Subfamily V member 4) that reduces cell stiffness via the regulation of actin cortex (Lee et al., 2016). MiRNAs may be involved in the regulation of invasive cytoskeleton during extravasation. In fact, the overexpression of MiR-25 in normal and highly metastatic prostate cancer cells strongly decreases their in vitro migration and in vivo extravasation capabilities. In particular, MiR-25 mechanism of action involves the downregulation of Integrin αv and Integrin α5 and the affection of cytoskeletal dynamics (Zoni et al., 2015). MiR-155, which is downregulated in metastatic breast cancer cell lines compared to non-metastatic cell lines, regulates extravasation and colonization to distant organs via alternative mechanisms, which include suppressing *ALDH1A1* (Aldehyde Dehydrogenase 1 family member A1), *PIR* (Pirin), and *PDCD4* (Thomsen et al., 2015). In endothelial cells, MiR-155 is induced by inflammatory cytokines during neuroinflammation, and this probably occurs also in brain metastasis. In this context, MiR-155 negatively regulates the blood barrier function by downregulating the junction proteins Annexin-2 and Claudin-1 and the focal adhesion components Syntenin-1 and DOCK-1 (Dedicator Of Cytokinesis) (Lopez-Ramirez et al., 2014).

6. MICROMETASTASIS FORMATION (FIG. 1D)

CTCs that successfully penetrate the endothelium and then infiltrate the target organ are called disseminating tumor cells (DTCs). The clinical observation that many cancer patients suffer from metastatic relapse months or years after initial diagnosis or surgical resection indicates the existence of a "lag time" between tumor dissemination and metastatic outgrowth (Giancotti, 2013). This period, referred to as metastatic dormancy, represents the time necessary for DTCs to adapt to the foreign microenvironment. As the adaptation to the newfound microenvironment delays, DTCs remain trapped in the target organ for protracted periods in the form of micrometastatic lesions, in which cell division and apoptosis occur at similar rates, or as single-infiltrated cells, that are blocked in a state of proliferative quiescence (Aguirre-Ghiso, 2007; Giancotti, 2013; Massagué and Obenauf, 2016). The process by which DTCs remain dormant in the target organ until their reactivation resembles the behavior of adult stem cells that reside into the organs in a state of proliferative quiescence whence they exit to sustain tissue homeostasis (Brabletz, 2012). In fact, several lines of evidence sustain the idea that tumor dormancy and stemness are two sides of the same coin: DTCs from various tumors have been found to display the traits of adult stem cells, such as negativity to Ki-67 staining (Pantel et al., 1993) and expression of stem cell surface markers, such as $CD44^{+}/CD24^{-/low}$ (Alix-Panabières et al., 2008; Balic et al., 2006), but most importantly the ability to proliferate upon the appropriate stimuli. Signals from the microenvironment can influence the behavior of DTCs, leading them to remain dormant, or conversely, to reactivate and give rise to metastatic outgrowth (Aguirre-Ghiso et al., 2003; Zhang et al., 2009).

MiRNAs play an important role in the switch of DTCs between dormant and reactivated state. To induce the dormancy of disseminated breast cancer cells, bone marrow mesenchymal stem cells secrete exosomes containing MiR-23b and these reach breast DTCs in which MiR-23b sustains a state of proliferative quiescence by targeting Marks (Myristoylated alanine-rich c-kinase substrate), a protein that promotes cell cycling and cell motility (Ono et al., 2014).

Reactivation of breast cancer DTCs in the target organ was studied by Gao and colleagues using an intriguing model consisting in a progression series of mouse mammary carcinoma cells arrested at defined steps of the metastatic cascade, as described in Aslakson and Miller (1992). 4T07 cells are a model of dormant mammary carcinoma cells, as they extravasate and invade the lungs without giving macroscopic metastasis. To identify MiRNAs involved in the exit of breast cancer from dormancy, the authors transduced a library of MiRNAs cloned in the retroviral vector pMSCV into dormant 4T07 cells and then injected these cells into mice. Interestingly, some of the injected mice developed lung metastases, which contained integrated MiR-138, MiR-223, MiR-340, and MiR-346. Among these MiRNAs, only MiR-138 and MiR-346 were confirmed to regulate metastatic reactivation (Gao et al., 2014).

To guarantee survival in a foreign microenvironment, DTCs generate appropriate "niches" for their engraftment in the target organ. The concept of metastatic niche originated over a century ago from the observations of the English surgeon Stephen Paget, who noted that the location of metastatic sites is not casual, as many cancers show a predilection to metastasize certain organs and not others. In his famous "seed and soil" theory (Ribatti et al., 2006), Paget stated that the susceptibility of some organs to be metastasized is not explicable only by vessels anatomy and blood flow physics, but an accommodating metastatic niche is required for malignant cells to engraft the distant sites and to establish metastasis. Different types of niches have been described (Massagué and Obenauf, 2016). In the case of "premetastatic niche," DTCs can prepare an appropriate niche for their engraftment prior to seeding, through the release of soluble molecules that travel in the circulation and reach the target organ. The concept of premetastatic niche is exemplified by the work of Kaplan and colleagues, who demonstrated that, when they are still at the primary tumor site, melanoma cells model a premetastatic niche in the bone marrow by secreting VEGF. VEGF, in turn, binds to VEGFR1(Vascular Endothelial Growth Factor Receptor 1) expressed by hematopoietic progenitors recruiting them to premetastatic sites where they form clusters awaiting the arrival of tumor cells (Kaplan et al., 2005). Another type of niche is the "perivascular niche" that exerts a supportive function for metastatic cells after their extravasation (Massagué and Obenauf, 2016). An interesting work by Ghajar and colleagues showed that disseminated breast cancer cells reside in a state of dormancy upon microvasculature of lung, brain, and bone marrow under the effect of endothelial-secreted Thrombospondin-1; curiously, this suppressive cue is completely lost in sprouting neovasculature that, conversely, supports breast cancer reactivation through endothelial-secreted TGF-β1 and Ostepontin (Ghajar et al., 2013).

7. METASTATIC COLONIZATION (FIG. 1D)

Proliferation of DTCs at the secondary tumor site and consequent metastatic colonization of the target organ are highly regulated events.

One of the key determinants of metastatic colonization is the capability of DTCs to adapt to the newfound microenvironment. In fact, we earlier described that tumor cells, when they are still at the primary tumor site, can model favorable premetastatic niches to gain the possibility to survive at distant sites. Several evidences indicate that MiRNAs are key regulators of tumor–stroma interactions that finally result in the remodeling of metastatic niche for tumor engraftment.

Breast, prostate, and other cancers preferentially metastasize to the bone marrow (Weilbaecher et al., 2011). One of the features of bone metastasis is bone resorption that is mediated by osteoclasts. As a strategy to prepare an appropriate niche in bone marrow, DTCs can induce formation and activation of osteoclasts, which in turn induce osteoporosis (Ell and Kang, 2013; Weilbaecher et al., 2011). Among the instruments used by DTCs to govern osteoclastogenesis, there are MiRNAs. In fact, the in vitro treatment of murine preosteoclasts with tumor-conditioned media from metastatic breast and bladder cancer cell lines not only induced the differentiation of preosteoclasts into mature osteoclasts (Sethi et al., 2011) but also identified a MiRNA signature in differentiated osteoclasts, including 42 upregulated and 45 downregulated MiRNAs. Importantly, the ectopic expression of two of the downregulated MiRNAs, namely MiR-141 and MiR-219, can impair osteoclast development and reduce bone metastasis burden through the downregulation of osteoclast genes that function at different steps of osteoclastogenesis, such as *MITF* (Microphthalmia-associated Transcription Factor), *CALCR* (Calcitonin receptor), and *TRAF* (Ell et al., 2013). Although its precise role in tumor–stroma interactions has not been fully understood, MiR-34a is a critical suppressor of osteoclastogenesis. In fact, MiR-34a is downregulated during osteoclast differentiation and MiR-34a-overexpressing transgenic mice show lower bone resorption and higher bone mass. Remarkably, bone metastases of breast and skin cancer are diminished in MiR-34a-overexpressing transgenic mice, as a consequence of MiR-34a-mediated downregulation of *TGIF2* (Transforming Growth-β-induced factor 2) (Krzeszinski et al., 2014). On the other hand, in prostate cancer cells, MiR-34a acts as a negative regulator of $CD44^{+}$ stem cell subpopulation, by inhibiting CD44 expression and thus prostate cancer self-renewal and metastasis (Li et al., 2015a; Liu et al., 2011). Similarly, in breast cancer, MiR-34a negatively affects stemness, metastasis, and chemosensitivity by downregulating *NOTCH1* (Kang et al., 2015). The Let-7 family of MiRNAs regulate self-renewal, tumorigenicity, EMT, and metastasis by targeting H-RAS and *HMGA2* (High-Mobility Group AT-hook 2) in breast cancer (Yu et al., 2007) and MUC-1 and Lin-28 in non-small cell lung cancer (Alam et al., 2015). In addition, Let-7 is induced by RKIP (Raf-Kinase Inhibitor Protein) and negatively regulates breast cancer invasion and bone metastasis directly, by targeting bone metastasis genes *MMP-1*, *CXCR4* (C-X-C Chemokine Receptor Type 4), and *OPN*, or indirectly, by targeting *BACH1* (BTB domain And CNC Homolog1) (Yun et al., 2011).

To establish an accommodating metastatic niche, DTCs need the support and contribution of other stromal cells, such as fibroblasts. Therefore, to adapt to the newfound stroma, DTCs can transform normal resident fibroblasts into Cancer-Associated Fibroblasts (CAFs) (Strell et al., 2012). CAFs, in turn, can support metastatic progression by producing growth factors, ECM-degrading enzymes, and ECM components (Vong and Kalluri, 2011). Epithelial ovarian cancer metastases are characterized by a stromal component in which CAFs are important constituents; moreover, these metastases typically concentrate in the abdominal cavity at the level of the omentum (Schauer et al., 2011). In 2012, Mitra and colleagues reported that, in ovarian cancer metastases, a pivotal role in the conversion of fibroblasts into CAFs is played by MiRNAs. The authors compared MiRNA expression profiles in normal omental fibroblasts and CAFs. CAFs were isolated from omental metastases (ex vivo approach) or were obtained coculturing normal omental fibroblasts with ovarian cancer cells (in vitro approach). Several MiRNAs that are differentially expressed in CAFs were identified with both approaches: the MiRNA most significantly upregulated was MiR-155, whereas the MiRNAs most significantly downregulated were MiR-214 and MiR-31 (Mitra et al., 2012). Consistently, the triple transfection of normal omental fibroblasts with anti-miR-214, anti-miR-31, and pre-miR-155 increased migration, invasion, and colony-forming capability of cocultured ovarian cancer cells, indicating the effective reprogramming of normal omental fibroblasts into CAFs; in this context, *CCL5* (Chemokine C–C motif Ligand 5) was identified as a direct target of MiR-214 (Mitra et al., 2012).

Alternatively, DTCs can directly modify the ECM to sustain metastatic growth. Breast cancer cells infiltrated into the lungs, in fact, support their own metastasis-initiating capability by producing Tenascin-C, an ECM glycoprotein typical of stem cells' niches. In turn, Tenascin-C nurses metastatic cells by promoting activation of Notch and Wnt/β-catenin signaling (Oskarsson et al., 2011). In breast cancer, MiR-335 loss affects cancer motility and invasion by repressing the expression of Tenascin-C and of the progenitor transcription factor Sox4, while MiR-126 loss correlates with tumor growth and proliferation at the secondary site (Tavazoie et al., 2008). DTCs can create a favorable niche by modifying niche-cell metabolism. For example, breast cancer cells increase metastatic colonization by reducing glucose uptake and thus increasing nutrient availability in the premetastatic niche. Intriguingly, this is possible via the secretion of MiR-122-containing extravesicles that act by targeting PKM2, an isoform of pyruvate kinase enzyme (Fong et al., 2015).

When delineating the mechanisms regulating DTCs proliferation at the secondary site, it must be mentioned the contribution of molecules directly involved in the regulation of proliferative and antiapoptotic pathways. In fact, activated proliferative pathways (MAPK, PI3K–AKT), positional–mechanical pathways (Hippo), or inflammatory pathways (NFKB, STAT1) all support metastatic cell growth (Massagué and Obenauf, 2016), and all of these pathways are also tuned by MiRNAs. For example, astrocyte-derived exosomes containing MiR-19a reversibly downregulate the levels of the tumor suppressor gene *PTEN* in the brain, which, in turn, leads to increase of CCL2 (Chemokine C–C motif Ligand 2) secretion and recruitment of myeloid cells that favor brain metastasis of breast cancer (Zhang et al., 2015a). Interestingly, MiR-19a, together with MiR-21 and MiR-17, was already demonstrated to be induced by PRL3 (Phosphates of Regenerating Liver-3)/STAT3 and to increase proliferation and invasiveness of colorectal cancer (Zhang et al., 2012).

Metastatic lesions of almost all carcinomas display epithelial features and a high degree of differentiation. Therefore, metastatic colonization of the target organ may require plasticity of tumor cells that switch from an EMT/dedifferentiated state, characterized by proliferative quiescence, to a MET/differentiated state, characterized by active proliferation (Brabletz, 2012). Overexpression of MiR-34 and MiR-200 family members induce MET, and consequently metastatic colonization, by repressing Snail-1 and Zeb-1 transcription factors, which in turn lead to the increase of E-cadherin expression levels (Dykxhoorn et al., 2009; Gregory et al., 2008; Siemens et al., 2011). Alternatively, MiR-200s can promote breast cancer colonization by targeting Sec23a, a protein that mediates the secretion of metastasis-suppressor proteins, such as Igfbp4 (Insulin-like growth factor-binding protein 4) and Tinagl1 (Tubulointerstitial nephritis antigen-like 1) (Korpal et al., 2011). Consistent with these in vitro observations, MiR-200c is upregulated in colorectal cancer lymph node and liver metastases (Drusco et al., 2014).

8. ANGIOGENESIS (FIG. 1E)

Angiogenesis is a physiological process in which new blood vessels grow from preexisting vessels, and is sustained by vasculature endothelial cells under the strict control of pro- and antiangiogenic factors. In tumors, this delicate balance is subverted, as tumor cells turn on the "angiogenic switch" (Tanaka et al., 2003), producing high amounts of proangiogenic factors and stimulating neovascularization to satisfy their growing demand for oxygen and metabolites. The rapid outburst of tumor cells usually creates a hypoxic environment, which, besides leading to tumor necrosis, induces cancer cells to adapt to this condition by engaging in HIF (Hypoxic-Induced Factor)-dependent survival pathways.

Several MiRNAs, such as MiR-26 family, MiR-107, and MiR-210, at least in part through HIF-1, are induced in response to low oxygen and sustain both tumor cell growth and survival (Kulshreshtha et al., 2006). Another MiRNA, MiR-382, is directly induced by HIF-1α and inhibits in vitro and in vivo tumor angiogenesis as well as reduces tumor growth by downregulating PTEN and AKT/mTOR signaling pathway (Seok et al., 2014). Moreover, HIF-1α can be directly targeted by MiRNAs; for example, p53-induced MiR-107 suppresses tumor angiogenesis, tumor growth, and VEGF expression by targeting *HIF1A* (Yamakuchi et al., 2010). Similarly, MiR-20b directly targets *HIF1A* and *STAT3* in breast cancer cells, thus modulating VEGF expression (Cascio et al., 2010).

During tumor progression, tumor cells can recruit quiescent endothelial cells and stimulate them to proliferate and organize to form new blood vessels that are characterized by weaker cell–cell junctions (Utoguchi et al., 1996). Neoangiogenesis is favored by various angiogenic factors, including VEGF. VEGF is induced in tumors by a wide range of effectors such as growth factors, cytokines, and hormones in addition to hypoxia and HIF-1 (Loureiro and Amore, 2005). VEGF is modulated, directly, by MiRNAs that are downregulated in hypoxic conditions, such as MiR-16, MiR-15a, MiR-20a, and MiR-20b (Cascio et al., 2010; Hua et al., 2006; Yamakuchi et al., 2010), and, indirectly, by MiRNAs that are over overexpressed hypoxic conditions, such as MiR-27a that targets *SP1*, *SP3*, and *SP4* (Specificity Protein 1, 3, and 4). In fact, MiR-27a, in breast cancer, leads to decreased expression of Sp-dependent survival and angiogenic genes, such as *BIRC5* (Survivin), *VEGF*, and *VEGFR* (Mertens-Talcott et al., 2007).

The oncogenic transcription factor c-Myc exerts proangiogenic functions by regulating the transcription of both protein-coding genes (VEGF that then controls TSP1) (Janz et al., 2000) and MiRNAs (i.e., MiR-17/92 cluster). In addition, the c-Myc-induced MiR-17/92 cluster participates in the creation of a more aggressive, richly perfused tumor phenotype. In particular, MiR-18 and MiR-19 are responsible for *CTGF* (Connective Tissue Growth Factor) and *TSP1* repression, respectively (Dews et al., 2006, 2010). Importantly, *TSP1* is downregulated also by other MiRNAs, namely MiR-27b and Let-7 (Kuehbacher et al., 2007), as well as MiR-194 that increases colorectal cancer angiogenesis in response to p53 (Sundaram et al., 2011).

Some MiRNAs are responsible for the modulation of endothelial functions. Clustered MiR-221 and MiR-222, which share a common seed sequence, act as negative regulators of the tyrosine kinase receptor c-Kit, repressing proliferative and angiogenic properties of its ligand Scf (Stem Cell Factor) (Poliseno et al., 2006). Moreover, MiR-221 and MiR-222 show proliferative and metastatic potential through the suppression of the cyclin-dependent kinase inhibitor p27 (Baldassarre et al., 2005; le Sage et al., 2007).

MiR-126 regulates the response of endothelial cells to VEGF and its repression results in the loss of vascular integrity during embryonic development. Interestingly, MiR-126 mechanism of action involves suppression of *SPRED1* (Sprouty-related protein) and *PK3R2*/p85 beta (Phosphoinositol-3 Kinase Regulatory subunit 2) (Fish et al., 2008). Consistently, MiR-126 inhibits breast cancer metastasis by suppressing metastatic endothelial cell recruitment, angiogenesis, and colonization through targeting of *IGFBP2* (Insulin-like Growth Factor-Binding Protein 2), *PITPNC1* (Phosphatidylinositol Transfer Protein Cytoplasmic 1), and *MERTK* (Protooncogene tyrosine-protein kinase MER) (Png et al., 2012).

Tumor cell-derived microvesicles containing MiR-9 promote endothelial cell migration and angiogenesis by reducing *SOCS5* (Suppressor of Cytokine Signaling 5) levels and by leading to activation of JAK/STAT signaling (Zhuang et al., 2012). MiR-1908, MiR-199a-3p, and MiR-199a-5p have been found to increase melanoma invasion, angiogenesis, and colonization. In particular, the overexpression of MiR-1908 and MiR-199a in parental melanoma cells increases in vitro metastatic endothelial recruitment, as well as in vivo endothelial density at metastatic nodules in the lungs (Pencheva et al., 2012).

The antiangiogenic *GAX* homeobox gene is highly expressed in quiescent endothelial cells, whereas it is rapidly downregulated after mitogenic and proangiogenic signals. MiR-130a, a MiRNA involved in megakaryocytic differentiation and expressed by endothelial cells, is triggered by angiogenic stimuli, in a reciprocal way compared to *GAX*. Accordingly, MiR-130a directly represses at the posttranscriptional level the expression of *GAX* and, to a lesser extent, the one of a similar antiangiogenic factor, such as HOXA5 (Homeobox protein HOX-A5), thereby granting the feat of the angiogenic process (Chen and Gorski, 2008). Other MiRNAs regulate tumor progression and angiogenesis by using alternative mechanisms. For example, MiR-214, whose loss is associated with hepatocellular carcinoma recurrence and worse prognosis. MiR-214 suppresses microvascularity of tumors and of their surrounding tissue by suppressing *HDGF* (Hepatocyte-Derived Growth Factor) (Shih et al., 2012). Instead, MiR-150 that is actively secreted in microvesicles by monocytes displays strong proangiogenic properties: MiR-150 increases in vitro capillary tube formation of endothelial cell lines and stimulates in vivo vascular invasion of sarcoma-implanted mice, by targeting the protooncogene *MYB* (Li et al., 2013a).

8.1 The Interaction Between MiRNAs and Other RNAs Species in the Regulation of Metastasis

As described earlier, MiRNAs can contribute to the regulation of metastatic cascade by controlling the expression of target protein-coding mRNAs. However, to give a complete picture of the role of MiRNAs in general and of MiRNA in metastasis regulation in particular, it must be mentioned that other RNA species, apart from mRNA, are capable of binding MiRNAs. These RNA species include transcribed pseudogenes, circular RNAs, and lncRNAs (long noncoding RNAs), which, together with mRNAs, are referred to as ceRNA (competing endogenous RNAs). ceRNAs share MiRNA response elements through which they compete for MiRNA binding, by sequestering MiRNAs and by preventing MiRNA-mediated control of mRNA translation (de Giorgio et al., 2013; Salmena et al., 2011).

Accumulating evidences indicate that ceRNAs have an important role in the regulation of cancer metastasis. In lung cancer, HMGA2 transcript acts as ceRNA for let-7 MiRNA family, by promoting invasion and dissemination in a manner independent of its protein-coding function (Kumar et al., 2014). In breast cancer, EMT and metastasis are inhibited by FOXO1 3′-UTR, which acts as ceRNA by sequestering MiR-9 (Yang et al., 2014).

Probably, the class of RNA molecules most frequently acting as ceRNAs in metastasis are lncRNAs (Ling et al., 2015; Xu et al., 2016). lncRNAs are noncoding transcripts longer than 200 nucleotides that are transcribed by RNA polymerase II from various genomic regions, including intergenic, intronic, and enhancer regions (Kung et al., 2013). For example, lncRNA H19 promotes pancreatic ductal adenocarcinoma migration and invasion by enhancing HMGA2-mediated EMT by competing for Let-7 (Ma et al., 2014). Moreover, lncRNA H19 is involved in a regulatory loop with MiR-141 in gastric cancer. In fact, in this cancer, lncRNA H19 promotes tumor cell proliferation and invasion, but these functions are repressed by the binding with MiR-141 (Zhou et al., 2015).

Another lncRNA, HOTAIR, promotes proliferation, migration, and invasion of gastric cancer by acting as ceRNA for MiR-331-3p, thus derepressing of HER2 (Liu et al., 2014; Wan et al., 2015). Moreover, HOTAIR promotes breast cancer metastasis by repressing epigenetically MiR-568, which targets *NFAT5* (Nuclear Factor of Activated T-cells 5) (Li et al., 2014).

Additional examples of lnRNA with ceRNA functions in metastasis are lncRNA ATB (lncRNA activated by TGF-β), which promotes hepatocellular carcinoma invasion by competitively binding to MiR-200 with subsequent upregulation of Zeb-1 and Zeb-2 EMT transcription factors (Yuan et al., 2014), and lncRNA-ROR, which similarly acts as a sponge for MiR-205 by preventing MiR-205-mediated Zeb-2 loss, with consequent promotion of EMT and metastasis in breast cancer (Hou et al., 2014).

8.2 MiRNA Signatures of Cancer Metastasis

As MiRNAs can distinguish between cancers of different origins (Volinia et al., 2006), MiRNA signatures can also distinguish primary cancers from their metastatic counterpart (Baffa et al., 2009). In fact, during tumor progression and metastasis establishment, either as innocent bystanders but more likely as driving forces as discussed earlier, MiRNAs are used by cancer cells to adapt to the changing microenvironment they must confront with. This information can be important in the development of metastasis-specific therapies; but, for metastasis prevention and for tailoring treatment, clinicians actually need to identify those patients who most likely will develop metastasis from those who won't.

The relatively small number of MiRNAs in our genome compared to protein-coding genes, the relative easiness of MiRNA laboratory detection methods (e.g., qRT-PCR), and most importantly the greater specificity of MiRNA based profiles compared to mRNA ones (Lu et al., 2005), make MiRNAs ideal prognostic markers.

One possible approach, which definitely needs systematic validation either by analyses of publicly available MiRNA expression data sets or by independent prospective validation studies, is the direct comparison between primary tumors coupled by stage, which differ in metastasis occurrence. The study by Bobowicz and colleagues addresses with a similar approach who, among stage pT2-T3 N0 colorectal cancer patients, might benefit from adjuvant chemotherapy after surgical resection vs who might not. They selected a cohort of 85 patients that received surgery only, and of these almost 50% ($n=40$) developed distant metastasis within a median range of 1.7 years. By comparing relapsed vs nonrelapsed patients, it was possible to identify five MiRNAs able to predict patients that were more likely to relapse. This signature warrants further validation in prospective studies, but if combined with T stage (70% of the relapsed were stage IIA), it could help clinicians decide whom could benefit from adjuvant chemotherapy avoiding unnecessary toxicities (i.e., to the 30% of patients' stage IIA that did not relapse) (Bobowicz et al., 2016).

Instead, in the analysis by Pizzini and colleagues, which was done on matched colorectal cancer samples (normal vs tumor vs metastasis either synchronous or metachronous), it was observed that most of MiRNA expression changes occurred in the transition from normal colon mucosa to primary tumor; whereas most of these MiRNAs remained stable in the liver metastasis, only few MiRNAs were confirmed to significantly change between tumor and associated liver metastasis (Pizzini et al., 2013). Yet, in the determination of this metastamiR signature, the miRNome profile of cancers that do not develop metastasis was not considered.

Indeed, the expression levels of specific MiRNAs, either alone or in combination, have been associated to metastasis occurrence in a cancer-specific manner (e.g., MiR-21 in breast cancer, MiR-122 in hepatocellular cancer, MiR-224 in colon cancer, MiR-106b in renal cancer view; see Table 2 for further examples). All these studies have usually analyzed retrospectively collected samples. For the same tumor type a plethora of studies of different sizes, most of them with independent validation cohorts, have addressed this issue, but few share common MiRNAs. It could be that collaborative prospective studies, for example, among international consortia (Søreide et al., 2016) reaching higher numbers of samples and most importantly sharing same sample preparation and molecular detection procedures, will offer more robust and reproducible signatures.

All of the studies reported in search of MiRNAs predicting metastasis look at the levels of MiRNAs inside the primary tumor or metastatic sample. But as mentioned earlier (Ostenfeld, 2014), MiRNAs can also be secreted actively by cancer cells to modulate, for example, the premetastatic niche and can be found in the circulation. For example, MiR-224 levels in primary colorectal cancer correlate with patients' stage and they can also be found in the patients' blood (Vega et al., 2013). Moreover, a signature of 21 plasma

Table 2 Examples of MiRNA Signatures in Different Cancer Types

Cancer Type	References	Technology	Training Set No. of Samples	Validation Set No. of Samples	MiR	Clinical Significance
Breast cancer	Pérez-Rivas et al. (2014)	Microarray qRT-PCR	71	—	MiR-149, MiR-10a, MiR-20b, MiR-30a-3p, and MiR-342-5p	Discriminates nonrelapsing vs early-relapsing patients
	Marino et al. (2014)	Microarray qRT-PCR	64	—	MiR-183, MiR-494, and MiR-21 in primary BC	Predicts risk of developing metastasis better than lobular histology, T2, N1 and HER2+
	Farazi et al. (2011), Volinia et al. (2012)	Deep sequencing	179[a]	—	MiR-423 and MiR-375 MiR-127-3p, MiR-210, MiR-185, MiR-143[b], let-7b, and MiR-210	MiR-423 and MiR-375 associate with metastasis occurrence 5-MiR signature predicts shorter distant metastasis-free survival
	Lerebours et al. (2013)	qRT-array	51	160	MiR-421, MiR-486, MiR-503, MiR-720, and MiR-1303	5-MiR signature predicts inflammatory breast cancer (IBC), and in non-IBC patients, it predicts poor metastasis-free survival
	Cascione et al. (2013)	Nanostring qRT-PCR	224	—	MiR-125b, MiR-655, MiR-421, MiR-16, MiR-374a/b, and MiR-497	7-MiR signature predicts shorter distant metastasis-free survival in triple-negative BC
	Qian et al. (2009)	qRT-PCR	—[c]	344	MiR-21	Predicts poor disease-free survival
	Yan et al. (2008)	Microarray qRT-PCR	8[b]	113	MiR-21	Predicts poor OS and is associated with advanced stage and lymph node metastasis

Continued

Table 2 Examples of MiRNA Signatures in Different Cancer Types—cont'd

Cancer Type	References	Technology	Training Set No. of Samples	Validation Set No. of Samples	MiR	Clinical Significance
Cervical cancer	Hu et al. (2010)	qRT-PCR	102		MiR-200a and MiR-9	
Colorectal cancer	Drusco et al. (2014)	Microarray TaqMan array qRT-PCR ISH	66	16	MiR-103, MiR-21, MiR-93, MiR-200c, and MiR-566	Association with liver or lymph node metastatic disease
	Hur et al. (2015)	Nanostring qRT-PCR ISH	9[b]	58[b] 84 175 169[d]	let-7i, MiR-10b, MiR-221, MiR-320a, and MiR-885-5p in met compared to primary CRC (pCRC)	Low let-7i in pCRC predicts poor prognosis and liver metastasis; high MiR-10b in pCRC predicts liver metastasis; high serum MiR-885-5p predicts poor prognosis and lymph node and liver metastasis
	Mudduluru et al. (2015)	Microarray qRT-PCR	8[b]		MiR-135b, MiR-210, and MiR-218	
	Bobowicz et al. (2016)	qRT-array (754 MiRs)	85	50[e]	1296, MiR-135b, MiR-539, MiR-572, and MiR-185	5-MiR signature for shorter distant metastasis-free survival
	Ellermeier et al. (2014)	TaqMan array	19[b]	—	MiR-210 and MiR-133b in liver mets	Reduced OS of metastatic CRC

	Pizzini et al. (2013)	Microarray qRT-PCR	46[b]	—	MiR-10b in liver mets	Reduced OS
	Vickers et al. (2012)	qRT-PCR	—[c]	34[b]	MiR-21, MiR-135a, MiR-335, MiR-206, and let-7a	Predicts in CRC between presence or absence of metastasis
	Chen et al. (2012b)	ISH	—[c]	188	MiR-103/107 in pCRC	Reduced DMFS and OS
Melanoma	Hanniford et al. (2015)		92	119 45	MiR-150-5p, MiR-15b-5p, MiR-16-5p, and MiR-374b-3p	
Nasopharyngeal carcinoma	Liu et al. (2012)	Microarray qRT-PCR	312	153	MiR-142-3p, MiR-29c, MiR-26a, MiR-30e, and MiR-93	Associates with distant metastasis-free survival
	Bruce et al. (2015)		125	121	4-MiRNA signature	Correlates with clinical stage
Renal clear cell	Heinzelmann et al. (2011)	Microarray qRT-PCR	30	—	MiR-451, MiR-221, MiR-30a, MiR-10b, and MiR-29a	Associates with metastasis and prognosis
	Wu et al. (2012)	Microarray qRT-PCR	28	50	MiR-10b, MiR-139-5p, MiR-130b, and MiR-199b-5p MiR-421, MiR-486, MiR-503, MiR-720, and MiR-1303	Associates with ccRCC metastasis and prognosis

Continued

Table 2 Examples of MiRNA Signatures in Different Cancer Types—cont'd

Cancer Type	References	Technology	Training Set No. of Samples	Validation Set No. of Samples	MiR	Clinical Significance
	Wotschofsky et al. (2016)	Microarray qRT-PCR	38	—	MiR-146a-5p, MiR-128a-3p, and MiR-17-5p	Lost in patients with metastasis
	Mlcochova et al. (2016)	TaqMan array qRT-PCR	104[f]	126	MiR-429 in EMT	Belongs to prognostic scoring system enabling identification of ccRCC patients at high-risk of cancer-related death
Hepatocellular carcinoma	Budhu et al. (2008)		241[b]	131 110	20 MiRNA signature	Predicts metastasis and/or recurrence

[a]Same data set analyzed by a different group with different inclusion criteria for patients.
[b]Matched samples (i.e., primary, metastatic, and normal tissue from the same patient).
[c]MiRNAs to be tested were selected by relevance in metastasis.
[d]Serum.
[e]GEO data set (GSE63119).
[f]Including discovery set.

MiRNAs identified 1–2 years before diagnosis predicts lung cancer development and aggressiveness, including the development of early metastatic tumors undetectable with spiral computed tomography (Boeri et al., 2011). Remarkably, tissue- and disease-specific MiRNA signatures can be found also in other body fluids, such as urine, spinal fluid, and saliva (Moldovan et al., 2014). For example, a specific urinary MiRNA signature can distinguish between recurrent and not recurrent bladder urothelial carcinoma (Sapre et al., 2016).

The possibility to recover in a relatively simple manner MiRNAs in body fluids renders MiRNA-based liquid biopsy an attractive approach to predict an individual cancer patients' potential to metastasize and represents a promising alternative to invasive procedures, such as tissue biopsies or imaging techniques, which commonly detect advanced stage disease. Further studies in this field are necessary to assess the feasibility of this approach in the clinical practice.

8.3 Exploiting MiRNA as Metastatic Targets

Given the high lethality in cancer patients due to metastasis, much of today's clinical effort in cancer treatment has shifted from primary cancer treatment toward the prevention of metastasis occurrence. In fact, systemic chemotherapies are mostly administered when the primary lesion has been completely removed (i.e., as adjuvant treatment). This strategy aims to eradicate eventual cancer cells that have left the primary tumor (CTCs and DTCs), are present in the blood, bone marrow, or other organs, but have still to thrive as overt metastasis being clinically undetectable (i.e., micrometastasis). Difficulty in prevention and treatment of metastasis lies also on the fact that CTCs and DTCs are intrinsically different from primary cancer cells and standard adjuvant cytotoxic agents may not be equally effective. It is only by understanding better and taking into consideration these differences that antimetastatic therapies can succeed.

Due to their potential to target at different levels a single pathway, MiRNAs are attractive targets and tools to take power over malignant cell behavior. From the first demonstration that reintroduction of deregulated MiRNAs in tumor xenografts is feasible and effective in controlling tumor growth (Esquela-Kerscher et al., 2008; Kota et al., 2009), MiRNA mimics or inhibitors (e.g., antagomiRs) are currently successfully used in preclinical settings to better understand MiRNAs in in vivo mechanism of action and to validate their efficacy as anticancer agents. Reasonably, the use of MiRNA mimics or inhibitors specific for the most vulnerable steps of the metastatic cascade could provide alternative mechanisms useful in the management of cancer metastasis, either prevention or treatment.

For example, in breast cancer preclinical models, silencing of MiR-10b by using 2′-O-methyl-group (OMe)-modified, cholesterol-conjugated antisense MiRNA can inhibit metastasis, without affecting primary tumor growth (Ma et al., 2010a,b). Yet, this is only true when breast cancer cells are present as a tumor xenograft at the primary site; whereas, if breast cancer cells are directly injected in the circulation, resembling the steps downstream of cancer cell intravasation, MiR-10b is not effective. Thus, if we consider the parallel progression model in which cancer cells disseminate early in the formation of the primary tumor, likely when it is not yet detectable, MiR-10b inhibition might not be useful in clinical settings as a metastatic prevention approach, but only ad hoc clinical studies will tell.

Another preclinical example of MiRNA used to inhibit cancer metastasis was achieved by the coupling of MiRNA synthetic mimics with atelocollagen in mouse models of prostate cancer. This approach was able to redirect MiR-16 delivery to the bone and in this manner to inhibit bone metastasis (Takeshita et al., 2010).

Several formulations and molecule types of MiRNA mimics or inhibitors in preclinical mouse models have been safely and effectively used (Zhang et al., 2013a), among others MiR-34a carried by liposome-based nanoparticles (Wiggins et al., 2010). This led to the design of a phase I safety clinical trial in selected tumors, started in 2013, using a liposome-based formulation of MiR-34a, MRX34 (Mirna Therapeutics, Inc.). Unfortunately, this study was prematurely terminated due to five immune-related serious adverse events (http://clinicaltrials.gov/ct2/show/NCT01829971). Instead, Miravirsen (Santaris Pharma, Horsholm, Denmark), a MiR-122 antisense locked nucleic acid, also reached clinical trials in human and passed phase I safety studies (http://clinicaltrials.gov/ct2/show/NCT01200420), but clinical trials testing its efficacy in cancer have still to be designed.

9. CONCLUSION AND FUTURE PERSPECTIVES

Disentangling the molecular mechanisms and regulatory networks that grant cancer cells the ability to win over tissue anatomical and homeostatic barriers, disseminate throughout the entire organism, and settle undisturbed in foreign organs, eventually conquering omnipresent immortality and indestructibility, remains a major challenge. MiRNAs for their particular mechanism of action could represent the rosetta stone for unsolving this riddle. In fact, this last decade has described their major involvement in controlling cancer metastasis. After this initial burst in information regarding their transversal functions, it is now time to draw conclusions and identify those mastermind MiRNAs clinically relevant. Many studies have already tried to address MiRNA potential use in the clinic as diagnostic, prognostic factor and therapeutic targets (see, for example, Pichler and Calin, 2015).

On the basis of our review work, we may speculate that the most promising MiRNA candidates exploitable as therapeutic targets to contrast cancer metastasis are the MiRNAs, which have a solid experimental evidence (in both in vitro and in vivo models) of involvement in metastasis regulation, but that at the same time present a counterpart in the clinical practice, being consistently up- or downregulated in metastatic cancer patients. When reasoning in this manner, however, one must take in consideration that: (1) experimental models may be unable to reflect the effective complexity that, in the patient, has led to metastasis formation. A MiRNA, which is important for a particular step of metastatic cascade, reproduced experimentally, may become undetectable in metastatic patients, simply because it serves in a precise phase of metastatic transformation, but once metastasis is established it disappears; (2) the pro- or antimetastatic function of MiRNAs can be context dependent.

Yet, we believe that some of the MiRNAs that we reviewed could have the potentiality to control cancer metastasis once we have developed the correct tools to modulate their expression in patients. For example, a promising approach would be inhibiting MiRNAs (e.g., using antagomiRs or RNA sponges acting as ceRNAs), which have been identified as promoters of metastasis. Among these, MiR-21 could be the most encouraging candidate. In fact, MiR-21 is a master regulator of metastasis, as it promotes tumor cell invasion, intravasation, as well as metastatic colonization in a wide range of cancers (Asangani et al., 2008; Gabriely et al., 2008; Meng et al., 2007; Song et al., 2010). At the same time, MiR-21 predicts poor overall survival and metastasis in breast cancer patients (Marino et al., 2014; Qian et al., 2009; Yan et al., 2008) and distinguishes between presence or absence of metastasis in colorectal cancer patients (Drusco et al., 2014). A molecule called RG-012, a single-stranded chemically modified oligonucleotide that binds and inhibits MiR-21, has already been developed and is currently used in a phase II clinical trial for the treatment of Alport syndrome, a genetic kidney disease (https://clinicaltrials.gov/ct2/show/NCT02855268; Regulus Therapeutics Inc, 2015). The design of ad hoc clinical trials using this novel approach could be exploited in the future to understand if actually MiRNA-based therapies can help in the treatment and prevention of cancer metastasis. Needless to say, the precise selection of cancer patients, homogeneous for tumor type, stage, and biological and molecular characteristics is required in order to demonstrate a concrete benefit of RG-012 use. Similarly, other promising MiRNA candidates that could be inhibited for metastasis control are MiR-135a/b. In fact, MiR-135a/b promote tumor cell invasion and metastatic colonization in prostate, lung, and colorectal cancers (Kroiss et al., 2015; Lin et al., 2013; Mudduluru et al., 2015), predict shorter distant metastasis-free survival and distinguish between presence or absence of metastasis in colorectal cancer patients (Bobowicz et al., 2016; Vickers et al., 2012).

Instead, for antimetastatic MiRNAs, which are lost during cancer metastasis, the proper therapeutic approach consists in reestablishing their expression (e.g., using MiRNA mimics or demethylating agents). This may be the case of MiR-30s, which inhibit tumor cell invasion and extravasation in various cancers (Kim et al., 2011; Kumarswamy et al., 2012; Ouzounova et al., 2013; Qi et al., 2015; Siemens et al., 2011; Ye et al., 2015) and associate with metastasis-free survival in nasopharyngeal carcinoma patients (Liu et al., 2012) while discriminating nonrelapsing vs early-relapsing breast cancer patients (Pérez-Rivas et al., 2014).

Although MiRNA-based therapies are still in their infancy, it is likely that the more we understand which MiRNA represents the best switch to act on and once we identify the best tolerable formulation for patients, MiRNA use in the clinic may become a reality.

REFERENCES

Aguirre-Ghiso, J.A., 2007. Models, mechanisms and clinical evidence for cancer dormancy. Nat. Rev. Cancer 7 (11), 834–846.

Aguirre-Ghiso, J.A., et al., 2003. ERK(MAPK) activity as a determinant of tumor growth and dormancy; regulation by p38SAPK. Cancer Res. 63 (7), 1684–1695.

Alam, M., et al., 2015. MUC1-C induces the LIN28B → LET-7 → HMGA2 axis to regulate self-renewal in NSCLC. Mol. Cancer Res. 13 (3), 449–460.

Alečković, M., Kang, Y., 2015. Welcoming treat: astrocyte-derived exosomes induce PTEN suppression to foster brain metastasis. Cancer Cell 28 (5), 554–556.

Alix-Panabières, C., Riethdorf, S., Pantel, K., 2008. Circulating tumor cells and bone marrow micrometastasis. Clin. Cancer Res. 14 (16), 5013–5021.

Almeida, M.I., Reis, R.M., Calin, G.A., 2011. MicroRNA history: discovery, recent applications, and next frontiers. Mutat. Res. Fundam. Mol. Mech. Mutagen. 717 (1–2), 1–8.

American Cancer Society, 2015. Global Cancer Facts & Figures, third ed. 800, American Cancer Society, Atlanta, GA, pp. 1–64.

Anderberg, C., et al., 2013. Deficiency for endoglin in tumor vasculature weakens the endothelial barrier to metastatic dissemination. J. Exp. Med. 210 (3), 563–579.
Asangani, I.A., et al., 2008. MicroRNA-21 (miR-21) post-transcriptionally downregulates tumor suppressor Pdcd4 and stimulates invasion, intravasation and metastasis in colorectal cancer. Oncogene 27 (15), 2128–2136.
Aslakson, C.J., Miller, F.R., 1992. Selective events in the metastatic process defined by analysis of the sequential dissemination of subpopulations of a mouse mammary tumor. Cancer Res. 52 (6), 1399–1405.
Baek, D., et al., 2008. The impact of microRNAs on protein output. Nature 455 (7209), 64–71.
Baffa, R., et al., 2009. MicroRNA expression profiling of human metastatic cancers identifies cancer gene targets. J. Pathol. 219 (2), 214–221.
Baldassarre, G., et al., 2005. p27Kip1-stathmin interaction influences sarcoma cell migration and invasion. Cancer Cell 7 (1), 51–63.
Balic, M., et al., 2006. Most early disseminated cancer cells detected in bone marrow of breast cancer patients have a putative breast cancer stem cell phenotype. Clin. Cancer Res. 12 (19), 5615–5621.
Barber, D.F., Faure, M., Long, E.O., 2004. LFA-1 contributes an early signal for NK cell cytotoxicity. J. Immunol. 173 (6), 3653–3659.
Bartel, D.P., 2004. MicroRNAs: genomics, biogenesis, mechanism, and function. Cell 116 (2), 281–297.
Bhaumik, D., et al., 2008. Expression of microRNA-146 suppresses NF-kappaB activity with reduction of metastatic potential in breast cancer cells. Oncogene 27, 5643–5647.
Bobowicz, M., et al., 2016. Prognostic value of 5-microRNA based signature in T2-T3N0 colon cancer. Clin. Exp. Metastasis 33 (8), 765–773.
Boeri, M., et al., 2011. MicroRNA signatures in tissues and plasma predict development and prognosis of computed tomography detected lung cancer. Proc. Natl. Acad. Sci. U.S.A. 108 (9), 3713–3718.
Borsig, L., et al., 2001. Heparin and cancer revisited: mechanistic connections involving platelets, P-selectin, carcinoma mucins, and tumor metastasis. Proc. Natl. Acad. Sci. U.S.A. 98, 3352–3357.
Brabletz, T., 2012. To differentiate or not—routes towards metastasis. Nat. Rev. Cancer 12 (6), 425–436.
Bruce, J.P., et al., 2015. Identification of a microRNA signature associated with risk of distant metastasis in nasopharyngeal carcinoma. Oncotarget 6 (6), 4537.
Budhu, A., et al., 2008. Identification of metastasis-related microRNAs in hepatocellular carcinoma. Hepatology 47 (3), 897–907.
Byles, V., et al., 2012. SIRT1 induces EMT by cooperating with EMT transcription factors and enhances prostate cancer cell migration and metastasis. Oncogene 31 (43), 4619–4629.
Cabrita, M.A., Christofori, G., 2008. Sprouty proteins, masterminds of receptor tyrosine kinase signaling. Angiogenesis 11 (1), 53–62.
Calin, G.A., et al., 2002. Frequent deletions and down-regulation of micro-RNA genes miR15 and miR16 at 13q14 in chronic lymphocytic leukemia. Proc. Natl. Acad. Sci. U.S.A. 99 (24), 15524–15529.
Calin, G.A., et al., 2004. Human microRNA genes are frequently located at fragile sites and genomic regions involved in cancers. Proc. Natl. Acad. Sci. U.S.A. 101 (9), 2999–3004.
Cancer Facts Sheet, 2017 at http://www.who.int/mediacentre/factsheets/fs297/en/.
Cascio, S., et al., 2010. miR-20b modulates VEGF expression by targeting HIF-1alpha and STAT3 in MCF-7 breast cancer cells. J. Cell. Physiol. 224 (1), 242–249.
Cascione, L., et al., 2013. Integrated MicroRNA and mRNA signatures associated with survival in triple negative breast cancer. PLoS One 8 (2), e55910.

Chang, T.-C., et al., 2008. Widespread microRNA repression by Myc contributes to tumorigenesis. Nat. Genet. 40 (1), 43–50.
Chen, Y., Gorski, D.H., 2008. Regulation of angiogenesis through a microRNA (miR-130a) that down-regulates antiangiogenic homeobox genes GAX and HOXA5. Blood 111 (3), 1217–1226.
Chen, D., et al., 2012a. LIFR is a breast cancer metastasis suppressor upstream of the Hippo-YAP pathway and a prognostic marker. Nat. Med. 18 (10), 1511–1517.
Chen, H.Y., et al., 2012b. MiR-103/107 promote metastasis of colorectal cancer by targeting the metastasis suppressors DAPK and KLF4. Cancer Res. 72 (14), 3631–3641.
Chen, D.L., et al., 2014. Identification of microRNA-214 as a negative regulator of colorectal cancer liver metastasis by way of regulation of fibroblast growth factor receptor 1 expression. Hepatology 60 (2), 598–609.
Chen, J., et al., 2016. Efficient extravasation of tumor-repopulating cells depends on cell deformability. Sci. Rep. 6, 19304.
Correia, N.C., et al., 2016. MiR-146b negatively regulates migration and delays progression of T-cell acute lymphoblastic leukemia. Sci. Rep. 6, 31894.
Costinean, S., et al., 2006. Pre-B cell proliferation and lymphoblastic leukemia/high-grade lymphoma in E(mu)-miR155 transgenic mice. Proc. Natl. Acad. Sci. U.S.A. 103 (18), 7024–7029.
Dangi-Garimella, S., et al., 2009. Raf kinase inhibitory protein suppresses a metastasis signalling cascade involving LIN28 and let-7. EMBO J. 28 (4), 347–358.
Dean, Z.S., Riahi, R., Wong, P.K., 2015. Spatiotemporal dynamics of microRNA during epithelial collective cell migration. Biomaterials 37, 156–163.
Deckers, M., et al., 2006. The tumor suppressor Smad4 is required for transforming growth factor β-induced epithelial to mesenchymal transition and bone metastasis of breast cancer cells. Cancer Res. 66 (4), 2202–2209.
de Giorgio, A., et al., 2013. Emerging roles of competing endogenous RNAs in cancer: insights from the regulation of PTEN. Mol. Cell. Biol. 33 (20), 3976–3982.
Deryugina, E.I., Quigley, J.P., 2015. Tumor angiogenesis: MMP-mediated induction of intravasation- and metastasis-sustaining neovasculature. Matrix Biol. 44–46, 94–112.
Dews, M., et al., 2006. Augmentation of tumor angiogenesis by a Myc-activated microRNA cluster. Nat. Genet. 38 (9), 1060–1065.
Dews, M., et al., 2010. The myc-miR-17∼92 axis blunts TGF{beta} signaling and production of multiple TGF{beta}-dependent antiangiogenic factors. Cancer Res. 70 (20), 8233–8246.
Dhillon, A.S., et al., 2007. MAP kinase signalling pathways in cancer. Oncogene 26 (22), 3279–3290.
Di Modica, M., et al., 2017. Breast cancer-secreted miR-939 downregulates VE-cadherin and destroys the barrier function of endothelial monolayers. Cancer Lett. 384, 94–100.
Drabsch, Y., Ten Dijke, P., 2011. TGF-B signaling in breast cancer cell invasion and bone metastasis. J. Mammary Gland Biol. Neoplasia 16 (2), 97–108.
Drusco, A., et al., 2014. MicroRNA profiles discriminate among colon cancer metastasis. PLoS One 9 (6), e96670.
Dykxhoorn, D.M., et al., 2009. miR-200 enhances mouse breast cancer cell colonization to form distant metastases. PLoS One 4 (9), e7181.
Ell, B., Kang, Y., 2013. Transcriptional control of cancer metastasis. Trends Cell Biol. 23 (12), 603–611.
Ell, B., et al., 2013. Tumor-induced osteoclast miRNA changes as regulators and biomarkers of osteolytic bone metastasis. Cancer Cell 24 (4), 542–556.
Ellermeier, C., et al., 2014. Prognostic microRNA expression signature from examination of colorectal primary and metastatic tumors. Anticancer Res. 34 (8), 3957–3967.

Elson-Schwab, I., Lorentzen, A., Marshall, C.J., 2010. MicroRNA-200 family members differentially regulate morphological plasticity and mode of melanoma cell invasion. PLoS One 5 (10), e13176.

Esquela-Kerscher, A., et al., 2008. The let-7 microRNA reduces tumor growth in mouse models of lung cancer. Cell Cycle 7 (6), 759–764.

Farazi, T.A., et al., 2011. MicroRNA sequence and expression analysis in breast tumors by deep sequencing. Cancer Res. 71 (13), 4443–4453.

Fidler, I.J., 2003. The pathogenesis of cancer metastasis: the "seed and soil" hypothesis revisited. Nat. Rev. Cancer 3 (6), 453–458.

Filipowicz, W., Bhattacharyya, S.N., Sonenberg, N., 2008. Mechanisms of post-transcriptional regulation by microRNAs: are the answers in sight? Nat. Rev. Genet. 9 (2), 102–114.

Fish, J.E., et al., 2008. miR-126 regulates angiogenic signaling and vascular integrity. Dev. Cell 15 (2), 272–284.

Fong, M.Y., et al., 2015. Breast-cancer-secreted miR-122 reprograms glucose metabolism in premetastatic niche to promote metastasis. Nat. Cell Biol. 17 (2), 183–194.

Freedland, S.J., Moul, J.W., 2007. Prostate specific antigen recurrence after definitive therapy. J. Urol. 177 (6), 1985–1991.

Friedl, P., Alexander, S., 2011. Cancer invasion and the microenvironment: plasticity and reciprocity. Cell 147 (5), 992–1009.

Friedl, P., Bröcker, E.B., 2000. The biology of cell locomotion within three-dimensional extracellular matrix. Cell. Mol. Life Sci. 57 (1), 41–64.

Friedl, P., Wolf, K., 2003. Tumour-cell invasion and migration: diversity and escape mechanisms. Nat. Rev. Cancer 3 (5), 362–374.

Gabriely, G., et al., 2008. MicroRNA 21 promotes glioma invasion by targeting matrix metalloproteinase regulators. Mol. Cell. Biol. 28 (17), 5369–5380.

Gao, H., et al., 2014. Forward genetic screens in mice uncover mediators and suppressors of metastatic reactivation. Proc. Natl. Acad. Sci. U.S.A. 111 (46), 16532–16537.

Ghajar, C.M., et al., 2013. The perivascular niche regulates breast tumour dormancy. Nat. Cell Biol. 15 (7), 807–817.

Giancotti, F.G., 2013. Mechanisms governing metastatic dormancy and reactivation. Cell 155 (4), 750–764.

Gilmore, A.P., 2005. Anoikis. Cell Death Differ. 12, 1473–1477.

Golden, R.J., et al., 2017. An Argonaute phosphorylation cycle promotes microRNA-mediated silencing. Nature 542 (7640), 197–202.

Gregory, P.A., et al., 2008. The miR-200 family and miR-205 regulate epithelial to mesenchymal transition by targeting ZEB1 and SIP1. Nat. Cell Biol. 10 (5), 593–601.

Gulhati, P., et al., 2011. mTORC1 and mTORC2 regulate EMT, motility, and metastasis of colorectal cancer via RhoA and Rac1 signaling pathways. Cancer Res. 71 (9), 3246–3256.

Guo, S.T., et al., 2012. MicroRNA-497 targets insulin-like growth factor 1 receptor and has a tumour suppressive role in human colorectal cancer. Oncogene 32 (15), 1910–1920.

Gupta, G.P., Massagué, J., 2006. Cancer metastasis: building a framework. Cell 127 (4), 679–695.

Ha, M., Kim, V.N., 2014. Regulation of microRNA biogenesis. Nat. Rev. Mol. Cell Biol. 15 (8), 509–524.

Hanna, N., 1982. Role of natural killer cells in control of cancer metastasis. Cancer Metastasis Rev. 1 (1), 45–64.

Hanniford, D., et al., 2015. A miRNA-based signature detected in primary melanoma tissue predicts development of brain metastasis. Clin. Cancer Res. 21 (21), 4903–4912.

Harris, T.A., et al., 2008. MicroRNA-126 regulates endothelial expression of vascular cell adhesion molecule 1. Proc. Natl. Acad. Sci. U.S.A. 105 (5), 1516–1521.
He, L., He, X., Lim, L.P., et al., 2007a. A microRNA component of the p53 tumour suppressor network. Nature 447 (7148), 1130–1134.
He, L., He, X., Lowe, S.W., et al., 2007b. microRNAs join the p53 network—another piece in the tumour-suppression puzzle. Nat. Rev. Cancer 7 (11), 819–822.
Heinzelmann, J., et al., 2011. Specific miRNA signatures are associated with metastasis and poor prognosis in clear cell renal cell carcinoma. World J. Urol. 29 (3), 367–373.
Hoeben, A., et al., 2004. Vascular endothelial growth factor and angiogenesis. Pharmacol. Rev. 56 (4), 549–580.
Hou, P., et al., 2014. LincRNA-ROR induces epithelial-to-mesenchymal transition and contributes to breast cancer tumorigenesis and metastasis. Cell Death Dis. 5, e1287.
Hu, M., et al., 2008. Regulation of in situ to invasive breast carcinoma transition. Cancer Cell 13 (5), 394–406.
Hu, X., et al., 2010. A microRNA expression signature for cervical cancer prognosis. Cancer Res. 70 (4), 1441–1448.
Hua, Z., et al., 2006. MiRNA-directed regulation of VEGF and other angiogenic under hypoxia. PLoS One 1 (1), e116.
Huang, Q., et al., 2008. The microRNAs miR-373 and miR-520c promote tumour invasion and metastasis. Nat. Cell Biol. 10 (2), 202–210.
Huang, Y., et al., 2009. Pulmonary vascular destabilization in the premetastatic phase facilitates lung metastasis. Cancer Res. 69 (19), 7529–7537.
Huang, B., et al., 2015. Modeling the transitions between collective and solitary migration phenotypes in cancer metastasis. Sci. Rep. 5, 17379.
Hudson, R.S., et al., 2011. MicroRNA-1 is a candidate tumor suppressor and prognostic marker in human prostate cancer. Nucleic Acids Res. 40 (8), 1–15.
Hunt, S., et al., 2011. MicroRNA-124 suppresses oral squamous cell carcinoma motility by targeting ITGB1. FEBS Lett. 585 (1), 187–192.
Hur, K., et al., 2015. Identification of a metastasis-specific microRNA signature in human colorectal cancer. J. Natl. Cancer Inst. 107 (3). pii:dju492.
Hurst, D.R., Edmonds, M.D., Welch, D.R., 2009. Metastamir: the field of metastasis-regulatory microRNA is spreading. Cancer Res. 69 (19), 7495–7498.
Janse van Rensburg, H.J., Yang, X., 2016. The roles of the Hippo pathway in cancer metastasis. Cell. Signal. 28 (11), 1761–1772.
Janz, A., et al., 2000. Activation of the myc oncoprotein leads to increased turnover of thrombospondin-1 mRNA. Nucleic Acids Res. 28 (11), 2268–2275.
Johnnidis, J.B., et al., 2008. Regulation of progenitor cell proliferation and granulocyte function by microRNA-223. Nature 451 (7182), 1125–1129.
Jurmeister, S., et al., 2012. MicroRNA-200c represses migration and invasion of breast cancer cells by targeting actin-regulatory proteins FHOD1 and PPM1F. Mol. Cell. Biol. 32 (3), 633–651.
Kang, L., et al., 2015. MicroRNA-34a suppresses the breast cancer stem cell-like characteristics by downregulating Notch1 pathway. Cancer Sci. 106 (6), 700–708.
Kaplan, R.N., et al., 2005. VEGFR1-positive haematopoietic bone marrow progenitors initiate the pre-metastatic niche. Nature 438 (7069), 820–827.
Karrison, T.G., Ferguson, D.J., Meier, P., 1999. Dormancy of mammary carcinoma after mastectomy. J. Natl. Cancer Inst. 91 (1), 80–85.
Kato, T., et al., 2014. TRIM27/MRTF-B-dependent integrin beta 1 expression defines leading cells in cancer cell collectives. Cell Rep. 7 (4), 1156–1167.
Keklikoglou, I., et al., 2012. MicroRNA-520/373 family functions as a tumor suppressor in estrogen receptor negative breast cancer by targeting NF-κB and TGF-β signaling pathways. Oncogene 31 (37), 4150–4163.

Kent, O.A., Mendell, J.T., 2006. A small piece in the cancer puzzle: microRNAs as tumor suppressors and oncogenes. Oncogene 25 (46), 6188–6196.
Khella, H.W.Z., et al., 2013. miR-192, miR-194 and miR-215: a convergent microRNA network suppressing tumor progression in renal cell carcinoma. Carcinogenesis 34 (10), 2231–2239.
Khew-Goodall, Y., Goodall, G.J., 2010. Myc-modulated miR-9 makes more metastases. Nat. Cell Biol. 12 (3), 209–211.
Kim, J., et al., 1998. Requirement for specific proteases in cancer cell intravasation as revealed by a novel semiquantitative PCR-based assay. Cell 94 (3), 353–362.
Kim, N.H., et al., 2011. A p53/miRNA-34 axis regulates Snail1-dependent cancer cell epithelial-mesenchymal transition. J. Cell Biol. 195 (3), 417–433.
Klein, C.A., 2009. Parallel progression of primary tumours and metastases. Nat. Rev. Cancer 9 (4), 302–312.
Kong, W., et al., 2008. MicroRNA-155 is regulated by the transforming growth factor beta/Smad pathway and contributes to epithelial cell plasticity by targeting RhoA. Mol. Cell. Biol. 28 (22), 6773–6784.
Konstantopoulos, K., Thomas, S.N., 2009. Cancer cells in transit: the vascular interactions of tumor cells. Annu. Rev. Biomed. Eng. 11 (1), 177–202.
Korpal, M., et al., 2011. Direct targeting of Sec23a by miR-200s influences cancer cell secretome and promotes metastatic colonization. Nat. Med. 17 (9), 1101–1108.
Koscielny, S., Tubiana, M., 2010. Parallel progression of tumour and metastases. Nat. Rev. Cancer 10 (2), 156.
Kota, J., et al., 2009. Therapeutic microRNA delivery suppresses tumorigenesis in a murine liver cancer model. Cell 137 (6), 1005–1017.
Krakhmal, N.V., et al., 2015. Cancer invasion: patterns and mechanisms. Acta Nat. 7 (2), 17–28.
Kroiss, A., et al., 2015. Androgen-regulated microRNA-135a decreases prostate cancer cell migration and invasion through downregulating ROCK1 and ROCK2. Oncogene 34 (22), 2846–2855.
Krzeszinski, J.Y., et al., 2014. miR-34a blocks osteoporosis and bone metastasis by inhibiting osteoclastogenesis and Tgif2. Nature 512 (7515), 431–435.
Kuehbacher, A., et al., 2007. Role of dicer and drosha for endothelial microRNA expression and angiogenesis. Circ. Res. 101 (1), 59–68.
Kulshreshtha, R., et al., 2006. A microRNA signature of hypoxia. Mol. Cell. Biol. 27 (5), 1859–1867.
Kumar, M.S., et al., 2014. HMGA2 functions as a competing endogenous RNA to promote lung cancer progression. Nature 505 (7482), 212–217.
Kumarswamy, R., et al., 2012. MicroRNA-30a inhibits epithelial-to-mesenchymal transition by targeting Snai1 and is downregulated in non-small cell lung cancer. Int. J. Cancer 130 (9), 2044–2053.
Kunej, T., et al., 2011. Epigenetic regulation of microRNAs in cancer: an integrated review of literature. Mutat. Res. Fundam. Mol. Mech. Mutagen. 717 (1–2), 77–84.
Kung, J.T.Y., Colognori, D., Lee, J.T., 2013. Long noncoding RNAs: past, present, and future. Genetics 193 (3), 651–669.
Kurrey, N.K., Amit, K., Bapat, S.A., 2005. Snail and Slug are major determinants of ovarian cancer invasiveness at the transcription level. Gynecol. Oncol. 97 (1), 155–165.
Leber, M.F., Efferth, T., 2009. Molecular principles of cancer invasion and metastasis (review). Int. J. Oncol. 34 (4), 881–895.
Lee, W.H., et al., 2016. TRPV4 regulates breast cancer cell extravasation, stiffness and actin cortex. Sci. Rep. 6, 27903.
Lerebours, F., et al., 2013. MiRNA expression profiling of inflammatory breast cancer identifies a 5-miRNA signature predictive of breast tumor aggressiveness. Int. J. Cancer 133 (7), 1614–1623.

le Sage, C., et al., 2007. Regulation of the p27(Kip1) tumor suppressor by miR-221 and miR-222 promotes cancer cell proliferation. EMBO J. 26 (15), 3699–3708.

Leupold, J.H., et al., 2007. Tumor suppressor Pdcd4 inhibits invasion/intravasation and regulates urokinase receptor (u-PAR) gene expression via Sp-transcription factors. Oncogene 26 (31), 4550–4562.

Li, J., et al., 2013a. Microvesicle-mediated transfer of microRNA-150 from monocytes to endothelial cells promotes angiogenesis. J. Biol. Chem. 288 (32), 23586–23596.

Li, X., et al., 2013b. C-MYC-regulated miR-23a/24-2/27a cluster promotes mammary carcinoma cell invasion and hepatic metastasis by targeting sprouty2. J. Biol. Chem. 288 (25), 18121–18133.

Li, Y., et al., 2013c. MiR-146b-5p inhibits glioma migration and invasion by targeting MMP16. Cancer Lett. 339 (2), 260–269.

Li, J.-T., et al., 2014. Nuclear factor of activated T cells 5 maintained by Hotair suppression of miR-568 upregulates S100 calcium binding protein A4 to promote breast cancer metastasis. Breast Cancer Res. 16 (5), 454.

Li, J., et al., 2015a. Registered report: The microRNA miR-34a inhibits prostate cancer stem cells and metastasis by directly repressing CD44. eLife 4, e06434.

Li, N., et al., 2015b. miR-215 promotes malignant progression of gastric cancer by targeting RUNX1. Oncotarget 5 (4), 4817–4828.

Liang, Y.J., et al., 2013. MiR-124 targets Slug to regulate epithelial-mesenchymal transition and metastasis of breast cancer. Carcinogenesis 34 (3), 713–722.

Lim, L.P., et al., 2005. Microarray analysis shows that some microRNAs downregulate large numbers of target mRNAs. Nature 433 (7027), 769–773.

Lin, S.-L., et al., 2008. Loss of mir-146a function in hormone-refractory prostate cancer. RNA 14 (3), 417–424.

Lin, C.-W., et al., 2013. MicroRNA-135b promotes lung cancer metastasis by regulating multiple targets in the Hippo pathway and LZTS1. Nat. Commun. 4, 1877.

Ling, H., et al., 2015. Junk DNA and the long non-coding RNA twist in cancer genetics. Oncogene 34, 5003–5011.

Ling, H., et al., 2016. The clinical and biological significance of MIR-224 expression in colorectal cancer metastasis. Gut. 65 (6), 977–989.

Liu, C., et al., 2011. Identification of miR-34a as a potent inhibitor of prostate cancer progenitor cells and metastasis by directly repressing CD44. Nat. Med. 17 (2), 211–215.

Liu, N., et al., 2012. Prognostic value of a microRNA signature in nasopharyngeal carcinoma: a microRNA expression analysis. Lancet Oncol. 13 (6), 633–641.

Liu, X., et al., 2013. MiRNA-296-3p-ICAM-1 axis promotes metastasis of prostate cancer by possible enhancing survival of natural killer cell-resistant circulating tumour cells. Cell Death Dis. 4 (11), e928.

Liu, X.H., et al., 2014. Lnc RNA HOTAIR functions as a competing endogenous RNA to regulate HER2 expression by sponging miR-331-3p in gastric cancer. Mol. Cancer 13 (1), 92.

Lopez-Ramirez, M.A., et al., 2014. MicroRNA-155 negatively affects blood-brain barrier function during neuroinflammation. FASEB J. 28 (6), 2551–2565.

Loureiro, R.M.B., Amore, P.A.D., 2005. Transcriptional regulation of vascular endothelial growth factor in cancer. Cytokine Growth Factor Rev. 16, 77–89.

Lu, L.F., Liston, A., 2009. MicroRNA in the immune system, microRNA as an immune system. Immunology 127 (3), 291–298.

Lu, J., et al., 2005. MicroRNA expression profiles classify human cancers. Nature 435 (7043), 834–838.

Ma, L., Teruya-Feldstein, J., Weinberg, R.A., 2007. Tumour invasion and metastasis initiated by microRNA-10b in breast cancer. Nature 449 (7163), 682–688.

Ma, L., et al., 2010a. MiR-9, a MYC/MYCN-activated microRNA, regulates E-cadherin and cancer metastasis. Nat. Cell Biol. 12 (3), 247–256.

Ma, L., et al., 2010b. Therapeutic silencing of miR-10b inhibits metastasis in a mouse mammary tumor model. Nat. Biotechnol. 28 (4), 341–347.
Ma, C., et al., 2014. H19 promotes pancreatic cancer metastasis by derepressing let-7's suppression on its target HMGA2-mediated EMT. Tumor Biol. 35 (9), 9163–9169.
Marino, A.L.F., et al., 2014. MicroRNA expression as risk biomarker of breast cancer metastasis: a pilot retrospective case-cohort study. BMC Cancer 14, 739.
Massagué, J., Obenauf, A.C., 2016. Metastatic colonization by circulating tumour cells. Nature 529 (7586), 298–306.
Meng, F., et al., 2007. MicroRNA-21 regulates expression of the PTEN tumor suppressor gene in human hepatocellular cancer. Gastroenterology 133 (2), 647–658.
Mertens-Talcott, S.U., et al., 2007. The oncogenic microRNA-27a targets genes that regulate specificity protein transcription factors and the G2-M checkpoint in MDA-MB-231 breast cancer cells. Cancer Res. 67 (22), 11001–11011.
Mitra, A.K., et al., 2012. MicroRNAs reprogram normal fibroblasts into cancer-associated fibroblasts in ovarian cancer. Cancer Discov. 2 (12), 1100–1108.
Mlcochova, H., et al., 2016. Epithelial-mesenchymal transition-associated microRNA/mRNA signature is linked to metastasis and prognosis in clear-cell renal cell carcinoma. Sci. Rep. 6, 31852.
Mohammadi-Yeganeh, S., et al., 2016. MicroRNA-340 inhibits the migration, invasion, and metastasis of breast cancer cells by targeting Wnt pathway. Tumor Biol. 37 (7), 8993–9000.
Moldovan, L., et al., 2014. Methodological challenges in utilizing miRNAs as circulating biomarkers. J. Cell. Mol. Med. 18 (3), 371–390.
Mudduluru, G., et al., 2015. A systematic approach to defining the microrna landscape in metastasis. Cancer Res. 75 (15), 3010–3019.
Muller, W.a., 2011. Mechanisms of leukocyte transendothelial migration. Annu. Rev. Pathol. 6, 323–344.
Nabeshima, K., 1999. Cohort migration of carcinoma cells: differentiated colorectal carcinoma cells move as coherent cell clusters or sheets. Histol. Histopathol. 14 (4), 1183–1197.
Napier, S.L., et al., 2007. Selectin ligand expression regulates the initial vascular interactions of colon carcinoma cells: the roles of CD44V and alternative sialofucosylated selectin ligands. J. Biol. Chem. 282 (6), 3433–3441.
Nicoloso, M.S., et al., 2009. MicroRNAs—the micro steering wheel of tumour metastases. Nat. Rev. Cancer 9 (4), 293–302.
Ochalek, T., et al., 1988. Correlation between cell deformability and metastatic potential in B16-F1 melanoma cell variants. Cancer Res. 48 (18), 5124–5128.
O'Donnell, K.A., et al., 2005. c-Myc-regulated microRNAs modulate E2F1 expression. Nature 435 (7043), 839–843.
Ono, M., et al., 2014. Exosomes from bone marrow mesenchymal stem cells contain a microRNA that promotes dormancy in metastatic breast cancer cells. Sci. Signal. 7 (332), ra63.
Orso, F., et al., 2016. Mir-214 and mir-148b targeting inhibits dissemination of melanoma and breast cancer. Cancer Res. 76 (17), 5151–5162.
Oskarsson, T., et al., 2011. Breast cancer cells produce tenascin C as a metastatic niche component to colonize the lungs. Nat. Med. 17 (7), 867–874.
Ostenfeld, M.S., et al., 2014. Cellular disposal of miR23b by RAB27-dependent exosome release is linked to acquisition of metastatic properties. Cancer Res. 74 (20), 5758–5771.
Ouzounova, M., et al., 2013. MicroRNA miR-30 family regulates non-attachment growth of breast cancer cells. BMC Genomics 14, 139.
Padua, D., et al., 2008. TGFβ primes breast tumors for lung metastasis seeding through angiopoietin-like 4. Cell 133 (1), 66–77.

Pantel, K., et al., 1993. Differential expression of proliferation-associated molecules in individual micrometastatic carcinoma cells. J. Natl. Cancer Inst. 85 (17), 1419–1424.
Paoli, P., Giannoni, E., Chiarugi, P., 2013. Anoikis molecular pathways and its role in cancer progression. Biochim. Biophys. Acta 1833 (12), 3481–3498.
Pencheva, N., et al., 2012. Convergent multi-miRNA targeting of ApoE grives LRP1/LRP8-dependent melanoma metastasis and angiogenesis. Cell 151 (5), 1068–1082.
Penna, E., et al., 2011. microRNA-214 contributes to melanoma tumour progression through suppression of TFAP2C. EMBO J. 30 (10), 1990–2007.
Penna, E., et al., 2013. MiR-214 coordinates melanoma progression by upregulating ALCAM through TFAP2 and miR-148b downmodulation. Cancer Res. 73 (13), 4098–4111.
Pérez-Rivas, L.G., et al., 2014. A microRNA signature associated with early recurrence in breast cancer. PLoS One 9 (3), e91884.
Pichler, M., Calin, G.A., 2015. MicroRNAs in cancer: from developmental genes in worms to their clinical application in patients. Br. J. Cancer 113 (4), 569–573.
Pichler, M., et al., 2017. Genome-wide miRNA analysis identifies miR-188-3p as a novel prognostic marker and molecular factor involved in colorectal carcinogenesis. Clin. Cancer Res. 23 (5), 1323–1333.
Pizzini, S., et al., 2013. Impact of microRNAs on regulatory networks and pathways in human colorectal carcinogenesis and development of metastasis. BMC Genomics 14 (1), 589.
Png, K.J., et al., 2012. A microRNA regulon that mediates endothelial recruitment and metastasis by cancer cells. Nature 481 (7380), 190–194.
Polette, M., et al., 1998. Association of fibroblastoid features with the invasive phenotype in human bronchial cancer cell lines. Clin. Exp. Metastasis 16 (2), 105–112.
Poliseno, L., et al., 2006. MicroRNAs modulate the angiogenic properties of HUVECs. Blood 108 (9), 3068–3071.
Qi, F., et al., 2015. The miR-30 family inhibits pulmonary vascular hyperpermeability in the premetastatic phase by direct targeting of Skp2. Clin. Cancer Res. 21 (13), 3071–3080.
Qian, B., et al., 2009. High miR-21 expression in breast cancer associated with poor disease-free survival in early stage disease and high TGF-beta1. Breast Cancer Res. Treat. 117 (1), 131–140.
Regulus Therapeutics Inc., 2015. Regulus initiates phase I clinical study of RG-012, a microRNA therapeutic in development for the treatment of Alport syndrome. In: Achieves Key 'Clinical Map Initiative' Goal for 2015 and Advances Orphan Disease Efforts. PR Newswire, La Jolla, CA, http://www.prnewswire.com/.
Reymond, N., D'Água, B.B., Ridley, A.J., 2013. Crossing the endothelial barrier during metastasis. Nat. Rev. Cancer 13 (12), 858–870.
Ribatti, D., Mangialardi, G., Vacca, A., 2006. Stephen Paget and the "seed and soil" theory of metastatic dissemination. Clin. Exp. Med. 6 (4), 145–149.
Rintoul, R.C., Sethi, T., 2001. The role of extracellular matrix in small-cell lung cancer. Lancet Oncol. 2 (7), 437–442.
Roybal, J.D., et al., 2011. miR-200 Inhibits lung adenocarcinoma cell invasion and metastasis by targeting Flt1/VEGFR1. Mol. Cancer Res. 9 (1), 25–35.
Sachdeva, M., et al., 2014. MicroRNA-182 drives metastasis of primary sarcomas by targeting multiple genes. J. Clin. Invest. 124 (10), 4305–4319.
Sahai, E., et al., 2007. Smurf1 regulates tumor cell plasticity and motility through degradation of RhoA leading to localized inhibition of contractility. J. Cell Biol. 176 (1), 35–42.
Salmena, L., et al., 2011. A ceRNA hypothesis: the rosetta stone of a hidden RNA language? Cell 146 (3), 353–358.
Sapre, N., et al., 2016. A urinary microRNA signature can predict the presence of bladder urothelial carcinoma in patients undergoing surveillance. Br. J. Cancer 114 (4), 454–462.

Sayed, D., et al., 2008. MicroRNA-21 targets Sprouty2 and promotes cellular outgrowths. Mol. Biol. Cell 19, 3272–3282.

Schauer, I.G., et al., 2011. Cancer-associated fibroblasts and their putative role in potentiating the initiation and development of epithelial ovarian cancer. Neoplasia 13 (5), 393–405.

Segura, M.F., et al., 2009. Aberrant miR-182 expression promotes melanoma metastasis by repressing FOXO3 and microphthalmia-associated transcription factor. Proc. Natl. Acad. Sci. 106 (6), 1814–1819.

Selbach, M., et al., 2008. Widespread changes in protein synthesis induced by microRNAs. Nature 455 (7209), 58–63.

Seok, J.K., et al., 2014. MicroRNA-382 induced by HIF-1 is an angiogenic miR targeting the tumor suppressor phosphatase and tensin homolog. Nucleic Acids Res. 42 (12), 8062–8072.

Sethi, N., et al., 2011. Tumor-derived jagged1 promotes osteolytic bone metastasis of breast cancer by engaging notch signaling in bone cells. Cancer Cell 19 (2), 192–205.

Shih, T.C., et al., 2012. MicroRNA-214 downregulation contributes to tumor angiogenesis by inducing secretion of the hepatoma-derived growth factor in human hepatoma. J. Hepatol. 57 (3), 584–591.

Siemens, H., et al., 2011. miR-34 and SNAIL form a double-negative feedback loop to regulate epithelial-mesenchymal transitions. Cell Cycle 10 (24), 4256–4271.

Song, B., et al., 2010. MicroRNA-21 regulates breast cancer invasion partly by targeting tissue inhibitor of metalloproteinase 3 expression. J. Exp. Clin. Cancer Res. 29, 29.

Søreide, K., et al., 2016. Assessment of clinically related outcomes and biomarker analysis for translational integration in colorectal cancer (ACROBATICC): study protocol for a population-based, consecutive cohort of surgically treated colorectal cancers and resected colorectal liver. J. Transl. Med. 14 (1), 1–9.

Sossey-Alaoui, K., Bialkowska, K., Plow, E.F., 2009. The miR200 family of microRNAs regulates WAVE3-dependent cancer cell invasion. J. Biol. Chem. 284 (48), 33019–33029.

Sotiropoulou, G., et al., 2009. Emerging roles of microRNAs as molecular switches in the integrated circuit of the cancer cell. RNA 15 (8), 1443–1461.

Strell, C., Rundqvist, H., Ostman, A., 2012. Fibroblasts—a key host cell type in tumor initiation, progression, and metastasis. Ups. J. Med. Sci. 117 (2), 187–195.

Suárez, Y., et al., 2010. Cutting edge: TNF-induced microRNAs regulate TNF-induced expression of E-selectin and intercellular adhesion molecule-1 on human endothelial cells: feedback control of inflammation. J. Immunol. 184 (1), 21–25.

Sugarbaker, E.V., 1979. Some characteristics of metastasis in man. Am. J. Pathol. 97 (3), 623–632.

Sundaram, P., et al., 2011. p53-responsive miR-194 inhibits thrombospondin-1 and promotes angiogenesis in colon cancers. Cancer Res. 71 (24), 7490–7501.

Taganov, K.D., et al., 2006. NF- B-dependent induction of microRNA miR-146, an inhibitor targeted to signaling proteins of innate immune responses. Proc. Natl. Acad. Sci. U.S.A. 103 (33), 12481–12486.

Takai, Y., Nakanishi, H., 2003. Nectin and afadin: novel organizers of intercellular junctions. J. Cell Sci. 116, 17–27.

Takeshita, F., et al., 2010. Systemic delivery of synthetic microRNA-16 inhibits the growth of metastatic prostate tumors via downregulation of multiple cell-cycle genes. Mol. Ther. 18 (1), 181–187.

Tan, X., et al., 2010. Pin1 expression contributes to lung cancer prognosis and carcinogenesis. Cancer Biol. Ther. 9 (2).

Tanaka, S., et al., 2003. Angiogenic switch as a molecular target of malignant tumors. J. Gastroenterol. 38 (Suppl. 1), 93–97.

Tavazoie, S.F., et al., 2008. Endogenous human microRNAs that suppress breast cancer metastasis. Nature 451 (7175), 147–152.
Taylor, M.A., et al., 2013. TGF-beta upregulates miR-181a expression to promote breast cancer metastasis. J. Clin. Invest. 123 (1), 150–163.
Thiery, J.P., et al., 2009. Epithelial-mesenchymal transitions in development and disease. Cell 139 (5), 871–890.
Thomsen, K.G., et al., 2015. miR-155, identified as anti-metastatic by global miRNA profiling of a metastasis model, inhibits cancer cell extravasation and colonization in vivo and causes significant signaling alterations. Oncotarget 6 (30), 29224–29239.
Thomson, J.M., et al., 2006. Extensive post-transcriptional regulation of microRNAs and its implications for cancer. Genes Dev. 20 (16), 2202–2207.
Tian, M., et al., 2011. Binding of RhoA by the C2 domain of E3 ligase Smurf1 is essential for Smurf1-regulated RhoA ubiquitination and cell protrusive activity. FEBS Lett. 585 (14), 2199–2204.
Tominaga, N., et al., 2015. Brain metastatic cancer cells release microRNA-181c-containing extracellular vesicles capable of destructing blood–brain barrier. Nat. Commun. 6, 6716.
Turner, N., Grose, R., 2010. Fibroblast growth factor signalling: from development to cancer. Nat. Rev. Cancer 10 (2), 116–129.
Utoguchi, N., et al., 1996. Effect of tumour cell-conditioned medium on endothelial macromolecular permeability and its correlation with collagen. Br. J. Cancer 73 (1), 24–28.
Valastyan, S., Weinberg, R.A., 2011. Tumor metastasis: molecular insights and evolving paradigms. Cell 147 (2), 275–292.
Vega, A.B., et al., 2013. microRNA expression profile in stage III colorectal cancer. Oncol. Rep. 30 (1), 320–326.
Verschueren, H., De Baetselier, P., Bereiter-Hahn, J., 1991. Dynamic morphology of metastatic mouse T-lymphoma cells invading through monolayers of 10T 1/2 cells. Cell Motil. Cytoskeleton 20 (3), 203–214.
Vickers, M.M., et al., 2012. Stage-dependent differential expression of microRNAs in colorectal cancer: potential role as markers of metastatic disease. Clin. Exp. Metastasis 29 (2), 123–132.
Vidigal, J.A., Ventura, A., 2015. The biological functions of miRNAs: lessons from in vivo studies. Trends Cell Biol. 25 (3), 137–147.
Volinia, S., et al., 2006. A microRNA expression signature of human solid tumors defines cancer gene targets. Proc. Natl. Acad. Sci. U.S.A. 103 (7), 2257–2261.
Volinia, S., et al., 2012. Breast cancer signatures for invasiveness and prognosis defined by deep sequencing of microRNA. Proc. Natl. Acad. Sci. U.S.A. 109 (8), 3024–3029.
Vong, S., Kalluri, R., 2011. The role of stromal myofibroblast and extracellular matrix in tumor angiogenesis. Genes Cancer 2, 1139–1145.
Voorhoeve, P.M., et al., 2006. A genetic screen implicates miRNA-372 and miRNA-373 as oncogenes in testicular germ cell tumors. Cell 124 (6), 1169–1181.
Wan, X., et al., 2015. The functional sites of miRNAs and lncRNAs in gastric carcinogenesis. Tumor Biol. 36 (2), 521–532.
Wang, W., et al., 2014. MicroRNA-497 inhibition of ovarian cancer cell migration and invasion through targeting of SMAD specific E3 ubiquitin protein ligase 1. Biochem. Biophys. Res. Commun. 449 (4), 432–437.
Wang, L., et al., 2015. MiR-573 inhibits prostate cancer metastasis by regulating epithelial-mesenchymal transition. Oncotarget 6 (34), 35978–35990.
Weber, B., et al., 2007. Methylation of human microRNA genes in normal and neoplastic cells. Cell Cycle 6 (9), 1001–1005.
Wei, F., et al., 2016a. miR-373 inhibits glioma Cell U251 migration and invasion by down-regulating CD44 and TGFBR2. Cell. Mol. Neurobiol. 36 (8), 1389–1397.

Wei, J.L., et al., 2016b. MiR-181a-5p promotes anoikis by suppressing autophagy during detachment induction in the mammary epithelial cell line MCF10A. Protein Cell 7 (4), 305–309.

Weilbaecher, K.N., Guise, T.A., McCauley, L.K., 2011. Cancer to bone: a fatal attraction. Nat. Rev. Cancer 11 (6), 411–425.

Weis, S., et al., 2004. Endothelial barrier disruption by VEGF-mediated Src activity potentiates tumor cell extravasation and metastasis. J. Cell Biol. 167 (2), 223–229.

Wiggins, J.F., et al., 2010. Development of a lung cancer therapeutic based on the tumor suppressor microRNA-34. Cancer Res. 70 (14), 5923–5930.

Wirtz, D., Konstantopoulos, K., Searson, P.C., 2011. The physics of cancer: the role of physical interactions and mechanical forces in metastasis. Nat. Rev. Cancer 11 (7), 512–522.

Wotschofsky, Z., et al., 2016. Integrated microrna and mRNA signature associated with the transition from the locally confined to themetastasized clear cell renal cell carcinoma exemplified by miR-146-5p. PLoS One 11 (2), 1–20.

Wu, X., et al., 2012. Identification of a 4-microRNA signature for clear cell renal cell carcinoma metastasis and prognosis. PLoS One 7 (5), e35661.

Xie, J., et al., 2014. Ovarian tumor-associated microRNA-20a decreases natural killer cell cytotoxicity by downregulating MICA/B expression. Cell. Mol. Immunol. 11 (5), 495–502.

Xie, X., et al., 2005. Systematic discovery of regulatory motifs in human promoters and 3′ UTRs by comparison of several mammals. Nature 434 (7031), 338–345.

Xu, H., et al., 2009. MicroRNA miR-29 modulates expression of immunoinhibitory molecule B7-H3: potential implications for immune based therapy of human solid tumors. Cancer Res. 69 (15), 6275–6281.

Xu, Q., et al., 2016. Long non-coding RNA regulation of epithelial–mesenchymal transition in cancer metastasis. Cell Death Dis. 7, 1–10.

Yamakuchi, M., et al., 2010. P53-induced microRNA-107 inhibits HIF-1 and tumor angiogenesis. Proc. Natl. Acad. Sci. U.S.A. 107 (14), 6334–6339.

Yan, L.-X., et al., 2008. MicroRNA miR-21 overexpression in human breast cancer is associated with advanced clinical stage, lymph node metastasis and patient poor prognosis. RNA 14 (11), 2348–2360.

Yang, J., Weinberg, R.A., 2008. Epithelial-mesenchymal transition: at the crossroads of development and tumor metastasis. Dev. Cell 14 (6), 818–829.

Yang, J., et al., 2014. FOXO1 3′UTR functions as a ceRNA in repressing the metastases of breast cancer cells via regulating miRNA activity. FEBS Lett. 588 (17), 3218–3224.

Ye, Z., et al., 2015. miR-30d blocked transforming growth factor β1-induced epithelial-mesenchymal transition by targeting snail in ovarian cancer cells. Int. J. Gynecol. Cancer 25 (9), 1574–1581.

Yilmaz, M., Christofori, G., 2010. Mechanisms of motility in metastasizing cells. Mol. Cancer Res. 8 (5), 629–642.

Yu, F., et al., 2007. let-7 regulates self renewal and tumorigenicity of breast cancer cells. Cell 131 (6), 1109–1123.

Yu, J., et al., 2010. MicroRNA, hsa-miR-200c, is an independent prognostic factor in pancreatic cancer and its upregulation inhibits pancreatic cancer invasion but increases cell proliferation. Mol. Cancer 9, 169.

Yun, J., et al., 2011. Signalling pathway for RKIP and Let-7 regulates and predicts metastatic breast cancer. EMBO J. 30 (21), 4500–4514.

Yuan, J., et al., 2014. A long noncoding RNA activated by TGF-β promotes the invasion-metastasis cascade in hepatocellular carcinoma. Cancer Cell 25 (5), 666–681.

Zhang, X.H.F., et al., 2009. Latent bone metastasis in breast cancer tied to Src-dependent survival signals. Cancer Cell 16 (1), 67–78.

Zhang, B., et al., 2010. Antimetastatic role of Smad4 signaling in colorectal cancer. Gastroenterology 138 (3), 969–980.

Zhang, J., et al., 2012. miR-21, miR-17 and miR-19a induced by phosphatase of regenerating liver-3 promote the proliferation and metastasis of colon cancer. Br. J. Cancer 107 (2), 352–359.

Zhang, Y., Wang, Z., Gemeinhart, R.A., 2013a. Progress in microRNA delivery. J. Control. Release 172 (3), 962–974.

Zhang, X., et al., 2013b. Regulation of the MicroRNA 200b (miRNA-200b) by transcriptional regulators PEA3 and ELK-1 protein affects expression of Pin1 protein to control anoikis. J. Biol. Chem. 288 (45), 32742–32752.

Zhang, L., et al., 2015a. Microenvironment-induced PTEN loss by exosomal microRNA primes brain metastasis outgrowth. Nature 527 (7576), 100–104.

Zhang, X., et al., 2015b. MicroRNA-26a promotes anoikis in human hepatocellular carcinoma cells by targeting alpha5 integrin. Oncotarget 6 (4), 2277–2289.

Zheng, F., et al., 2012. The putative tumour suppressor microRNA-124 modulates hepatocellular carcinoma cell aggressiveness by repressing ROCK2 and EZH2. Gut 61 (2), 278–289.

Zhou, W., et al., 2014. Cancer-Secreted miR-105 destroys vascular endothelial barriers to promote metastasis. Cancer Cell 25 (4), 501–515.

Zhou, X., et al., 2015. The interaction between MiR-141 and lncRNA-H19 in regulating cell proliferation and migration in gastric cancer. Cell. Physiol. Biochem. 36 (4), 1440–1452.

Zhu, S., et al., 2008. MicroRNA-21 targets tumor suppressor genes in invasion and metastasis. Cell Res. 18 (3), 350–359.

Zhuang, G., et al., 2012. Tumour-secreted miR-9 promotes endothelial cell migration and angiogenesis by activating the JAK-STAT pathway. EMBO J. 31 (17), 3513–3523.

Zoni, E., et al., 2015. miR-25 modulates invasiveness and dissemination of human prostate cancer cells via regulation of αv- and α6-integrin expression. Cancer Res. 75 (11), 2326–2336.

CHAPTER SIX

MicroRNAs in Oncogenesis and Tumor Suppression

Simona Barbato*, Giancarlo Solaini*, Muller Fabbri†,1

*Laboratory of Biochemistry and Mitochondrial Pathophysiology, University of Bologna, Bologna, Italy
†Children's Center for Cancer and Blood Diseases and The Saban Research Institute, Norris Comprehensive Cancer Center, Keck School of Medicine, University of Southern California, Children's Hospital Los Angeles, Los Angeles, CA, United States
[1]Corresponding author: e-mail address: mfabbri@chla.usc.edu

Contents

Abstract

MicroRNAs (MiRNAs) have emerged in the last 15 years as central players in the biology of cancer. Increasing lines of evidence have supported their regulatory role in the expression of both oncogenes and tumor-suppressor genes, progressively clarifying which genes are modulated by specific MiRNAs dysregulated in cancer. Intriguingly, a "target-specific" understanding of MiRNA function in oncology has been replaced by a more "pathway-specific" vision of their involvement in cancer biology. This work provides a state-of-the-art knowledge of the role of MiRNAs in the most frequently altered signaling pathways in cancer cells and provides an updated overview on some of the most relevant findings trying to decode the complex molecular mechanisms of cancer.

International Review of Cell and Molecular Biology, Volume 333
ISSN 1937-6448
http://dx.doi.org/10.1016/bs.ircmb.2017.05.001

1. INTRODUCTION

MicroRNAs (MiRNAs) are 19–24-nucleotide-long noncoding RNAs (ncRNAs) whose regulatory function was first discovered in the nematode *Caenorhabditis elegans* in 1993 (Lee et al., 1993). Specifically, it was discovered that the tiny ncRNA let-7 could bind to the 3′-untranslated region (3′-UTR) of the lin-4 RNA regulating its expression during the nematode development. Further studies clarified that such tiny ncRNAs (in the meantime defined as MiRNAs) are present also in vertebrates and their function was not limited to heterochronic regulation of developmental timing (Ambros and Lee, 2004; Lagos-Quintana et al., 2001; Lau et al., 2001; Lee and Ambros, 2001). The biogenesis of MiRNAs involves specific and well-regulated steps that are shared by the small interfering RNA (siRNA) and RNA interference machinery of the cell, and it has been extensively described in other excellent papers (Grishok et al., 2001; Hammond et al., 2000; Hutvagner et al., 2001; Ketting et al., 2001; Reinhart et al., 2002). The first involvement of MiRNAs in human cancer was provided by Dr. Croce's lab, which identified the MiR-15a/16-1 cluster of MiRNAs as frequently downregulated in chronic lymphocytic leukemia (CLL) (Calin et al., 2002). Moreover, Calin et al. showed that the loci of the genes encoding for MiRNAs are not randomly located in the human genome, but more frequently map in cancer-associated genomic regions that are regions of chromosomal instability typically deleted or amplified in cancer cells (Calin et al., 2004). Frank Slack's group identified the RAS oncogene as a direct target of let-7, providing the first mechanistic evidence that MiRNAs can act as tumor suppressors by silencing driving cancer promoting genes (Johnson et al., 2005). At the same time, Cimmino et al., identified BCL2 as a direct target of MiR-15a and MiR-16 and showed that restoration of these MiRNA expressions could promote apoptosis in leukemic cells (Cimmino et al., 2005). Since these initial discoveries, a plethora of studies have developed, leading to the identification of several MiRNAs dysregulated in all types of cancers (both solid and hematological malignancies) and classifying these MiRNAs as oncogenes (if they were silencing tumor-suppressor genes) or, vice versa, as tumor-suppressor MiRNAs. In the following years, it has become clearer that classifying MiRNAs purely as oncogenes or tumor suppressors is not only inaccurate but actually misleading. Indeed, it has been shown that one cluster of two MiRNAs (the MiR-15a/16-1 cluster) alone is able to regulate (directly or indirectly) the expression of about 14% of the total human genome (Calin et al., 2008). Obviously, such an impact does not exclusively involve pro- or antitumoral genes, but a mixture of them (Fabbri et al., 2007b). Therefore, it is now accepted that MiRNAs have hybrid targeting effects both on oncogenes and on tumor-suppressor genes, but the overall phenotype can be more pro- or antitumoral depending on their concentration, on the type of tumor, and sometimes even on the species in which their function is studied.

Several groups have clearly and convincingly shown that MiRNAs can be successfully used as cancer biomarkers and as therapeutics (Ling et al., 2013). The ultimate translational dimension of the MiRNA research has been recently achieved by introducing MiRNAs in clinical trials. The first trial (Clinical Trial #: NCT01200420) aimed at testing antagomiR-122 in the treatment of HCV patients (Jopling et al., 2005; Luna et al., 2015). This study began in Denmark in 2010 and has currently completed phase II. Currently, serum levels of MiR-122 are investigated as a real-time detection biomarker of drug-induced liver injury by chemotherapy in Clinical Trial # NCT03039062 that started recruiting patients in July 2016. One of the reasons for the rapid progression of this treatment in the clinical setting can be attributed to the specificity of MiR-122 expression in the liver and to the preclinical research clearly showing significant upregulation of MiR-122 in the liver of HCV-infected patients (Jopling, 2008). The first clinical trial in cancer patients started in April 2013 and studied the use of MiR-34a mimic (MRX34) in the treatment of patients with primary liver cancer or other selected solid tumors or hematologic malignancies (Clinical Trial # NCT01829971). This phase I study has been terminated in September 2016 due to five immune-related serious adverse events. Unfortunately, these concerning results led to withdraw Clinical Trial # NCT02862145 that aimed studying the pharmacodynamics of MRX34 liposomal injection in melanoma patients with biopsy accessible lesions. The failure of these clinical trials rose important questions on the specificity of MiR-34a targeting effects. Further preclinical studies addressing how the pattern of target genes of a given MiRNA is affected by the concentration of administered MiRNA are warranted. When dealing with MiRNAs as therapeutics, one of the most important and frequently asked questions is what is the benefit of using a MiRNA to silence let's say, a key oncogene, compared to the use of a siRNA specifically designed against that oncogene. Certainly, the degree of silencing achieved by the siRNA approach is higher than the one elicited by a MiRNA. However, the siRNA silencing is limited to the target gene it is designed against, whereas a single MiRNA can silence several different genes at the same time, despite reducing their expression by about 40%–50% at best. Interestingly, the lack of complete silencing achieved by MiRNAs can be compensated by the number and type of targeted genes: specifically, a single MiRNA could silence several genes involved in one common (and strictly related) signal transduction molecular pathway resulting in a profound effect in terms of pathway activation and pathway-dependent phenotype. This is probably the most important difference between a siRNA- and a MiRNA-based therapy. As we are learning from the clinical experience, the promiscuity of MiRNA target genes represents a double-edge sword, in the sense that it could also lead to unwanted side effects. In order to exploit the multiplicity of MiRNA targets to develop more effective therapeutics, it is imperative that a comprehensive and detailed analysis of the "targetome" of a given MiRNA (meaning the totality of the genome modulated by a MiRNA) is completed in the preclinical setting, to avoid undesired side effects. A first step in this direction is to focus on the impact of MiRNA regulation on signaling pathways instead of on a specific target gene. This is the approach that we decided to use in this work.

2. MiRNAs IN ONCOGENIC PATHWAYS

MiRNA dysregulation in cancer was first demonstrated in 2002 (Calin et al., 2002), opening a new branch of investigation over cancer-related issues. Since then, increasing studies have shown that MiRNAs play important roles in tumor growth, driving tumor cell reprogramming by regulating oncogenes and tumor-suppressor genes (Calin and Croce, 2006), and by modulating cancer metabolism (Cairns et al., 2011). As a result, cancer cells undergo uncontrolled cell division and escape cell death mechanisms, ensuring cells a high potential to spread and metastasize in other parts of the body. Albeit all genes are potentially affected by the resultant increased rate of mutation, alterations in cancer genes that directly control cell cycle and cell proliferation can affect net cell growth, thereby conferring a selective advantage to the mutant cell. In this sense, MiRNA-associated activation of driver oncogenes can result in rapid and sustained tumor progression and plenty of data in the literature already highlighted the tight connection existing between certain MiRNAs and some crucial oncogenes, but most of them mainly focused on a specific target or on a specific MiRNA. Therefore, in the first part of the review, we decided to offer a broader overview on MiRNA-driven regulation of Myc, RAS, NOTCH, BRAF, and the Sonic-Hedgehog (Shh) pathway, since these pathways have a well-established role in human carcinogenesis and a lot of efforts have been geared toward understanding the role of MiRNAs in these biological pathways.

2.1 Myc

Myc is an immediate early response gene downstream of many ligand–membrane receptor complexes (Armelin et al., 1984; Kelly et al., 1983) binding to 10%–15% of genomic loci in mammals. In this way, Myc governs fundamental cellular functions, including energy metabolism, cell cycle progression, and apoptosis.

Being at the crossroads of many growth promoting signal transduction pathways, Myc dysregulation has been linked to carcinogenesis and several studies showed that Myc activation was sufficient to provoke diverse types of cancers (Adams et al., 1985; Leder et al., 1986), including Burkitt's lymphoma, multiple myeloma, myeloid leukemia, and plasma cell lymphoma. Moreover, mutations, overexpression, rearrangement, and translocation of this gene have also been reported in breast, prostate, gastrointestinal cancers, and melanomas (Nesbit et al., 1999). Notably, Myc aberrant expression and dysfunctional activity have often been associated with tumor aggressiveness and poor clinical outcome (Gamberi et al., 1998; Kawate et al., 1999), underlying the pivotal role of this human proto-oncogene in tumor development and rendering Myc one of the most widely investigated human transcription factors (TFs). Indeed, several studies report that the transformed state of many cancer cells critically relies on Myc (Felsher and Bishop, 1999; Pelengaris et al., 1999). For instance, studies with animal models showed that the activation of MYC in lymphocyte precursor drives lymphomagenesis, whereas its inactivation was associated with a rapid, complete, and sustained tumor regression (Li et al., 2014b).

Myc family proteins (c-Myc, N-Myc, L-Myc, and B-Myc) are basic helix–loop–helix leucine zipper TFs, and their expression is tightly regulated at multiple levels, including transcription, translation, and protein stability. As posttranscriptional modulators of gene expression, MiRNAs play a crucial role in controlling cell fate and often cooperate with TFs to regulate the same set of protein-coding genes. Notably, recent studies have demonstrated that Myc exerts its oncogenic role by regulating the expression of a wide repertoire of MiRNAs that simultaneously modulate complex genetic networks, including various oncogenes and tumor-suppressor genes—thereby affecting cell death and proliferation—by inhibiting translation of their mRNAs (Gao et al., 2009; Ma et al., 2010; Polioudakis et al., 2013).

More specifically, further investigations singled out the MiR-17–92 as a Myc-related cluster of MiRNAs and the large amount of data reported in the literature show that the MiR-17–92 family is one of the most well-investigated clusters of MiRNAs, whose action involves both protumoral and tumor-suppressive functions. As for its oncogenic activity, a recurrent focal amplification of the cluster in diffuse large B-cell lymphomas was first pointed out in 2004 by Ota et al. (2004). Since this first evidence, a plethora of data have then highlighted that MiR-17–92 cluster is a direct transcriptional target of c-Myc (Dews et al., 2006; Li et al., 2014a; O'Donnell et al., 2005), and that MiR-17–92 levels are correlated with c-Myc expression levels in human lymphomas (He et al., 2005), chronic myeloid leukemia (Venturini et al., 2007), mantle cell lymphoma (Rinaldi et al., 2007), and adenocarcinoma (Diosdado et al., 2009), among others. Notably, N-Myc was also demonstrated to increase MiR-17–92 in neuroblastoma (Schulte et al., 2008; Fig. 1).

The MiR-17–92 cluster is located in a nonprotein-coding region on chromosome 13 (13q31.3), and its primary polycistronic MiRNA transcript (pri-miR-17–92, ~0.8 kb-long) contains six tandem-loop hairpin structures, ultimately generating the six mature MiRNAs: MiR-17 (MiR-17-5p and MiR-17-3p), MiR-18a, MiR-19a, MiR-19b, MiR-20a, and MiR-92a-1 (Ota et al., 2004). Acting independently and/or coordinately to target various mRNAs, the six members of the cluster control various cellular processes that generally promote oncogenic transformation by antagonizing senescence, escaping apoptosis, and evading tumor suppressing mechanisms (Dews et al., 2010; He et al., 2005; O'Donnell et al., 2005; Ventura et al., 2008). Indeed, this genomic locus is amplified in hematopoietic malignancies (Ota et al., 2004), and high levels of MiR-17–92 expression are reported in many types of cancer, including breast, lung, colon, pancreas prostate, thyroid, and lymphoma (Mogilyansky and Rigoutsos, 2013; Volinia et al., 2006). Indeed, consistent with the proliferative functions associated with c-Myc, MiR-17–92 has been associated with cell survival and proliferation by activating the E2F family, an important regulator of cell proliferation that coordinates the expression of genes in the G1/S transition (DeGregori and Johnson, 2006) and by inducing the cyclin-dependent kinase inhibitor (CDKN1A) or inhibiting BCL2L11/BIM and p21 (Frediani and Fabbri, 2016; Umezu et al., 2013). Furthermore, a series of studies have also shown that the MiR-17–92 cluster promotes angiogenesis by directly downregulating antiangiogenic factors

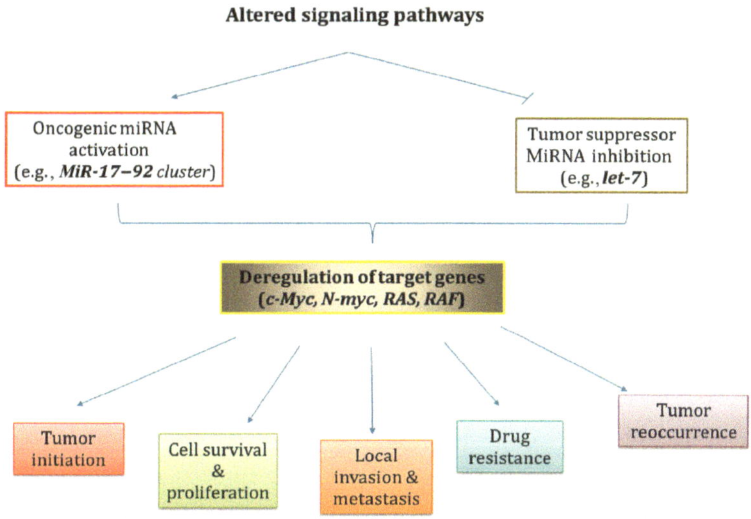

Fig. 1 Cancer formation and progression critically rely on MiRNAs aberrant expression. MiRNAs finely regulate fundamental cellular processes and have recently emerged as critical players in cancer formation and/or progression. Indeed, the alteration of many signaling pathways may induce the expression of MiRNAs (also referred to as oncogenic MiRNAs or OncomiRs) that actively promote cancer by enhancing the activity of driving oncogenes, such as Myc, RAS, and RAF, among others. Concomitantly, a reduced expression of tumor-suppressor MiRNAs (silencing these key oncogenes) may occur. As a result, an important interplay among oncogenic and tumor-suppressor MiRNAs is established. This, in turn, modulates the expression of different MiRNAs and of target genes that directly, or indirectly, drive cancer progression, leading to cell death escape, uncontrolled cell division, resistance to drug therapies, and the ability to disseminate in other parts of the body.

thrombospondin-1 and connective tissue growth factor—by MiR-18a and MiR-19, respectively (Dews et al., 2006)—as well as cell invasion, by inducing p-AKT (Lin et al., 2013). Additionally, it has been reported that MiR-17–92 downregulation induced apoptosis and reduced cell proliferation upon Myc silencing, while deletion of MiR-17–92 significantly impeded the ability of Myc to maintain a tumorigenic state (Li et al., 2014a).

Intriguingly, the role of Myc as "orchestrator" of multiple biological functions is reflected in the plurality of Myc-regulated MiRNAs. It has been reported that in addition to activating the MiR-17–92 cluster in human and mouse models of B-cell lymphoma, Myc also promotes downregulation of other MiRNAs including the let-7 family, MiR-22, and MiR-34a (Chang et al., 2008; O'Donnell et al., 2005). Moreover, recent studies showed that the biogenesis of these MiRNAs is mediated by Lin-28 that directly binds to let-7 stem loops, thereby inhibiting both Drosha- and Dicer-mediated cleavage and degradation of let-7 precursors in mammalian cells.

Lin-28 is an RNA-binding protein that coordinates both proliferative growth and metabolism, playing an important role in tumorigenesis (Thornton et al., 2012) by regulating the let-7 family, essential for promoting cellular differentiation and for tumor suppression (Esquela-Kerscher and Slack, 2006). Despite its early discover in the early 1980s (Ambros and Horvitz, 1984; Horvitz and Sulston, 1980; Sulston and Horvitz, 1981), the functional role of Lin-28 in mammals has only recently been elucidated. Particularly, it has been suggested that increased levels of Lin-28 in cancer could limit the amount of mature let-7, preventing its action as an effective tumor suppressor. By modulating the expression of various regulators of cell cycle, including Myc, Ras, and different types of cyclins, among others, Lin-28 overexpression enables unchecked cellular growth and promotes a less-differentiated state. Incidentally, multiple studies in hepatocellular and nasopharyngeal carcinoma cells revealed that the inhibition of Myc inhibited proliferation, elucidating the close interaction between Lin-28/let-7 and Myc for tumor progression (Lan et al., 2011; Wong et al., 2011). Posttranscriptional regulation of MiRNAs by Lin-28 is currently specific to the let-7 family and MiR-128; however, this role of Lin-28 as a posttranscriptional regulator of the let-7 family needs to be further investigated.

More recent studies showed that Myc regulates MiR-200c, a member of the MiR-200 family that has been positively associated with human malignancies (Iorio et al., 2007). A crucial role of MiR-200c in Myc-mediated cellular transformation, partially due to the inhibition of the tumor suppressor PTEN and the subsequent activation of the AKT signaling pathway was also reported (Chen et al., 2016).

In support to these data, the downregulation of Myc-suppressed MiRNAs, including let-7, MiR-34a-c, MiR-15/16, and MiR-29, has been reported and confirmed in several primary tumors. These findings, together with the upregulation of Myc-induced oncogenic MiRNAs (such as the MiR-17–92 cluster), underline the central role of Myc in orchestrating pro- and antitumoral functions by modulating specific MiRNAs.

2.2 RAS

The RAS–mitogen-activated protein kinase (MAPK) signaling pathway integrates extracellular signals and coordinates a complex network of numerous downstream effector pathways, such as the well-characterized

RAF–MEK–ERK pathway and phosphoinositide 3-kinase (PI3K)–AKT–mTOR kinase cascade, to regulate many fundamental cellular processes, including cell differentiation, survival, and proliferation.

Specifically, the human genome harbors three different RAS genes (HRAS, KRAS, and NRAS, located on chromosomes 11, 12, and 1, respectively), encoding for a family of membrane-bound small GTPases, all sharing an overall 82%–90% amino acid sequence identity: Harvey-Ras (HRAS), neuroblastoma-Ras (NRAS), and the most studied, Kirsten-Ras (KRAS).

Normally, RAS proteins are positively regulated by guanine nucleotide exchange factors (GEFs) to exchange GDP for GTP, thereby promoting the active state of the pathway, whereas the GTPase-activating proteins (GAPs) catalyze the hydrolysis of GTP back to GDP to switch the signaling off. Nevertheless, oncogenic mutations of these proteins typically result in a constitutive active signaling cascade, rendering RAS proteins relatively independent from the GEFs and/or insensitive to RAS modulation by the GAPs, thereby promoting carcinogenesis and proliferation via its many downstream effectors.

Indeed, RAS aberrant activation was first implicated in the development of colorectal cancer described by Vogelstein et al. (1988) and is a major and highly prevalent oncogenic event in many human cancers, taking place in approximately 30% of all tumor types. Moreover, RAS–MAPK signaling pathway is altered in T-cell precursor acute lymphoblastic leukemia and lung adenocarcinomas (Knight and Irving, 2014) and activating mutations in NRAS are a genetic hallmark of non-BRAF-related melanomas, while KRAS somatic mutations were identified in more than 90% of pancreatic ductal adenocarcinoma (PDAC) (Tuveson and Neoptolemos, 2012).

Notably, mutations in these genes can actively modify the energy metabolism of tumor cells, since KRAS mutations have been consistently correlated with an enhanced expression of the glucose transporter GLUT1 and increased glycolytic rate (Yun et al., 2009). Studies in T fibroblasts showed that KRAS mutation resulted in a dramatic loss of activity and content of respiratory complex I (NADH-CoQ oxidoreductase), with a consequent decrease of both NAD-linked respiration and ATP synthesis rate (Baracca et al., 2010), and this phenotype appears to be sufficient to explain the impairment of mitochondrial function and the Warburg effect in transformed cells.

Being KRAS mutation a critical event in the onset of different malignant phenotypes, RAS activity must be optimally regulated to induce and maintain the tumorigenic state and its expression is differentially regulated by amplification, mRNA stability, and altered translation. The observation that RAS is directly regulated by MiRNAs added a new facet to the regulation of RAS. Among MiRNAs, let-7 has been demonstrated to be a direct regulator of KRAS expression in human cells: as Johnson and collaborators showed in 2005, all three human RAS oncogenes have let-7 complementary sites in their 3′-UTRs (Johnson et al., 2005), enabling this MiRNA to regulate RAS expression. let-7 is a heterochronic switch gene that causes reiterations of larval fates in the adult when lost and the premature expression of adult fates during larval stages when overexpressed (Reinhart et al., 2000).

The expression of let-7 members is significantly reduced and inversely correlated to that of KRAS in human cancers and in cancer stem cells (CSCs) (Jin et al., 2016; Schwarzenbacher et al., 2013). Reduced expression of let-7 has been often associated with poor prognosis in lung cancer patients (Takamizawa et al., 2004), suggesting let-7 to be a bona fide tumor-suppressor gene, at least in lung carcinogenesis. In this sense, targeting let-7 and the above-mentioned Lin-28 (that regulates let-7 biogenesis) might offer viable approaches in the treatment of cancers with activated RAS signaling cascade.

Although various strategies have been proposed in the treatment of RAS mutation-related nonsmall cell lung cancer (NSCLC), surgical resection still remains the most effective approach. Nevertheless, late-stage diagnosis and metastatic lesions, along with drug resistance, still remain the main obstacles to be overcome. Remarkably, let-7 and MiR-34 are often the most dysregulated MiRNAs in NSCLC (Gallardo et al., 2009; Johnson et al., 2005; Lodygin et al., 2008; Takamizawa et al., 2004), controlling the expression of multiple target oncogenes, and therefore rendering them attractive therapeutic agents. As such, various studies showed that exogenous let-7 can both prevent and treat KRAS-driven lung tumors (Esquela-Kerscher et al., 2008; Kumar et al., 2008; Trang et al., 2010). Additionally, it has been shown that MiR-34 prevents the initiation and progression of $KRas^{G12D+/}$;$p53^{R172H/+}$ lung tumors in human NSCLC tumor xenografts (Kasinski and Slack, 2012; Trang et al., 2011), suggesting a central role of MiR-34 in KRAS-dependent NSCLC, despite a direct targeting of KRAS by MiR-34 has not been established. Likewise, the combinatorial therapeutic approach of let-7 and MiR-34 MiRNAs proposed by Kasinski and collaborators resulted in an even wider repression of key oncogenes, enhancing the individual effect of each individual tumor-suppressor MiRNA (Kasinski et al., 2015). Notably, recent data show that the inactivation of both Myc and KRAS inhibits carcinogenesis of lung cancer/lymphomas. Remarkably, since the expression of both Myc and KRAS is suppressed by let-7, it has been suggested that reintroducing let-7 or its mimics may be a useful approach in the treatment of malignancies carrying mutations of Myc and KRAS.

Additionally, several epidemiological cancer and nutrition studies highlighted the connection between the risk of cancer development or progression and a specific dietary pattern. For this reason, an increasing attention has been recently given to diet and nutrients to provide new therapeutic perspectives and ameliorate the effect of current therapies. Notably, it has been reported that a combinatorial subministration of dietary agents including sulforaphane, quercetin, and a mixture of green tea catechins can mediate the induction of let-7a expression, which in turn inhibits KRAS expression and CSC features in pancreatic cancer (Appari et al., 2014), showing the broad range and widespread effect of MiRNAs in mediating the transformed and tumorigenic state.

RAS expression is also regulated by a number of other MiRNAs. Among these, MiR-181 was demonstrated to directly target KRAS, and it was found to be downregulated in oral squamous cell carcinomas (Shin et al., 2011), whereas MiR-126 was shown to be strongly downregulated in PDAC (Jiao et al., 2012). Similarly, other members of the MiR-181 family have also been described to directly target KRAS mRNA, including MiR-181c in gastric carcinoma (Hashimoto et al., 2010) and MiR-181d in human gliomas (Wang et al., 2012).

Reports also suggest that MiR-193a-mediated targeting of KRAS and plasminogen activator urokinase factors can arrest growth and invasiveness of human colon adenocarcinomas, in which both these factors are highly expressed (Deng et al., 2014). Additionally, MiR-134 was shown to directly target the KRAS. Its overexpression was linked to inhibited migration and invasiveness and suppressed proliferation in renal cell carcinoma (Liu et al., 2015). In addition to KRAS, MiR-134 can also target DHHC9 mRNA, which is involved in posttranslational modifications required for HRAS activity (Chai et al., 2013). Therefore, cancer-related downregulation of MiR-134 could impact the RAS–MAK pathway activation, through dual KRAS and HRAS targeting.

2.3 Notch

The Notch signaling pathway is a highly conserved cell signaling system that promotes proliferation during neurogenesis, maintains tissue homeostasis, and plays crucial roles in carcinogenesis.

Notch and most of its ligands are transmembrane proteins and four different single-pass transmembrane Notch receptor paralogues (Notch1–4) have been described in mammals (Andersson and Lendahl, 2014; Yin et al., 2010). When the ligands engage the Notch receptors, these are cleaved to release active intracellular Notch domains that are consequently internalized into the cell nucleus to activate the expression of Notch signaling target genes. Because of its transmembrane nature, the receptor is normally triggered via direct cell-to-cell contact, whereas the activity of this pathway is blocked by Numb. However, in case of Notch receptor-activating mutations, an accumulation of Notch intracellular domain (NICD) occurs into the nucleus, as reported in adult T-cell acute lymphoblastic leukemia (T-ALL) and lymphoma (Gu et al., 2016).

An intronic MiRNA encoded on chromosome 8q (a frequent altered region in cancers), MiR-151, was confirmed to be a potent inducer of the Notch1 pathway and a p53 suppressor (Hsu et al., 2016). According to this recent study on gastric carcinogenesis, Notch1 pathway mediates tumor progression through MiR-151-5p, FAK, and p53 (Hsu et al., 2016), which may serve as potential diagnostic markers and therapeutic targets of gastric cancer (GC). Possibly, and as suggested by this study, the subministration of MiR-151-5p antagonists into cancer cells could reduce the GC progression by restoring the tumor-suppressor activity of p53.

Another MiRNA, MiR-34a inhibits the Notch signaling by directly targeting Notch receptors, as demonstrated by de Sousa et al. (2011). Indeed, MiR-34a-lowered expression enhanced Notch signaling cascade, eventually preserving CSC formation (Nguyen et al., 2012). Conversely, when MiR-34a was overexpressed, the Notch signaling cascade was impaired.

In addition, it has been reported that the Notch signaling pathway plays a crucial role in the regulation of MiR-17–92 levels. Besides the well-established (and already described) induction of the MiR-17–92 cluster by Myc, studies in papillary thyroid cancer demonstrated that Notch1 knockdown reduced MiR-19 levels (Fuziwara and Kimura, 2014), and that the activation of NICD induced substantial MiR-17–92 expression in normal thyroid follicular cells. Furthermore, an interplay between MiR-19 and Notch1 activation was also observed in T-ALL (Mavrakis et al., 2010), where the role of the oncogenic Myc is fundamental, rendering the Notch1-Myc regulatory circuit an emerging attractive therapeutic target for the treatment of T-ALL (Sanchez-Martin and Ferrando, 2017).

2.4 BRAF

In metazoans, the major effectors of RAS are RAF kinases, regulating the mitogenic cascade, one of the MAP kinase cascades. The RAF family members include ARAF, BRAF, and CRAF, all three sharing two conserved regions, CR1 and CR2, at the N-terminus, whereas a third conserved region, CR3, encodes the kinase domain at the C-terminus. Among these, the orthologue of DRAF—identified in *Drosophila melanogaster*—BRAF is the most studied member of the RAF family. Specifically, research over the last decade showed that multiple genetic alterations of BRAF, ultimately involved in cell growth, proliferation, and survival, are a genetic hallmark of different types of tumors and play a role during the progression of the disease. Specifically, mutations in BRAF isoforms have been described in approximately 30% of all human cancers and have therefore become a target of interest for molecular therapy, with promising results emerging from clinical trials.

Mutations in BRAF commonly occur in the same cancer types that harbor RAS mutations, in a mutually exclusive fashion. Like RAS, BRAF promotes carcinogenesis through a constitutive activation of the MAP kinase pathway. As a result, these mutations can cause unintended and overactive signaling inside the cell (e.g., in the absence of incoming signals), leading to increased and uncontrolled cell proliferation and resistance to apoptosis (Wellbrock and Hurlstone, 2010). Specifically, one of the most common mutations of BRAF in human carcinogenesis is the point mutation BRAF (V600E), present in more than 50% of metastatic melanomas (Davies et al., 2002), an average of 45% of papillary thyroid cancers and 10% of colorectal cancers, involving a substitution of a glutamate with a valine in position 600, that causes a constitutively active protein serine kinase. This mutation is reflected in the constantly activated BRAF → MEK1/2 → ERK1/2 MAP kinase pathway, critically implicated in the regulation of gene expression, cell proliferation, and survival, all involved in the initiation and progression of the disease (Smalley, 2003).

Gene expression profiling studies revealed that MiRNAs are involved in BRAF-mediated effects in human carcinogenesis. In thyroid cancer, the induction of the BRAF oncogene induces the overexpression of members of the MiR-17–92 cluster (Fuziwara and Kimura, 2015). In addition, data in the literature report the upregulation of MiR-146b, MiR-221, MiR-222, MiR-181b, and MiR-21 as a characteristic feature of papillary thyroid carcinoma, whereas MiR-31-5p is reported to be the most upregulated

MiRNA in CRCs with mutated BRAF (V600E) (Choi et al., 2016; Nosho et al., 2014).

In a comparative study of BRAF-mutated vs KRAS-mutated CRCs patients, a list of 10 MiRNAs was found to be significantly dysregulated in BRAF-mutated cancers. Among these, MiR-31-5p, MiR-877-5p, and MiR-1237-5p were significantly upregulated, whereas the expression of let-7f-1-3p, MiR-363-5p, MiR-425-3p, MiR-500a-5p, MiR-532-5p, MiR-660-5p, and MiR-1825 was significantly diminished (Choi et al., 2016).

Moreover, MiR-34a expression is downregulated in many tumor types by promoter methylation and this MiRNA is regarded as a tumor suppressor, regulating hundreds of transcripts, including ARAF expression and other RAS–MAPK pathway members (Masliah-Planchon et al., 2016).

2.5 Sonic Hedgehog

Normal development during embryogenesis requires a number of complex signaling cascades, including the Hedgehog (Hh) pathway, a key mediator of many fundamental processes in vertebrate embryonic development that controls cell fate, patterning, proliferation, survival, and differentiation of many different regions (Fig. 2).

First described in 1980 by Nusslein-Volhard and Wieschaus (1980), in their genetic analysis of the fruit fly *D. melanogaster*, the name Hh originates from the short and "spiked" phenotype of the cuticle of the Hh mutant *Drosophila* larvae, which resembled the spikes of a hedgehog (Varjosalo and Taipale, 2008).

In vertebrates, the canonical Hh pathway contains several key components, including HH glycoproteins *Desert Hedgehog*, *Indian Hedgehog*, and Shh, three Hh gene homologues discovered in the early 1990s (Echelard et al., 1993; Krauss et al., 1993; Roelink et al., 1994). Among these, Shh is the most potent of these ligands and the most widely expressed mammalian Hh signaling molecule, inducing cell proliferation in a tissue-specific manner.

In humans, the Hh cascade in the target cell is triggered by the Hh ligand binding to the Patched (Ptch) receptor, a 12-span transmembrane protein that exists as two isoforms in vertebrates, Ptc1 and Ptc2. In the absence of the ligand, Ptch inhibits downstream signaling by catalytically inactivation of the seven-transmembrane-span-receptor-like protein, Smoothened (Smo) (Taipale et al., 2002), affecting its localization to the cell surface, or by eliciting endogenous Smo agonists to prevent its binding. Conversely, upon secretion of Shh glycoproteins, Ptch is inactivated and Smo is

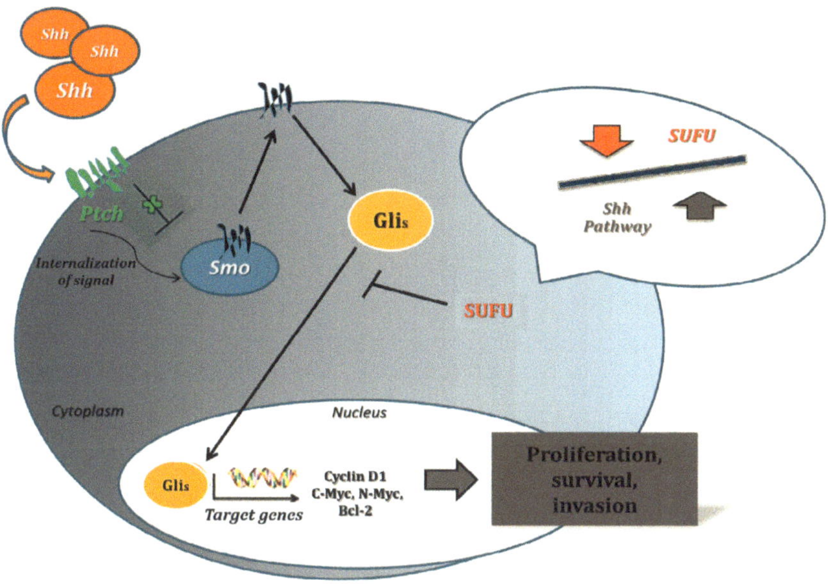

Fig. 2 The Sonic Hedgehog signaling pathway. Upon Shh ligand binding to its receptor Patched (Ptch), the repression of Smoothened (Smo) is relieved, resulting in the movement of Smo from the intracellular vesicles to the primary cilium. Activated Smo induces the Gli proteins, which enter the nucleus and promote transcription of the target genes, enhancing proliferation. This process is negatively regulated by the internal protein SUFU: in the absence of ligand, SUFU directly binds to GLI transcription factors and anchors them in the cytoplasm, thus facilitating their degradation (in addition, SUFU forms a repressor complex suppressing GLI1-induced gene expression inside the nucleus).

activated, resulting in the transduction of the Hh signaling to the cytoplasm and the nuclear localization of the terminal effectors of the Shh signaling, the glioma-associated (GLI) transcription factors. Once in the nucleus, GLI transcription factors can activate target genes involved in Hh pathway feedback (e.g., GLI1, Ptc1), proliferation (e.g., cyclin D1, MYC), apoptosis (e.g., Bcl-2), angiogenesis (e.g., ANG1/2), epithelial-to-mesenchymal transition (e.g., SNAIL), and stem cell self-renewal (e.g., NANOG, SOX2) (Hui and Angers, 2011; Scales and de Sauvage, 2009; Stecca and Ruiz, 2010).

Hh signaling is crucial in development, as evinced by the dramatic consequences in human fetuses with inherited or sporadic mutations in the Shh signaling pathway, resulting in a number of human birth defects, like holoprosencephaly, microencephaly, cyclopia, absent nose, or cleft palate (Belloni et al., 1996; Roessler et al., 1996). Moreover, abnormal activation of the Shh pathway has been recently shown to affect carcinogenesis, apoptosis, migration, and angiogenesis of various cancers that usually originate from endoderm, enclosing gastric, pancreatic, colon, lung, prostate, and hepatocellular cancer (Lum and Beachy, 2004).

Specifically, it has been proposed that a ligand-independent Hh pathway-activating mutation was responsible for aberrant Shh pathway activity, leading to the development of basal cell carcinomas and medulloblastomas (Kimonis et al., 1997). Furthermore, Hh might also promote carcinogenesis or tumor growth acceleration in a wide variety of tissues, by secreting Shh ligands from the tumor to the surrounding stroma, thereby stimulating the production of secondary growth factors that drive tumor angiogenesis, tumor cell proliferation, and survival (Yauch et al., 2008) in multiple tumor types, including lung (Watkins et al., 2003), pancreatic, colorectal (Varnat et al., 2009), prostate (Karhadkar et al., 2004) metastatic carcinomas, and gliomas (Carpenter et al., 2015), or in CSCs.

Despite the growing evidence that MiRNAs are crucial posttranscriptional regulators of gene expression and their implication in tumor formation, little is still known on the mechanism by which MiRNAs target specific developmental pathways, including the Shh pathway. In an attempt to investigate whether the expression of MiRNAs correlates with the functional status of Hh signaling pathway in glioblastomas (GBMs), Wentao and collaborators compared the expression of different MiRNAs in GLI1-high- or low-expressing GBMs, and this comparison revealed a differential expression of 17 MiRNAs among the two groups, including MiR-125b-1 and MiR-144 (Gu et al., 2014). Since the expression of these MiRNAs (MiR-125b-1 and MiR-144) correlated with different levels of Shh signaling in human GBMs, it was hypothesized that these MiRNAs may be responsible for Shh regulation, and therefore had an effect on cell proliferation (Gu et al., 2014).

The oncogene N-MYC is a target of the activation of the Shh/Ptch signaling cascade in medulloblastoma (Northcott et al., 2009) and regulates the expression of the MiR-17–92 cluster, which has been found frequently overexpressed in this type of tumor, especially when associated with a constitutive active Shh pathway. Among the MiRNA cluster members, MiR-19 is the highest expressed in Shh-dependent medulloblastoma, and the intravenous administration of MiR-17 and MiR-19 antagonists diminished tumor cell proliferation in vitro and inhibited xenograft tumor growth in vivo, as demonstrated by Murphy BL, who reported a prolonged survival of mice upon MiR-17–92 inhibition to treat Shh medulloblastomas (Murphy et al., 2013). A role for MiR-17–92 in medulloblastoma was further confirmed by Uziel and colleagues through in vivo studies in Ink4c-homozygous-deleted-mice and in Ptch1-heterozygous-deleted-mice (Uziel et al., 2009).

Shh has been reported to play a critical role in the control of chondrogenic differentiation (Tavella et al., 2004), and the physiological role of MiRNAs in the maintenance of joint homeostasis and degeneration in osteoarthrosis (OA) is an active area of investigation. Recent findings on this topic revealed a reduced expression of MiR-602 and MiR-608 (both targeting the Shh pathway) in damaged OA cartilage compared to smooth cartilage, whereas their overexpression inhibited the SHH pathway and protein expression in human OA chondrocytes (Akhtar et al., 2015).

3. MiRNAs IN TUMOR-SUPPRESSOR PATHWAYS

MiRNAs silencing oncogenes are considered tumor-suppressor MiRNAs, and the role of epigenetic repression of tumor-suppressor genes in cancer is well documented in the literature. Indeed, the number of human genes repressed by the binding of a MiRNA to the 3′-UTR of their mRNAs has been growing rapidly (Calin and Croce, 2006; Chen and Lodish, 2005), and large attention has been given to the involvement of MiRNAs in mediating the reduced expression of tumor-suppressor genes. Therefore, in this second part of the review, we will summarize the main studies on the role of MiRNAs in the regulation of the TP53, PTEN, TGFβ, and Wnt signaling pathways.

As discussed earlier, however, it is important to underline that MiRNAs have a wide pleiotropic activity and that the same types of MiRNAs can affect both pro- and antitumoral genes. Additionally, recent discoveries have also shed light on the spatiotemporal plasticity of MiRNAs, that may shift from oncogenic to tumor-suppressive activity during cancer progression, as discussed by Bonaldi and Mihailovich on the dual role of the above-mentioned MiR-17–92 cluster (Bonaldi and Mihailovich, 2016; Fig. 3).

3.1 TP53

TP53 has been one of the most investigated genes in cancer in the last decades, and the importance of its encoded product p53 in tumor suppression is unequivocal. As many data in the literature report, p53 is a cellular stress sensor involved in cell cycle regulation that acts to negatively regulate cell division, by controlling a set of genes required for this process. Induced by diverse stimuli, including DNA damage, hypoxia, oxidative stress, hyperproliferative signals, and nutrient starvation, p53 promotes growth arrest,

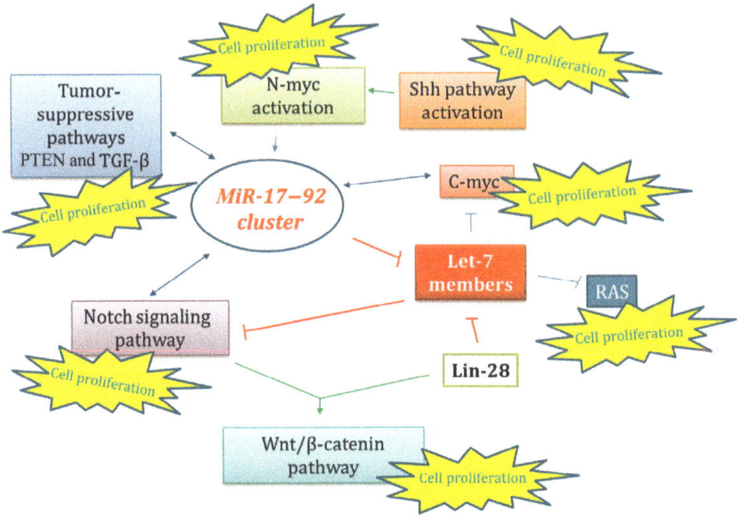

Fig. 3 The complex network activity of some of the discussed MiRNAs. MiRNAs have been widely connected to malignant processes and it is largely known that they exert a pleiotropic activity inside the cells. As an example, the oncogenic MiR-17–92 cluster of MiRNAs, which is upregulated in most types of cancer, is directly involved to modulate the expression of various signaling pathways and target genes, ultimately resulting in the uncontrolled proliferation of cancer cells.

senescence, and/or apoptosis, to ensure an appropriate and controlled cellular growth.

Mutations in TP53, either sporadic or inherited, are one of the most frequent and recurrent events during cancer formation, ultimately resulting in a partial or complete p53 functional deficiency in most types of cancer. Most of the TP53 mutations in human tumors localize in the DNA-binding domain of the protein, suggesting that p53 mainly acts as a TF to induce transient G1 cell cycle arrest in response to DNA-damage signals, allowing cells to repair their genomes before proceeding through the cell cycle, thereby acting as a "guardian of the genome" to limit cell propagation of potential oncogenic mutation (el-Deiry et al., 1993; Lane, 1992). Depending on the tumor type, this event can enhance the initiation or progression of cancer, and tumor-lacking p53 is commonly characterized by a more aggressive and malignant phenotype, with increased invasiveness and metastatic potential (Malkin et al., 1990; Song et al., 2007).

Moreover, new functions of p53 as a tumor suppressor have emerged in recent years, including p53-mediated regulation of cellular metabolism, stem cell functions, invasiveness and metastasis, as well as cell–cell communication within the tumor microenvironment (Vousden and Prives, 2009).

Notably, Leung and Sharp proposed that MiRNAs have presumably evolved to allow organisms to respond to various stimuli and overcome stressing conditions (Leung and Sharp, 2007, 2010). According to this idea, TP53 is one of the major downstream targets of MiRNAs and the p53 stress-response pathway is largely interconnected with MiRNAs, since a number of MiRNAs contribute to tumor suppression by mediating the down-regulation of many mRNAs and protein expression occurring after p53 activation. Consistently, Tavakoli and coworkers reported a significant decrease in tumor-suppressor transcripts, upon transduction of the above-mentioned MiR-17–92 cluster of MiRNAs (which has abundantly been connected with an increased cell proliferation and suppression of apoptosis, as previously discussed), with a prominent effect on p53 (Tavakoli et al., 2017).

To identify MiRNAs potentially regulated by p53, Suzuki and collaborators measured the expression levels of a selected group of MiRNAs after treatment with doxorubicin, a potent DNA-damaging agent and p53 inducer, revealing an increase in MiR-34a, MiR-15a, MiR-16-1, MiR-23a, MiR-26a, MiR-103, MiR-143, MiR-145, MiR-203, and MiR-206 expression (Suzuki et al., 2009). Among these, the MiR-34 family (composed of MiR-34a, -34b, and -34c) has been the first group of MiRNAs directly transactivated by p53 (Chang et al., 2007; He et al., 2007; Tarasov et al., 2007), with the most pronounced p53-regulatory activity exerted for MiR-34a.

Considering that p53 loss of function is a typical event occurring in cancer, the link between p53 and MiR-34 family members is noteworthy and deserves more attention and further explanation. As reported, p53 stimulates the transcription of MiR-34 family members, eventually resulting in the induction of apoptosis and/or senescence. Conversely, p53 downregulation correlates with reduced levels of MiR-34 levels in most p53-mutated ovarian cancer patients, as reported by Corney et al. (2007). Furthermore, different studies reported that the ectopic expression of MiR-34a and MiR-34b/c resulted in multiple tumor-suppressive activities (Hermeking, 2007), including cells arrest in G1 phase, signs of premature senescence (He et al., 2007; Lodygin et al., 2008), apoptosis (Chang et al., 2007; Tarasov et al., 2007), and the epithelial–mesenchymal transition (EMT). The latter event occurs through the induction of the TF SNAIL (also known as SNAI1) (Kim et al., 2011), that can itself modulate the expression of these MiRNAs, forming a double-negative feedback loop (Brabletz, 2012). Concomitantly, the stem cell markers CD133, CD44, and BMI1 were found to be downregulated, upon the ectopic expression of MiR-34a (Siemens et al., 2011). Moreover, numerous cell cycle regulators involved in the p53-controlled transition between G1 and S phases have also been validated to be direct targets of MiR-34, such as cyclin-dependent kinase 4 (CDK4), CDK6, cyclin E2, E2F3, Myc, and N-Myc (Zhu et al., 2011), being CDK4 and CDK6 and both Myc and N-Myc also regulated by MiR-145, as reported by Sachdeva et al. (2009).

Besides regulating the expression of MiRNA-coding genes, p53 may also affect the processing of MiRNAs, as p53 interacts with the Drosha-processing complex through the association with DEAD-box RNA helicase p68 (also known as DDX5), whereas transcriptionally inactive p53 mutants interfere with the functional assembly between Drosha complex and DDX5, thereby hampering the maturation of primary MiRNAs to precursor MiRNAs, as reported for MiR-16-1, MiR-143, and MiR-145, that exert a growth-suppressive function in response to DNA damage (Suzuki et al., 2009).

A p53 role in mediating and checking the proper processing of MiRNAs was further confirmed in DICER1-deficient cells, that displayed an enhanced expression of p53 and ARF tumor suppressors, followed by reduced proliferation and induced premature senescence in cells with an incomplete maturation of MiRNAs, that was prevented by suppressing ARF and p53 formation, further confirming that p53 plays an important role to ensure a proper processing of MiRNAs inside the cells (Mudhasani et al., 2008). Additionally, DICER1 was shown to be a direct target of MiR-107 that enhances migration and promotes metastatic dissemination upon its ectopic expression, whereas its inhibition can reverse migration and metastatic dissemination of cancer cells (Martello et al., 2010).

As discussed earlier, the expression of MiR-15a and MiR-16-1 is induced by p53 activation and the deletion/downregulation of these MiRNAs has been detected in several tumor types, including NSCLC, ovarian, and prostate cancer (Bandi and Vassella, 2011). In CLL, Fabbri et al. identified a feedback loop in which TP53 (located on chromosome 17) transactivates the MiR-15a/16-1 cluster (mapping on chromosome 13) and it is targeted by these two MiRNAs, explaining, at least in part, the prognostic implications of recurrent chromosomal aberrations in human CLL (Fabbri et al., 2011). Novel targets of these MiRNAs include important regulators of the cell cycle and cell proliferation, such as KRAS (as a target of MiR-143), CDK6 (as a target of MiR-16, MiR-26a, MiR-107, MiR-145, and MiR-206), and cyclin E, presumably playing an important role in p53-mediated cell cycle arrest in G1/S phase (Suzuki et al., 2009). Conversely, G2/M phase transition MiR-192 appears to be mediated by other p53-

induced MiRNAs, like MiR-192 and MiR-215, targeting CDC7 and MAD2 transition factors (Georges et al., 2008). Very recently, Van Roosbroeck provided the first evidence (corroborated on a huge cohort of primary samples) that MiR-155, a MiRNA frequently upregulated in several types of human cancers (Volinia et al., 2006), directly targets TP53 and showed that silencing of MiR-155 elicits an antitumoral effect both in vitro and in xenograft murine models (Van Roosbroeck et al., 2016).

Thus far, p53 may be considered a central regulator of tumor suppression, exerting its crucial role by interacting with a large variety of cancer-related genes, proteins, and MiRNAs. Nevertheless, the exact mechanisms underlying p53 loss of functions as well as the p53-induced MiRNA modulation in cancer still remain to be completely elucidated. Noteworthy, newer approaches and techniques involving the combined deletion and modulation of different MiRNAs may be useful to better understand their relevance to p53 function.

3.2 PTEN

Phosphatase and tensin homologue deleted on chromosome ten (PTEN) is a general controlling factor in cell proliferation and neuronal growth during development with a phosphatase activity against phosphatidylinositol (3,4,5)-trisphosphate that is metabolized into PI(4,5)P_2 (Worby and Dixon, 2014). Alterations of these direct actions of PTEN at synapses may lead to synaptic dysfunction with behavioral and cognitive consequences and may even participate in neurodegenerative processes and cancer.

Mutations in the evolutionarily conserved phosphatase domain of PTEN, consisting of a central five-stranded β-sheets included within two and four α-helices at the terminal sides, are often associated with cancer. Indeed, the aberrant activation of the PI3K signaling network derived from PTEN loss of function is a common event in the development of tumor, including prostate and breast cancer (Torbett et al., 2008; Wee et al., 2008), and PTEN is also reported to be involved in the self-renewing activity of normal and leukemic hematopoietic stem cells (Yilmaz et al., 2006; Zhang et al., 2006). Additionally, further confirmations of its tumor-suppressive role derived, more than a decade ago, from animal studies in PTEN-knockout mice and yet the mechanistic basis for this link still need to be completely elucidated (Di Cristofano et al., 1998; Podsypanina et al., 1999).

PTEN is constitutively expressed in normal tissues, but epigenetic silencing of PTEN is a common feature in cancer: for this reason, large attention is now given to those MiRNAs that may mediate the activation of signaling through PTEN and the PI3K network. As an example, PTEN was among the first validated targets of the MiR-17–92 cluster (Sylvestre et al., 2007; Ventura et al., 2008), and recent findings suggest that MiR-148a directly binds the PTEN mRNA 3′-UTR sequences and that PTEN expression can be alternatively suppressed or enhanced upon MiR-148a overexpression or inhibition, respectively (Zhang et al., 2016).

Intriguingly, numerous other MiRNAs have been demonstrated to influence PTEN. As reported, increased expression of MiR-221/MiR-222 can actually suppress tumor/metastasis-suppressor proteins (such as PTEN and TIMP3) in human NSCLC and hepatocellular carcinoma (Garofalo et al., 2009), whereas the reduced expression of the same MiRNAs correlates with the inhibition of carcinogenesis, migration, and invasiveness through upregulation of PTEN, along with the increased expression of CDKN1C and TIMP3 (Boominathan, 2010).

Noteworthy, some studies highlight the influence of Myc in oncogenic transformation, by reprogramming the expression of different MiRNAs that directly target PTEN, casting a new light in the Myc–PTEN interactions as mediated by MiRNAs. For example, it has been reported that the induction of MiR-17–92 cluster of MiRNAs reduces the expression of PTEN, as well as it occurs upon the induction of MiR-19 (Mu et al., 2009) or MiR-200c (Chen et al., 2016). Coherently, high levels of Myc bind to tumor-suppressor MiRNAs such as let-7, MiR-29, MiR-15/16a, MiR-34, MiR-26, MiR-30, MiR-150, MiR-146a, and MiR-22, thereby suppressing their expression (Chang et al., 2008). This, in turn, may promote tumorigenesis, EMT, angiogenesis, invasiveness, and CSC proliferation (Dews et al., 2006; Mendell, 2008).

PTEN expression has also been shown to be regulated by p53, which is a transcriptional activator of PTEN. Interestingly, experimental evidence linked apoptotic events induced by p53 to the activation of p53-mediated PTEN transcription (Stambolic et al., 2001) and many p53-induced MiRNAs, like MiR-23, MiR-26, MiR-29, MiR-25, MiR-32, MiR-92, MiR-200, and MiR-130, appear to target PTEN. In line with these findings, the tumor suppressor p53 homologues, TA-p73, and p63 have been shown to function as tumor suppressors and are able to transactivate transcriptional targets of p53, including MiR-34, MiR-192/215, MiR-107, and MiR-145, that play a key role in controlling tumor progression, angiogenesis, and metastasis. Intriguingly, p53 increases the transcription of Notch1—also reported to regulate the transcription of PTEN either positively, by the C-repeat binding factor 1, CBF1, or negatively, by the induction of Myc (Whelan et al., 2007)—while it decreases its expression at the posttranscriptional level through its target MiR-34 (Ji et al., 2009).

Collectively, these data suggest a complex layer of interactions between p53/PTEN/Myc pathways.

3.3 TGFβ

Transforming growth factor-β is a multifunctional polypeptide cytokine involved in various physiologic processes (Fan et al., 2013). The TGFβ superfamily consists of highly pleiotropic molecules involved in regulation of inflammation, fibrosis, cell apoptosis, and cell proliferation. Among these molecules, morphogenic proteins, growth differentiation factors, and glial-derived neurotrophic factors, activins, and inhibins are described to initiate the signaling cascade inside the cells.

Activated TGFβ stimulated the overexpression of many extracellular matrix proteins and suppressed their degradation by matrix metalloproteinases through the upregulation of tissue inhibitor of metalloproteinases (TIMP) (Gressner et al., 2002). Once TGFβ binds to its transmembrane receptors (TGFβR-I and TGFβR-II), the TGFβ pathway is activated and signals through the phosphorylation of receptor-associated Smads downstream transducers (Smad2 and Smad3; R-Smads), which then form a complex with Smad4. The latter complex enters the nucleus and binds to the TGFβ promoter to regulate its expression (Wang et al., 2008).

TGFβ pathway dysregulation has been reported in several diseases, including cancer (Han et al., 2015). However, the role of TGFβ in human carcinogenesis is still somewhat controversial, since its induction can either exert tumor-suppressive or tumor-promoting functions, depending on the stage of the disease and cellular context. As an example, TGFβR-II was downregulated in multiple cancer types (Han et al., 2015) and the canonical Smad2/3 signaling activity induced by TGFβ inhibited the growth of malignant cells and promoted apoptosis (Padua and Massague, 2009) in epithelial cells and during tumor initiation. However, in later stages of tumor progression, TGFβ induced EMT and cancer cells activated the TGFβ pathway to enhance their malignant properties, especially in the presence of oncogenic events and epigenetic perturbations (Tang et al., 2003). The mechanisms of such decreased sensitivity to TGFβ-mediated growth inhibition remain unclear and the stromal-derived enhancer and suppressor molecules that can directly modulate TGFβ increase the level of complexity of this pathway.

A possible explanation of this dual role in carcinogenesis resides in the hypothesis that MiRNAs may be involved. In particular, it has been shown that MiR-17 and MiR-20a directly target TGFβR-II, whereas MiR-18a targets Smad2 and Smad4 (Dews et al., 2010). Additionally, recent evidence shows that TGFβ regulates the expression of several MiRNAs, either positively or negatively regulated by the TGFβ signaling. Among these, MiR-216a, MiR-217, MiR-192, and MiR-193 were shown to be upregulated in the TGFβ-treated mouse model MC (Kato et al., 2009). Of note, MiR-216 and MiR-217 have been reported to activate Akt (Ser-473 phosphorylation) and the same effect was observed upon the induction of TGFβ. Interestingly, since PTEN is a potential target of both MiR-216a and MiR-217 and its alteration can enhance Akt activity in certain cancer cells, it has been proposed that PTEN might be a link between TGFβ and Akt activation, further confirming the hypothesis that PTEN-mediated downregulation of certain MiRNAs may modulate the TGFβ pathway (Kato et al., 2009).

Taylor and collaborators also reported that the expression of MiR-181a is upregulated by TGFβ and this, in turn, promotes breast cancer metastasis (Taylor et al., 2013). Moreover, it has been reported that several MiRNAs are posttranscriptionally induced by TGFβ, suggesting that the rapid modulation of MiRNA levels could play an important role in cellular response to cytokine signaling (Hata and Davis, 2009). For instance, treatment with either TGFβ or its family member, bone morphogenetic protein 4, increases the expression of pre- and mature MiR-21, by facilitating the Drosha-processing step (Davis et al., 2008). Interestingly, the oncogenic MiR-21 has been recently reported to mediate the attenuation of TGFβ signaling in prostate cancer cells by directly interacting with the androgen receptor (AR). Indeed, TGFβR-II is downregulated through a positive feedback loop, with the consequent reduction in the TGFβ-mediated inhibitory growth responses in prostate cancer (Mishra et al., 2014). Moreover, the expression of TGFβR-II was also found significantly suppressed in GC tissues and the interaction with MiR-130 has been recently investigated in the progression of GC. Notably, MiR-130 can directly bind to the 3′-UTR of the mRNA of TGFβR-II and this receptor's expression is inversely correlated with the MiR-130 expression level (Duan et al., 2016).

The EMT has been reported to play a crucial role in cancer progression and metastasis, and this transition involves the regulation of many TFs, cytoskeleton, and MiRNAs (Zaravinos, 2015). Thus far, a growing number of

MiRNAs were reported to modulate tumor metastasis by regulating EMT and many evidences indicate that TGFβ is a critical player in the EMT induction. As recently reported, MiR-132 and MiR-212, collectively named as the MiR-132/212 cluster, inhibit the TGFβ-induced EMT and their levels diminish upon TGFβ stimulation in prostate cancer cells, providing another example of a MiRNA-pathway feedback regulatory loop (Fu et al., 2016).

3.4 Wnt

The Wnt signaling pathway is an ancient and evolutionarily conserved pathway that regulates crucial aspects of cell fate determination, cell migration, cell polarity, neural patterning, and organogenesis during embryonic development.

Wnt proteins are secreted glycoproteins that bind to the N-terminal extracellular cysteine-rich domain of the Frizzled receptor family, a seven-transmembrane-span protein with topological homology to G-protein-coupled receptors (Schulte and Bryja, 2007), together with the low-density-lipoprotein-related protein5/6 (LRP5/6), as coreceptors. This trimeric complex stimulates several intracellular signal transduction cascades, eventually resulting in the stabilization and consequent accumulation of β-catenin in the cytoplasm. Accumulated β-catenin then translocates into the nucleus and acts as a transcriptional coactivator of TFs, driving the activation of target genes, such as Myc, c-Jun, Axin2, EphB/ephrin-B, and cyclin D1, all playing crucial roles in cell proliferation, cell division, stem cell maintenance, and regulation of various stages of embryogenesis (ten Berge et al., 2011), among others.

Conversely, in the absence of Wnt ligands, the accumulation of β-catenin is prevented by a destruction complex, composed of a scaffold combination of Axin, adenomatosis polyposis coli (APC), protein phosphatase 2A, casein kinase1, and glycogen synthase kinase 3β (GSK-3β) (Minde et al., 2013). This complex targets β-catenin for ubiquitination through a series of phosphorylating events (to a set of threonine and serine residues) that recruit APC and expose β-catenin to E3 ubiquitin ligase complex for degradation.

The Wnt signaling pathway controls numerous stages of animal development and tissue homeostasis, and aberrant Wnt/β-catenin signaling cascade, with the subsequent activation of Wnt target genes, may result in a series of diseases, including developmental disorders and cancer. Besides the contribution of mutations in key genes of the Wnt signaling cascade such as CTNNB1 (the gene for β-catenin), Axin, APC, and GSK-3β, several studies indicate that MiRNAs are powerful regulators of the Wnt pathway. As an example, MiR-410 has been connected to increased growth, migration, and invasion of NSCLC cells by activating the canonical Wnt signaling in vitro (Peng et al., 2016). Moreover, studies on the overexpression of MiR-374a revealed the stabilization and accumulation of nuclear β-catenin in 4T1 and MCF7 breast cancer cell lines (Aslakson and Miller, 1992), suggesting a role of MiR-374a in degrading APC, an essential component of the β-catenin-degrading complex.

The Wnt/β-catenin signaling pathway can also be negatively regulated by MiRNAs, which directly or indirectly mediate Wnt repression or β-catenin degradation. For instance, MiR-200a has been implicated in targeting and degrading the mRNA of E-cadherin repressors Zeb1 and Zeb2, leading E-cadherin to form a complex with β-catenin and promoting cell–cell adhesion (Su et al., 2011). This complex is beneficial for the cells, since it reduces the cytoplasmic levels of β-catenin, therefore avoiding its nuclear translocation and activation of Wnt target genes. Similarly, MiR-34 inhibits the canonical Wnt signaling, as demonstrated by deletion of functional MiR-34 gene, resulting in the increased expression of Wnt1 and β-catenin mRNA in A549 and MCF7 cancer cell lines (Wang et al., 2010). Furthermore, let-7 has been implicated in regulating cell proliferation, since the overexpression of let-7a in Wnt activated MDA-MB-231 cells inhibits β-catenin-activated cell growth and colony formation (Sun et al., 2015). Notably, Lin-28 is crucial to mediate let-7 repression and promote cell proliferation through the Wnt/β-catenin pathway (Zhang et al., 2007).

4. CONCLUSIONS

The last 15 years have witnessed a progressively increasing importance of MiRNAs in the biology of cancer. A better definition of the most widely dysregulated MiRNAs in cancer has been followed by a comprehensive (yet not exhaustive) identification of their target genes, shifting the interest from the single gene to the targeted pathway. It is extremely remarkable that several groups are constantly expanding the panel of newly identified MiRNA target genes, and it is also becoming clearer and clearer that MiRNAs are very versatile molecules with more than one mechanism of action. First, Vasudevan et al. showed that MiRNAs not only silence the expression of their target genes but can also increase the translation of target mRNAs into proteins (Vasudevan et al., 2007). Then, Perotti's lab provided the first evidence that MiRNAs can directly bind to proteins and affect their ability to regulate gene expression (Eiring et al., 2010). We identified the first group of "epi-miRNAs," meaning MiRNAs affecting the expression of epigenetically regulated genes by regulating the expression of effectors of the epigenetic machinery such as DNA methyltranferases, when targeted by the MiR-29 family (Fabbri et al., 2007a). More recently, a seminal paper by Lotvall's group showed that MiRNAs are contained in cell-derived nanovesicles called "exosomes" and can be functionally transferred from one cell to another (Valadi et al., 2007), opening the a new layer of intercellular interactions within the tumor microenvironment. Finally, in 2012, we discovered that MiRNAs can function as "hormones" by binding to a miRceptor (meaning a receptor for MiRNAs), namely human toll-like receptor 8 (TLR8) in tumor-associated macrophages, leading to activation of the TLR8-triggered NF-κB pathway which increases the secretion of protumoral inflammatory cytokines (Fabbri et al., 2012), and of the protumoral MiR-155 which increases the resistance of cancer cells to chemotherapy (Challagundla et al., 2015). The identification of MiRNAs in the blood and the fact that different circulating MiRNAs can be detected in cancer patients compared to the healthy counterpart suggests that circulating MiRNAs can be used as cancer biomarkers. The discovery of MiRNAs inside of exosomes and the fact that they can be functionally transferred from one cell to another introduced two important concepts: first, the need to assess whether a "naked" circulating MiRNA is a better biomarker than a MiRNA circulating inside of an extracellular vesicle, and second, whether MiRNAs inside of extracellular vesicles can be exploited for therapeutic options. Currently, several groups are focusing their endeavors on discovering and validating MiRNAs (intra- and extravesicles) as cancer biomarkers and in creating engineered extracellular vesicles enriched for specific MiRNAs to achieve a better therapeutic outcome.

Clearly, these studies highlight the importance of exploring "out of the box" options in the mechanism of action of MiRNAs and highlight the versatility of these fascinating molecules in the biology of cancer and other human diseases. Although there is still a pervasive need to fully clarify the entire "targetome" of MiRNAs, the full spectrum of their mechanisms of action (within the cancer cell and the tumor microenvironment), and assess how different MiRNAs cooperate in the regulation of key cancer signaling pathways, a series of impressive achievements have already been achieved in the field, leading to convincing translational studies on the role of MiRNAs as cancer biomarkers and therapeutics. For these reasons, it can be safely anticipated that the successful trajectory of the scientific discoveries in the MiRNA field will lead to the development of new strategies for a more effective early diagnosis of cancer and for the treatment of the disease, ultimately improving the clinical outcome for cancer patients.

ACKNOWLEDGMENTS

Dr. Fabbri is a St. Baldrick Foundation's Scholar and is supported by Hyundai "Hope on Wheels," Alex's Lemonade, the Jean Perkins Foundation, the Nautica Malibu Triathlon Funds, STOP Cancer, the Hugh and Audy Lou Colvin Foundation, the William Lawrence & Blanche Hughes Foundation, and the Award No. P30CA014089 from the National Cancer Institute.

REFERENCES

Adams, J.M., Harris, A.W., Pinkert, C.A., Corcoran, L.M., Alexander, W.S., Cory, S., Palmiter, R.D., Brinster, R.L., 1985. The c-myc oncogene driven by immunoglobulin enhancers induces lymphoid malignancy in transgenic mice. Nature 318, 533–538.

Akhtar, N., Makki, M.S., Haqqi, T.M., 2015. MicroRNA-602 and microRNA-608 regulate sonic hedgehog expression via target sites in the coding region in human chondrocytes. Arthritis Rheumatol. 67, 423–434.

Ambros, V., Horvitz, H.R., 1984. Heterochronic mutants of the nematode Caenorhabditis elegans. Science 226, 409–416.

Ambros, V., Lee, R.C., 2004. Identification of microRNAs and other tiny noncoding RNAs by cDNA cloning. Methods Mol. Biol. 265, 131–158.

Andersson, E.R., Lendahl, U., 2014. Therapeutic modulation of notch signalling—are we there yet? Nat. Rev. Drug Discov. 13, 357–378.

Appari, M., Babu, K.R., Kaczorowski, A., Gross, W., Herr, I., 2014. Sulforaphane, quercetin and catechins complement each other in elimination of advanced pancreatic cancer by miR-let-7 induction and K-ras inhibition. Int. J. Oncol. 45, 1391–1400.

Armelin, H.A., Armelin, M.C., Kelly, K., Stewart, T., Leder, P., Cochran, B.H., Stiles, C.D., 1984. Functional role for c-myc in mitogenic response to platelet-derived growth factor. Nature 310, 655–660.

Aslakson, C.J., Miller, F.R., 1992. Selective events in the metastatic process defined by analysis of the sequential dissemination of subpopulations of a mouse mammary tumor. Cancer Res. 52, 1399–1405.

Bandi, N., Vassella, E., 2011. miR-34a and miR-15a/16 are co-regulated in non-small cell lung cancer and control cell cycle progression in a synergistic and Rb-dependent manner. Mol. Cancer 10, 55.

Baracca, A., Chiaradonna, F., Sgarbi, G., Solaini, G., Alberghina, L., Lenaz, G., 2010. Mitochondrial complex I decrease is responsible for bioenergetic dysfunction in K-ras transformed cells. Biochim. Biophys. Acta 1797, 314–323.

Belloni, E., Muenke, M., Roessler, E., Traverso, G., Siegel-Bartelt, J., Frumkin, A., Mitchell, H.F., Donis-Keller, H., Helms, C., Hing, A.V., Heng, H.H., Koop, B., Martindale, D., Rommens, J.M., Tsui, L.C., Scherer, S.W., 1996. Identification of Sonic hedgehog as a candidate gene responsible for holoprosencephaly. Nat. Genet. 14, 353–356.

Bonaldi, T., Mihailovich, M., 2016. Spatiotemporal plasticity of miRNAs functions: the miR-17-92 case. Mol. Cell. Oncol. 3, e1156216.
Boominathan, L., 2010. The guardians of the genome (p53, TA-p73, and TA-p63) are regulators of tumor suppressor miRNAs network. Cancer Metastasis Rev. 29, 613–639.
Brabletz, T., 2012. MiR-34 and SNAIL: another double-negative feedback loop controlling cellular plasticity/EMT governed by p53. Cell Cycle 11, 215–216.
Cairns, R.A., Harris, I.S., Mak, T.W., 2011. Regulation of cancer cell metabolism. Nat. Rev. Cancer 11, 85–95.
Calin, G.A., Croce, C.M., 2006. MicroRNA signatures in human cancers. Nat. Rev. Cancer 6, 857–866.
Calin, G.A., Dumitru, C.D., Shimizu, M., Bichi, R., Zupo, S., Noch, E., Aldler, H., Rattan, S., Keating, M., Rai, K., Rassenti, L., Kipps, T., Negrini, M., Bullrich, F., Croce, C.M., 2002. Frequent deletions and down-regulation of micro- RNA genes miR15 and miR16 at 13q14 in chronic lymphocytic leukemia. Proc. Natl. Acad. Sci. U.S.A. 99, 15524–15529.
Calin, G.A., Liu, C.G., Sevignani, C., Ferracin, M., Felli, N., Dumitru, C.D., Shimizu, M., Cimmino, A., Zupo, S., Dono, M., Dell'aquila, M.L., Alder, H., Rassenti, L., Kipps, T.J., Bullrich, F., Negrini, M., Croce, C.M., 2004. MicroRNA profiling reveals distinct signatures in B cell chronic lymphocytic leukemias. Proc. Natl. Acad. Sci. U.S.A. 101, 11755–11760.
Calin, G.A., Cimmino, A., Fabbri, M., Ferracin, M., Wojcik, S.E., Shimizu, M., Taccioli, C., Zanesi, N., Garzon, R., Aqeilan, R.I., Alder, H., Volinia, S., Rassenti, L., Liu, X., Liu, C.G., Kipps, T.J., Negrini, M., Croce, C.M., 2008. MiR-15a and miR-16-1 cluster functions in human leukemia. Proc. Natl. Acad. Sci. U.S.A. 105, 5166–5171.
Carpenter, R.L., Paw, I., Zhu, H., Sirkisoon, S., Xing, F., Watabe, K., Debinski, W., Lo, H.W., 2015. The gain-of-function GLI1 transcription factor TGLI1 enhances expression of VEGF-C and TEM7 to promote glioblastoma angiogenesis. Oncotarget 6, 22653–22665.
Chai, S., Cambronne, X.A., Eichhorn, S.W., Goodman, R.H., 2013. MicroRNA-134 activity in somatostatin interneurons regulates H-Ras localization by repressing the palmitoylation enzyme, DHHC9. Proc. Natl. Acad. Sci. U.S.A. 110, 17898–17903.
Challagundla, K.B., Wise, P.M., Neviani, P., Chava, H., Murtadha, M., Xu, T., Kennedy, R., Ivan, C., Zhang, X., Vannini, I., Fanini, F., Amadori, D., Calin, G.A., Hadjidaniel, M., Shimada, H., Jong, A., Seeger, R.C., Asgharzadeh, S., Goldkorn, A., Fabbri, M., 2015. Exosome-mediated transfer of microRNAs within the tumor microenvironment and neuroblastoma resistance to chemotherapy. J. Natl. Cancer Inst. 107 (7), pii: djv135. http://dx.doi.org/10.1093/jnci/djv135.
Chang, T.C., Wentzel, E.A., Kent, O.A., Ramachandran, K., Mullendore, M., Lee, K.H., Feldmann, G., Yamakuchi, M., Ferlito, M., Lowenstein, C.J., Arking, D.E., Beer, M.A., Maitra, A., Mendell, J.T., 2007. Transactivation of miR-34a by p53 broadly influences gene expression and promotes apoptosis. Mol. Cell 26, 745–752.
Chang, T.C., Yu, D., Lee, Y.S., Wentzel, E.A., Arking, D.E., West, K.M., Dang, C.V., Thomas-Tikhonenko, A., Mendell, J.T., 2008. Widespread microRNA repression by Myc contributes to tumorigenesis. Nat. Genet. 40, 43–50.
Chen, C.Z., Lodish, H.F., 2005. MicroRNAs as regulators of mammalian hematopoiesis. Semin. Immunol. 17, 155–165.
Chen, P., Guo, X., Zhang, L., Zhang, W., Zhou, Q., Tian, Z., 2016. MiR-200c is a cMyc-activated miRNA that promotes nasopharingeal carcinoma by downregulating PTEN. Oncotarget, 1–13.
Choi, Y.W., Song, Y.S., Lee, H., Yi, K., Kim, Y.B., Suh, K.W., Lee, D., 2016. MicroRNA expression signatures associated with BRAF-mutated versus KRAS-mutated colorectal cancers. Medicine (Baltimore) 95, e3321.

Cimmino, A., Calin, G.A., Fabbri, M., Iorio, M.V., Ferracin, M., Shimizu, M., Wojcik, S.E., Aqeilan, R.I., Zupo, S., Dono, M., Rassenti, L., Alder, H., Volinia, S., Liu, C.G., Kipps, T.J., Negrini, M., Croce, C.M., 2005. miR-15 and miR-16 induce apoptosis by targeting BCL2. Proc. Natl. Acad. Sci. U.S.A. 102, 13944–13949.

Corney, D.C., Flesken-Nikitin, A., Godwin, A.K., Wang, W., Nikitin, A.Y., 2007. MicroRNA-34b and MicroRNA-34c are targets of p53 and cooperate in control of cell proliferation and adhesion-independent growth. Cancer Res. 67, 8433–8438.

Davies, H., Bignell, G.R., Cox, C., Stephens, P., Edkins, S., Clegg, S., Teague, J., Woffendin, H., Garnett, M.J., Bottomley, W., Davis, N., Dicks, E., Ewing, R., Floyd, Y., Gray, K., Hall, S., Hawes, R., Hughes, J., Kosmidou, V., Menzies, A., Mould, C., Parker, A., Stevens, C., Watt, S., Hooper, S., Wilson, R., Jayatilake, H., Gusterson, B.A., Cooper, C., Shipley, J., Hargrave, D., Pritchard-Jones, K., Maitland, N., Chenevix-Trench, G., Riggins, G.J., Bigner, D.D., Palmieri, G., Cossu, A., Flanagan, A., Nicholson, A., Ho, J.W., Leung, S.Y., Yuen, S.T., Weber, B.L., Seigler, H.F., Darrow, T.L., Paterson, H., Marais, R., Marshall, C.J., Wooster, R., Stratton, M.R., Futreal, P.A., 2002. Mutations of the BRAF gene in human cancer. Nature 417, 949–954.

Davis, B.N., Hilyard, A.C., Lagna, G., Hata, A., 2008. Smad proteins control DROSHA-mediated microRNA maturation. Nature 454, 56–61.

DeGregori, J., Johnson, D.G., 2006. Distinct and overlapping roles for E2F family members in transcription, proliferation and apoptosis. Curr. Mol. Med. 6, 739–748.

Deng, H., Lv, L., Li, Y., Zhang, C., Meng, F., Pu, Y., Xiao, J., Qian, L., Zhao, W., Liu, Q., Zhang, D., Wang, Y., Zhang, H., He, Y., Zhu, J., 2014. miR-193a-3p regulates the multi-drug resistance of bladder cancer by targeting the LOXL4 gene and the oxidative stress pathway. Mol. Cancer 13, 234.

de Sousa, E.M., Vermeulen, L., Richel, D., Medema, J.P., 2011. Targeting Wnt signaling in colon cancer stem cells. Clin. Cancer Res. 17, 647–653.

Dews, M., Homayouni, A., Yu, D., Murphy, D., Sevignani, C., Wentzel, E., Furth, E.E., Lee, W.M., Enders, G.H., Mendell, J.T., Thomas-Tikhonenko, A., 2006. Augmentation of tumor angiogenesis by a Myc-activated microRNA cluster. Nat. Genet. 38, 1060–1065.

Dews, M., Fox, J.L., Hultine, S., Sundaram, P., Wang, W., Liu, Y.Y., Furth, E., Enders, G.H., El-Deiry, W., Schelter, J.M., Cleary, M.A., Thomas-Tikhonenko, A., 2010. The myc-miR-17∼92 axis blunts TGF{beta} signaling and production of multiple TGF{beta}-dependent antiangiogenic factors. Cancer Res. 70, 8233–8246.

Di Cristofano, A., Pesce, B., Cordon-Cardo, C., Pandolfi, P.P., 1998. Pten is essential for embryonic development and tumour suppression. Nat. Genet. 19, 348–355.

Diosdado, B., Van De Wiel, M.A., Terhaar Sive Droste, J.S., Mongera, S., Postma, C., Meijerink, W.J., Carvalho, B., Meijer, G.A., 2009. MiR-17-92 cluster is associated with 13q gain and c-myc expression during colorectal adenoma to adenocarcinoma progression. Br. J. Cancer 101, 707–714.

Duan, J., Zhang, H., Qu, Y., Deng, T., Huang, D., Liu, R., Zhang, L., Bai, M., Zhou, L., Ying, G., Ba, Y., 2016. Onco-miR-130 promotes cell proliferation and migration by targeting TGFbetaR2 in gastric cancer. Oncotarget 7, 44522–44533.

Echelard, Y., Epstein, D.J., St-Jacques, B., Shen, L., Mohler, J., McMahon, J.A., McMahon, A.P., 1993. Sonic hedgehog, a member of a family of putative signaling molecules, is implicated in the regulation of CNS polarity. Cell 75, 1417–1430.

Eiring, A.M., Harb, J.G., Neviani, P., Garton, C., Oaks, J.J., Spizzo, R., Liu, S., Schwind, S., Santhanam, R., Hickey, C.J., Becker, H., Chandler, J.C., Andino, R., Cortes, J., Hokland, P., Huettner, C.S., Bhatia, R., Roy, D.C., Liebhaber, S.A., Caligiuri, M.A., Marcucci, G., Garzon, R., Croce, C.M., Calin, G.A., Perrotti, D., 2010. miR-328 functions as an RNA decoy to modulate hnRNP E2 regulation of mRNA translation in leukemic blasts. Cell 140, 652–665.

el-Deiry, W.S., Tokino, T., Velculescu, V.E., Levy, D.B., Parsons, R., Trent, J.M., Lin, D., Mercer, W.E., Kinzler, K.W., Vogelstein, B., 1993. WAF1, a potential mediator of p53 tumor suppression. Cell 75, 817–825.
Esquela-Kerscher, A., Slack, F.J., 2006. Oncomirs—microRNAs with a role in cancer. Nat. Rev. Cancer 6, 259–269.
Esquela-Kerscher, A., Trang, P., Wiggins, J.F., Patrawala, L., Cheng, A., Ford, L., Weidhaas, J.B., Brown, D., Bader, A.G., Slack, F.J., 2008. The let-7 microRNA reduces tumor growth in mouse models of lung cancer. Cell Cycle 7, 759–764.
Fabbri, M., Garzon, R., Cimmino, A., Liu, Z., Zanesi, N., Callegari, E., Liu, S., Alder, H., Costinean, S., Fernandez-Cymering, C., Volinia, S., Guler, G., Morrison, C.D., Chan, K.K., Marcucci, G., Calin, G.A., Huebner, K., Croce, C.M., 2007a. MicroRNA-29 family reverts aberrant methylation in lung cancer by targeting DNA methyltransferases 3A and 3B. Proc. Natl. Acad. Sci. U.S.A. 104, 15805–15810.
Fabbri, M., Ivan, M., Cimmino, A., Negrini, M., Calin, G.A., 2007b. Regulatory mechanisms of microRNAs involvement in cancer. Expert Opin. Biol. Ther. 7, 1009–1019.
Fabbri, M., Bottoni, A., Shimizu, M., Spizzo, R., Nicoloso, M.S., Rossi, S., Barbarotto, E., Cimmino, A., Adair, B., Wojcik, S.E., Valeri, N., Calore, F., Sampath, D., Fanini, F., Vannini, I., Musuraca, G., Dell'aquila, M., Alder, H., Davuluri, R.V., Rassenti, L.Z., Negrini, M., Nakamura, T., Amadori, D., Kay, N.E., Rai, K.R., Keating, M.J., Kipps, T.J., Calin, G.A., Croce, C.M., 2011. Association of a microRNA/TP53 feedback circuitry with pathogenesis and outcome of B-cell chronic lymphocytic leukemia. JAMA 305, 59–67.
Fabbri, M., Paone, A., Calore, F., Galli, R., Gaudio, E., Santhanam, R., Lovat, F., Fadda, P., Mao, C., Nuovo, G.J., Zanesi, N., Crawford, M., Ozer, G.H., Wernicke, D., Alder, H., Caligiuri, M.A., Nana-Sinkam, P., Perrotti, D., Croce, C.M., 2012. MicroRNAs bind to toll-like receptors to induce prometastatic inflammatory response. Proc. Natl. Acad. Sci. U.S.A. 109, E2110–E2116.
Fan, X., Zhang, Q., Li, S., Lv, Y., Su, H., Jiang, H., Hao, Z., 2013. Attenuation of CCl4-induced hepatic fibrosis in mice by vaccinating against TGF-beta1. PLoS One 8, e82190.
Felsher, D.W., Bishop, J.M., 1999. Transient excess of MYC activity can elicit genomic instability and tumorigenesis. Proc. Natl. Acad. Sci. U.S.A. 96, 3940–3944.
Frediani, J.N., Fabbri, M., 2016. Essential role of miRNAs in orchestrating the biology of the tumor microenvironment. Mol. Cancer 15, 42.
Fu, W., Tao, T., Qi, M., Wang, L., Hu, J., Li, X., Xing, N., Du, R., Han, B., 2016. MicroRNA-132/212 upregulation inhibits TGF-beta-mediated epithelial-mesenchymal transition of prostate cancer cells by targeting SOX4. Prostate 76, 1560–1570.
Fuziwara, C.S., Kimura, E.T., 2014. High iodine blocks a Notch/miR-19 loop activated by the Braf(V600E) oncoprotein and restores the response to TGFβ in thyroid follicular cells. Thyroid 24, 453–462.
Fuziwara, C.S., Kimura, E.T., 2015. Insights into regulation of the miR-17-92 cluster of miRNAs in cancer. Front. Med. (Lausanne) 2, 64.
Gallardo, E., Navarro, A., Vinolas, N., Marrades, R.M., Diaz, T., Gel, B., Quera, A., Bandres, E., Garcia-Foncillas, J., Ramirez, J., Monzo, M., 2009. miR-34a as a prognostic marker of relapse in surgically resected non-small-cell lung cancer. Carcinogenesis 30, 1903–1909.
Gamberi, G., Benassi, M.S., Bohling, T., Ragazzini, P., Molendini, L., Sollazzo, M.R., Pompetti, F., Merli, M., Magagnoli, G., Balladelli, A., Picci, P., 1998. C-myc and c-fos in human osteosarcoma: prognostic value of mRNA and protein expression. Oncology 55, 556–563.
Gao, P., Tchernyshyov, I., Chang, T.C., Lee, Y.S., Kita, K., Ochi, T., Zeller, K.I., De Marzo, A.M., Van Eyk, J.E., Mendell, J.T., Dang, C.V., 2009. c-Myc suppression of miR-23a/b enhances mitochondrial glutaminase expression and glutamine metabolism. Nature 458, 762–765.

Garofalo, M., Di Leva, G., Romano, G., Nuovo, G., Suh, S.S., Ngankeu, A., Taccioli, C., Pichiorri, F., Alder, H., Secchiero, P., Gasparini, P., Gonelli, A., Costinean, S., Acunzo, M., Condorelli, G., Croce, C.M., 2009. miR-221&222 regulate TRAIL resistance and enhance tumorigenicity through PTEN and TIMP3 downregulation. Cancer Cell 16, 498–509.
Georges, S.A., Biery, M.C., Kim, S.Y., Schelter, J.M., Guo, J., Chang, A.N., Jackson, A.L., Carleton, M.O., Linsley, P.S., Cleary, M.A., Chau, B.N., 2008. Coordinated regulation of cell cycle transcripts by p53-inducible microRNAs, miR-192 and miR-215. Cancer Res. 68, 10105–10112.
Gressner, A.M., Weiskirchen, R., Breitkopf, K., Dooley, S., 2002. Roles of TGF-beta in hepatic fibrosis. Front. Biosci. 7, d793–d807.
Grishok, A., Pasquinelli, A.E., Conte, D., Li, N., Parrish, S., Ha, I., Baillie, D.L., Fire, A., Ruvkun, G., Mello, C.C., 2001. Genes and mechanisms related to RNA interference regulate expression of the small temporal RNAs that control C. elegans developmental timing. Cell 106, 23–34.
Gu, W., Shou, J., Gu, S., Sun, B., Che, X., 2014. Identifying hedgehog signaling specific microRNAs in glioblastomas. Int. J. Med. Sci. 11, 488–493.
Gu, Y., Masiero, M., Banham, A.H., 2016. Notch signaling: its roles and therapeutic potential in hematological malignancies. Oncotarget 7, 29804–29823.
Hammond, S.M., Bernstein, E., Beach, D., Hannon, G.J., 2000. An RNA-directed nuclease mediates post-transcriptional gene silencing in Drosophila cells. Nature 404, 293–296.
Han, Y., Jia, C., Cong, X., Yu, F., Cai, H., Fang, S., Cai, L., Yang, H., Sun, Y., Li, D., Liu, J., Xie, R., Yuan, X., Zhong, X., Li, M., Wei, Q., Lv, Z., Fu, D., Ma, Y., 2015. Increased expression of TGFbetaR2 is associated with the clinical outcome of non-small cell lung cancer patients treated with chemotherapy. PLoS One 10, e0134682.
Hashimoto, Y., Akiyama, Y., Otsubo, T., Shimada, S., Yuasa, Y., 2010. Involvement of epigenetically silenced microRNA-181c in gastric carcinogenesis. Carcinogenesis 31, 777–784.
Hata, A., Davis, B.N., 2009. Control of microRNA biogenesis by TGFbeta signaling pathway—a novel role of Smads in the nucleus. Cytokine Growth Factor Rev. 20, 517–521.
He, L., Thomson, J.M., Hemann, M.T., Hernando-Monge, E., Mu, D., Goodson, S., Powers, S., Cordon-Cardo, C., Lowe, S.W., Hannon, G.J., Hammond, S.M., 2005. A microRNA polycistron as a potential human oncogene. Nature 435, 828–833.
He, L., He, X., Lim, L.P., De Stanchina, E., Xuan, Z., Liang, Y., Xue, W., Zender, L., Magnus, J., Ridzon, D., Jackson, A.L., Linsley, P.S., Chen, C., Lowe, S.W., Cleary, M.A., Hannon, G.J., 2007. A microRNA component of the p53 tumour suppressor network. Nature 447, 1130–1134.
Hermeking, H., 2007. p53 enters the microRNA world. Cancer Cell 12, 414–418.
Horvitz, H.R., Sulston, J.E., 1980. Isolation and genetic characterization of cell-lineage mutants of the nematode Caenorhabditis elegans. Genetics 96, 435–454.
Hsu, K.W., Fang, W.L., Huang, K.H., Huang, T.T., Lee, H.C., Hsieh, R.H., Chi, C.W., Yeh, T.S., 2016. Notch1 pathway-mediated microRNA-151-5p promotes gastric cancer progression. Oncotarget 7, 38036–38051.
Hui, C.C., Angers, S., 2011. Gli proteins in development and disease. Annu. Rev. Cell Dev. Biol. 27, 513–537.
Hutvagner, G., McLachlan, J., Pasquinelli, A.E., Balint, E., Tuschl, T., Zamore, P.D., 2001. A cellular function for the RNA-interference enzyme dicer in the maturation of the let-7 small temporal RNA. Science 293, 834–838.
Iorio, M.V., Visone, R., Di Leva, G., Donati, V., Petrocca, F., Casalini, P., Taccioli, C., Volinia, S., Liu, C.G., Alder, H., Calin, G.A., MÉnard, S., Croce, C.M., 2007. MicroRNA signatures in human ovarian cancer. Cancer Res. 67, 8699–8707.

Ji, Q., Hao, X., Zhang, M., Tang, W., Yang, M., Li, L., Xiang, D., Desano, J.T., Bommer, G.T., Fan, D., Fearon, E.R., Lawrence, T.S., Xu, L., 2009. MicroRNA miR-34 inhibits human pancreatic cancer tumor-initiating cells. PLoS One 4, e6816.
Jiao, L.R., Frampton, A.E., Jacob, J., Pellegrino, L., Krell, J., Giamas, G., Tsim, N., Vlavianos, P., Cohen, P., Ahmad, R., Keller, A., Habib, N.A., Stebbing, J., Castellano, L., 2012. MicroRNAs targeting oncogenes are down-regulated in pancreatic malignant transformation from benign tumors. PLoS One 7, e32068.
Jin, B., Wang, W., Meng, X.X., Du, G., Li, J., Zhang, S.Z., Zhou, B.H., Fu, Z.H., 2016. Let-7 inhibits self-renewal of hepatocellular cancer stem-like cells through regulating the epithelial-mesenchymal transition and the Wnt signaling pathway. BMC Cancer 16, 863.
Johnson, S.M., Grosshans, H., Shingara, J., Byrom, M., Jarvis, R., Cheng, A., Labourier, E., Reinert, K.L., Brown, D., Slack, F.J., 2005. Ras is regulated by the let-7 microRNA family. Cell 120, 635–647.
Jopling, C.L., 2008. Regulation of hepatitis C virus by microRNA-122. Biochem. Soc. Trans. 36, 1220–1223.
Jopling, C.L., Yi, M., Lancaster, A.M., Lemon, S.M., Sarnow, P., 2005. Modulation of hepatitis C virus RNA abundance by a liver-specific microRNA. Science 309, 1577–1581.
Karhadkar, S.S., Bova, G.S., Abdallah, N., Dhara, S., Gardner, D., Maitra, A., Isaacs, J.T., Berman, D.M., Beachy, P.A., 2004. Hedgehog signalling in prostate regeneration, neoplasia and metastasis. Nature 431, 707–712.
Kasinski, A.L., Slack, F.J., 2012. miRNA-34 prevents cancer initiation and progression in a therapeutically resistant K-ras and p53-induced mouse model of lung adenocarcinoma. Cancer Res. 72, 5576–5587.
Kasinski, A.L., Kelnar, K., Stahlhut, C., Orellana, E., Zhao, J., Shimer, E., Dysart, S., Chen, X., Bader, A.G., Slack, F.J., 2015. A combinatorial microrna therapeutics approach to suppressing non-small cell lung cancer. Oncogene 34, 3547–3555.
Kato, M., Putta, S., Wang, M., Yuan, H., Lanting, L., Nair, I., Gunn, A., Nakagawa, Y., Shimano, H., Todorov, I., Rossi, J.J., Natarajan, R., 2009. TGF-beta activates Akt kinase through a microRNA-dependent amplifying circuit targeting PTEN. Nat. Cell Biol. 11, 881–889.
Kawate, S., Fukusato, T., Ohwada, S., Watanuki, A., Morishita, Y., 1999. Amplification of c-myc in hepatocellular carcinoma: correlation with clinicopathologic features, proliferative activity and p53 overexpression. Oncology 57, 157–163.
Kelly, K., Cochran, B.H., Stiles, C.D., Leder, P., 1983. Cell-specific regulation of the c-myc gene by lymphocyte mitogens and platelet-derived growth factor. Cell 35, 603–610.
Ketting, R.F., Fischer, S.E., Bernstein, E., Sijen, T., Hannon, G.J., Plasterk, R.H., 2001. Dicer functions in RNA interference and in synthesis of small RNA involved in developmental timing in C. elegans. Genes Dev. 15, 2654–2659.
Kim, N.H., Kim, H.S., Li, X.Y., Lee, I., Choi, H.S., Kang, S.E., Cha, S.Y., Ryu, J.K., Yoon, D., Fearon, E.R., Rowe, R.G., Lee, S., Maher, C.A., Weiss, S.J., Yook, J.I., 2011. A p53/miRNA-34 axis regulates Snail1-dependent cancer cell epithelial-mesenchymal transition. J. Cell Biol. 195, 417–433.
Kimonis, V.E., Goldstein, A.M., Pastakia, B., Yang, M.L., Kase, R., Digiovanna, J.J., Bale, A.E., Bale, S.J., 1997. Clinical manifestations in 105 persons with nevoid basal cell carcinoma syndrome. Am. J. Med. Genet. 69, 299–308.
Knight, T., Irving, J.A., 2014. Ras/Raf/MEK/ERK pathway activation in childhood acute lymphoblastic leukemia and its therapeutic targeting. Front. Oncol. 4, 160.
Krauss, S., Concordet, J.P., Ingham, P.W., 1993. A functionally conserved homolog of the Drosophila segment polarity gene hh is expressed in tissues with polarizing activity in zebrafish embryos. Cell 75, 1431–1444.

Kumar, M.S., Erkeland, S.J., Pester, R.E., Chen, C.Y., Ebert, M.S., Sharp, P.A., Jacks, T., 2008. Suppression of non-small cell lung tumor development by the let-7 microRNA family. Proc. Natl. Acad. Sci. U.S.A. 105, 3903–3908.

Lagos-Quintana, M., Rauhut, R., Lendeckel, W., Tuschl, T., 2001. Identification of novel genes coding for small expressed RNAs. Science 294, 853–858.

Lan, F.F., Wang, H., Chen, Y.C., Chan, C.Y., Ng, S.S., Li, K., Xie, D., He, M.L., Lin, M.C., Kung, H.F., 2011. Hsa-let-7g inhibits proliferation of hepatocellular carcinoma cells by downregulation of c-Myc and upregulation of p16(Ink4A). Int. J. Cancer 128, 319–331.

Lane, D.P., 1992. Cancer. p53, guardian of the genome. Nature 358, 15–16.

Lau, N.C., Lim, L.P., Weinstein, E.G., Bartel, D.P., 2001. An abundant class of tiny RNAs with probable regulatory roles in Caenorhabditis elegans. Science 294, 858–862.

Leder, A., Pattengale, P.K., Kuo, A., Stewart, T.A., Leder, P., 1986. Consequences of widespread deregulation of the c-myc gene in transgenic mice: multiple neoplasms and normal development. Cell 45, 485–495.

Lee, R.C., Ambros, V., 2001. An extensive class of small RNAs in Caenorhabditis elegans. Science 294, 862–864.

Lee, R.C., Feinbaum, R.L., Ambros, V., 1993. The C. elegans heterochronic gene lin-4 encodes small RNAs with antisense complementarity to lin-14. Cell 75, 843–854.

Leung, A.K., Sharp, P.A., 2007. MicroRNAs: a safeguard against turmoil? Cell 130, 581–585.

Leung, A.K., Sharp, P.A., 2010. MicroRNA functions in stress responses. Mol. Cell 40, 205–215.

Li, Y., Casey, S.C., Choi, P.S., Felsher, D.W., 2014a. miR-17-92 explains Myc oncogene addiction. Mol. Cell. Oncol. 1, e970092.

Li, Y., Casey, S.C., Felsher, D.W., 2014b. Inactivation of MYC reverses tumorigenesis. J. Intern. Med. 276, 52–60.

Lin, Y.H., Liao, C.J., Huang, Y.H., Wu, M.H., Chi, H.C., Wu, S.M., Chen, C.Y., Tseng, Y.H., Tsai, C.Y., Chung, I.H., Wu, T.I., Tsai, M.M., Lin, C.D., Lin, K.H., 2013. Thyroid hormone receptor represses miR-17 expression to enhance tumor metastasis in human hepatoma cells. Oncogene 32, 4509–4518.

Ling, H., Fabbri, M., Calin, G.A., 2013. MicroRNAs and other non-coding RNAs as targets for anticancer drug development. Nat. Rev. Drug Discov. 12, 847–865.

Liu, Y., Zhang, M., Qian, J., Bao, M., Meng, X., Zhang, S., Zhang, L., Zhao, R., Li, S., Cao, Q., Li, P., Ju, X., Lu, Q., Li, J., Shao, P., Qin, C., Yin, C., 2015. miR-134 functions as a tumor suppressor in cell proliferation and epithelial-to-mesenchymal Transition by targeting KRAS in renal cell carcinoma cells. DNA Cell Biol. 34, 429–436.

Lodygin, D., Tarasov, V., Epanchintsev, A., Berking, C., Knyazeva, T., KÖrner, H., Knyazev, P., Diebold, J., Hermeking, H., 2008. Inactivation of miR-34a by aberrant CpG methylation in multiple types of cancer. Cell Cycle 7, 2591–2600.

Lum, L., Beachy, P.A., 2004. The Hedgehog response network: sensors, switches, and routers. Science 304, 1755–1759.

Luna, J.M., Scheel, T.K., Danino, T., Shaw, K.S., Mele, A., Fak, J.J., Nishiuchi, E., Takacs, C.N., Catanese, M.T., De Jong, Y.P., Jacobson, I.M., Rice, C.M., Darnell, R.B., 2015. Hepatitis C virus RNA functionally sequesters miR-122. Cell 160, 1099–1110.

Ma, L., Young, J., Prabhala, H., Pan, E., Mestdagh, P., Muth, D., Teruya-Feldstein, J., Reinhardt, F., Onder, T.T., Valastyan, S., Westermann, F., Speleman, F., Vandesompele, J., Weinberg, R.A., 2010. miR-9, a MYC/MYCN-activated microRNA, regulates E-cadherin and cancer metastasis. Nat. Cell Biol. 12, 247–256.

Malkin, D., Li, F.P., Strong, L.C., Fraumeni Jr., J.F., Nelson, C.E., Kim, D.H., Kassel, J., Gryka, M.A., Bischoff, F.Z., Tainsky, M.A., et al., 1990. Germ line p53 mutations in a familial syndrome of breast cancer, sarcomas, and other neoplasms. Science 250, 1233–1238.
Martello, G., Rosato, A., Ferrari, F., Manfrin, A., Cordenonsi, M., DuPont, S., Enzo, E., Guzzardo, V., Rondina, M., Spruce, T., Parenti, A.R., Daidone, M.G., Bicciato, S., Piccolo, S., 2010. A microRNA targeting dicer for metastasis control. Cell 141, 1195–1207.
Masliah-Planchon, J., Garinet, S., Pasmant, E., 2016. RAS-MAPK pathway epigenetic activation in cancer: miRNAs in action. Oncotarget 7, 38892–38907.
Mavrakis, K.J., Wolfe, A.L., Oricchio, E., Palomero, T., De Keersmaecker, K., Mcjunkin, K., Zuber, J., James, T., Khan, A.A., Leslie, C.S., Parker, J.S., Paddison, P.J., Tam, W., Ferrando, A., Wendel, H.G., 2010. Genome-wide RNA-mediated interference screen identifies miR-19 targets in Notch-induced T-cell acute lymphoblastic leukaemia. Nat. Cell Biol. 12, 372–379.
Mendell, J.T., 2008. miRiad roles for the miR-17-92 cluster in development and disease. Cell 133, 217–222.
Minde, D.P., Radli, M., Forneris, F., Maurice, M.M., Rudiger, S.G., 2013. Large extent of disorder in adenomatous polyposis coli offers a strategy to guard Wnt signalling against point mutations. PLoS One 8, e77257.
Mishra, S., Deng, J.J., Gowda, P.S., Rao, M.K., Lin, C.L., Chen, C.L., Huang, T., Sun, L.Z., 2014. Androgen receptor and microRNA-21 axis downregulates transforming growth factor beta receptor II (TGFBR2) expression in prostate cancer. Oncogene 33, 4097–4106.
Mogilyansky, E., Rigoutsos, I., 2013. The miR-17/92 cluster: a comprehensive update on its genomics, genetics, functions and increasingly important and numerous roles in health and disease. Cell Death Differ. 20, 1603–1614.
Mu, P., Han, Y.C., Betel, D., Yao, E., Squatrito, M., Ogrodowski, P., De Stanchina, E., D'andrea, A., Sander, C., Ventura, A., 2009. Genetic dissection of the miR-17∼92 cluster of microRNAs in Myc-induced B-cell lymphomas. Genes Dev. 23, 2806–2811.
Mudhasani, R., Zhu, Z., Hutvagner, G., Eischen, C.M., Lyle, S., Hall, L.L., Lawrence, J.B., Imbalzano, A.N., Jones, S.N., 2008. Loss of miRNA biogenesis induces p19Arf-p53 signaling and senescence in primary cells. J. Cell Biol. 181, 1055–1063.
Murphy, B.L., Obad, S., Bihannic, L., Ayrault, O., Zindy, F., Kauppinen, S., Roussel, M.F., 2013. Silencing of the miR-17∼92 cluster family inhibits medulloblastoma progression. Cancer Res. 73, 7068–7078.
Nesbit, C.E., Tersak, J.M., Prochownik, E.V., 1999. MYC oncogenes and human neoplastic disease. Oncogene 18, 3004–3016.
Nguyen, L.V., Vanner, R., Dirks, P., Eaves, C.J., 2012. Cancer stem cells: an evolving concept. Nat. Rev. Cancer 12, 133–143.
Northcott, P.A., Fernandez, L.A., Hagan, J.P., Ellison, D.W., Grajkowska, W., Gillespie, Y., Grundy, R., Van Meter, T., Rutka, J.T., Croce, C.M., Kenney, A.M., Taylor, M.D., 2009. The miR-17/92 polycistron is up-regulated in sonic hedgehog-driven medulloblastomas and induced by N-myc in sonic hedgehog-treated cerebellar neural precursors. Cancer Res. 69, 3249–3255.
Nosho, K., Igarashi, H., Nojima, M., Ito, M., Maruyama, R., Yoshii, S., Naito, T., Sukawa, Y., Mikami, M., Sumioka, W., Yamamoto, E., Kurokawa, S., Adachi, Y., Takahashi, H., Okuda, H., Kusumi, T., Hosokawa, M., Fujita, M., Hasegawa, T., Okita, K., Hirata, K., Suzuki, H., Yamamoto, H., Shinomura, Y., 2014. Association of microRNA-31 with BRAF mutation, colorectal cancer survival and serrated pathway. Carcinogenesis 35, 776–783.

Nusslein-Volhard, C., Wieschaus, E., 1980. Mutations affecting segment number and polarity in Drosophila. Nature 287, 795–801.

O'Donnell, K.A., Wentzel, E.A., Zeller, K.I., Dang, C.V., Mendell, J.T., 2005. c-Myc-regulated microRNAs modulate E2F1 expression. Nature 435, 839–843.

Ota, A., Tagawa, H., Karnan, S., Tsuzuki, S., Karpas, A., Kira, S., Yoshida, Y., Seto, M., 2004. Identification and characterization of a novel gene, C13orf25, as a target for 13q31-q32 amplification in malignant lymphoma. Cancer Res. 64, 3087–3095.

Padua, D., Massague, J., 2009. Roles of TGFbeta in metastasis. Cell Res. 19, 89–102.

Pelengaris, S., Littlewood, T., Khan, M., Elia, G., Evan, G., 1999. Reversible activation of c-Myc in skin: induction of a complex neoplastic phenotype by a single oncogenic lesion. Mol. Cell 3, 565–577.

Peng, Y., Zhang, X., Feng, X., Fan, X., Jin, Z., 2016. The crosstalk between microRNAs and the Wnt/beta-catenin signaling pathway in cancer. Oncotarget.

Podsypanina, K., Ellenson, L.H., Nemes, A., Gu, J., Tamura, M., Yamada, K.M., Cordon-Cardo, C., Catoretti, G., Fisher, P.E., Parsons, R., 1999. Mutation of Pten/Mmac1 in mice causes neoplasia in multiple organ systems. Proc. Natl. Acad. Sci. U.S.A. 96, 1563–1568.

Polioudakis, D., Bhinge, A.A., Killion, P.J., Lee, B.K., Abell, N.S., Iyer, V.R., 2013. A Myc-microRNA network promotes exit from quiescence by suppressing the interferon response and cell-cycle arrest genes. Nucleic Acids Res. 41, 2239–2254.

Reinhart, B.J., Slack, F.J., Basson, M., Pasquinelli, A.E., Bettinger, J.C., Rougvie, A.E., Horvitz, H.R., Ruvkun, G., 2000. The 21-nucleotide let-7 RNA regulates developmental timing in Caenorhabditis elegans. Nature 403, 901–906.

Reinhart, B.J., Weinstein, E.G., Rhoades, M.W., Bartel, B., Bartel, D.P., 2002. MicroRNAs in plants. Genes Dev. 16, 1616–1626.

Rinaldi, A., Poretti, G., Kwee, I., Zucca, E., Catapano, C.V., Tibiletti, M.G., Bertoni, F., 2007. Concomitant MYC and microRNA cluster miR-17-92 (C13orf25) amplification in human mantle cell lymphoma. Leuk. Lymphoma 48, 410–412.

Roelink, H., Augsburger, A., Heemskerk, J., Korzh, V., Norlin, S., Ruiz I Altaba, A., Tanabe, Y., Placzek, M., Edlund, T., Jessell, T.M., et al., 1994. Floor plate and motor neuron induction by vhh-1, a vertebrate homolog of hedgehog expressed by the notochord. Cell 76, 761–775.

Roessler, E., Belloni, E., Gaudenz, K., Jay, P., Berta, P., Scherer, S.W., Tsui, L.C., Muenke, M., 1996. Mutations in the human Sonic Hedgehog gene cause holoprosencephaly. Nat. Genet. 14, 357–360.

Sachdeva, M., Zhu, S., Wu, F., Wu, H., Walia, V., Kumar, S., Elble, R., Watabe, K., Mo, Y.Y., 2009. p53 represses c-Myc through induction of the tumor suppressor miR-145. Proc. Natl. Acad. Sci. U.S.A. 106, 3207–3212.

Sanchez-Martin, M., Ferrando, A., 2017. The NOTCH1-MYC highway towards T-cell acute lymphoblastic leukemia. Blood 129, 1124–1133.

Scales, S.J., de Sauvage, F.J., 2009. Mechanisms of Hedgehog pathway activation in cancer and implications for therapy. Trends Pharmacol. Sci. 30, 303–312.

Schulte, G., Bryja, V., 2007. The Frizzled family of unconventional G-protein-coupled receptors. Trends Pharmacol. Sci. 28, 518–525.

Schulte, J.H., Horn, S., Otto, T., Samans, B., Heukamp, L.C., Eilers, U.C., Krause, M., Astrahantseff, K., Klein-Hitpass, L., Buettner, R., Schramm, A., Christiansen, H., Eilers, M., Eggert, A., Berwanger, B., 2008. Mycn regulates oncogenic microRNAs in neuroblastoma. Int. J. Cancer 122, 699–704.

Schwarzenbacher, D., Balic, M., Pichler, M., 2013. The role of microRNAs in breast cancer stem cells. Int. J. Mol. Sci. 14, 14712–14723.

Shin, K.H., Bae, S.D., Hong, H.S., Kim, R.H., Kang, M.K., Park, N.H., 2011. miR-181a shows tumor suppressive effect against oral squamous cell carcinoma cells by downregulating K-ras. Biochem. Biophys. Res. Commun. 404, 896–902.

Siemens, H., Jackstadt, R., HÜnten, S., Kaller, M., Menssen, A., GÖtz, U., Hermeking, H., 2011. miR-34 and SNAIL form a double-negative feedback loop to regulate epithelial-mesenchymal transitions. Cell Cycle 10, 4256–4271.
Smalley, K.S., 2003. A pivotal role for ERK in the oncogenic behaviour of malignant melanoma? Int. J. Cancer 104, 527–532.
Song, H., Hollstein, M., Xu, Y., 2007. p53 gain-of-function cancer mutants induce genetic instability by inactivating ATM. Nat. Cell Biol. 9, 573–580.
Stambolic, V., Macpherson, D., Sas, D., Lin, Y., Snow, B., Jang, Y., Benchimol, S., Mak, T.W., 2001. Regulation of PTEN transcription by p53. Mol. Cell 8, 317–325.
Stecca, B., Ruiz, I.A.A., 2010. Context-dependent regulation of the GLI code in cancer by HEDGEHOG and non-HEDGEHOG signals. J. Mol. Cell Biol. 2, 84–95.
Su, W.L., Kleinhanz, R.R., Schadt, E.E., 2011. Characterizing the role of miRNAs within gene regulatory networks using integrative genomics techniques. Mol. Syst. Biol. 7, 490.
Sulston, J.E., Horvitz, H.R., 1981. Abnormal cell lineages in mutants of the nematode Caenorhabditis elegans. Dev. Biol. 82, 41–55.
Sun, X., Jiang, S., Liu, J., Wang, H., Zhang, Y., Tang, S.C., Wang, J., Du, N., Xu, C., Wang, C., Qin, S., Zhang, J., Liu, D., Zhang, Y., LI, X., Wang, J., Dong, J., Wang, X., Xu, S., Tao, Z., Xu, F., Zhou, J., Wang, T., Ren, H., 2015. MiR-208a stimulates the cocktail of SOX2 and beta-catenin to inhibit the let-7 induction of self-renewal repression of breast cancer stem cells and formed miR208a/let-7 feedback loop via LIN28 and DICER1. Oncotarget 6, 32944–32954.
Suzuki, H.I., Yamagata, K., Sugimoto, K., Iwamoto, T., Kato, S., Miyazono, K., 2009. Modulation of microRNA processing by p53. Nature 460, 529–533.
Sylvestre, Y., De Guire, V., Querido, E., Mukhopadhyay, U.K., Bourdeau, V., Major, F., Ferbeyre, G., Chartrand, P., 2007. An E2F/miR-20a autoregulatory feedback loop. J. Biol. Chem. 282, 2135–2143.
Taipale, J., Cooper, M.K., Maiti, T., Beachy, P.A., 2002. Patched acts catalytically to suppress the activity of Smoothened. Nature 418, 892–897.
Takamizawa, J., Konishi, H., Yanagisawa, K., Tomida, S., Osada, H., Endoh, H., Harano, T., Yatabe, Y., Nagino, M., Nimura, Y., Mitsudomi, T., Takahashi, T., 2004. Reduced expression of the let-7 microRNAs in human lung cancers in association with shortened postoperative survival. Cancer Res. 64, 3753–3756.
Tang, B., Vu, M., Booker, T., Santner, S.J., Miller, F.R., Anver, M.R., Wakefield, L.M., 2003. TGF-beta switches from tumor suppressor to prometastatic factor in a model of breast cancer progression. J. Clin. Invest. 112, 1116–1124.
Tarasov, V., Jung, P., Verdoodt, B., Lodygin, D., Epanchintsev, A., Menssen, A., Meister, G., Hermeking, H., 2007. Differential regulation of microRNAs by p53 revealed by massively parallel sequencing: miR-34a is a p53 target that induces apoptosis and G1-arrest. Cell Cycle 6, 1586–1593.
Tavakoli, R., Vakilian, S., Langroudi, L., Arefian, E., Sahmani, M., Soleimani, M., Jamshidi-Adegani, F., 2017. The role of miR-17-92 cluster in the expression of tumor suppressor genes in unrestricted somatic stem cells. Biologicals 46, 143–147.
Tavella, S., Biticchi, R., Schito, A., Minina, E., DI Martino, D., Pagano, A., Vortkamp, A., Horton, W.A., Cancedda, R., Garofalo, S., 2004. Targeted expression of SHH affects chondrocyte differentiation, growth plate organization, and Sox9 expression. J. Bone Miner. Res. 19, 1678–1688.
Taylor, M.A., Sossey-Alaoui, K., Thompson, C.L., Danielpour, D., Schiemann, W.P., 2013. TGF-β upregulates miR-181a expression to promote breast cancer metastasis. J. Clin. Invest. 123, 150–163.
ten Berge, D., Kurek, D., Blauwkamp, T., Koole, W., Maas, A., Eroglu, E., Siu, R.K., Nusse, R., 2011. Embryonic stem cells require Wnt proteins to prevent differentiation to epiblast stem cells. Nat. Cell Biol. 13, 1070–1075.

Thornton, J.E., Chang, H.M., Piskounova, E., Gregory, R.I., 2012. Lin28-mediated control of let-7 microRNA expression by alternative TUTases Zcchc11 (TUT4) and Zcchc6 (TUT7). RNA 18, 1875–1885.

Torbett, N.E., Luna-Moran, A., Knight, Z.A., Houk, A., Moasser, M., Weiss, W., Shokat, K.M., Stokoe, D., 2008. A chemical screen in diverse breast cancer cell lines reveals genetic enhancers and suppressors of sensitivity to PI3K isoform-selective inhibition. Biochem. J. 415, 97–110.

Trang, P., Medina, P.P., Wiggins, J.F., Ruffino, L., Kelnar, K., Omotola, M., Homer, R., Brown, D., Bader, A.G., Weidhaas, J.B., Slack, F.J., 2010. Regression of murine lung tumors by the let-7 microRNA. Oncogene 29, 1580–1587.

Trang, P., Wiggins, J.F., Daige, C.L., Cho, C., Omotola, M., Brown, D., Weidhaas, J.B., Bader, A.G., Slack, F.J., 2011. Systemic delivery of tumor suppressor microRNA mimics using a neutral lipid emulsion inhibits lung tumors in mice. Mol. Ther. 19, 1116–1122.

Tuveson, D.A., Neoptolemos, J.P., 2012. Understanding metastasis in pancreatic cancer: a call for new clinical approaches. Cell 148, 21–23.

Umezu, T., Ohyashiki, K., Kuroda, M., Ohyashiki, J.H., 2013. Leukemia cell to endothelial cell communication via exosomal miRNAs. Oncogene 32, 2747–2755.

Uziel, T., Karginov, F.V., Xie, S., Parker, J.S., Wang, Y.D., Gajjar, A., He, L., Ellison, D., Gilbertson, R.J., Hannon, G., Roussel, M.F., 2009. The miR-17∼92 cluster collaborates with the Sonic Hedgehog pathway in medulloblastoma. Proc. Natl. Acad. Sci. U.S.A. 106, 2812–2817.

Valadi, H., Ekström, K., Bossios, A., Sjöstrand, M., Lee, J.J., Lötvall, J.O., 2007. Exosome-mediated transfer of mRNAs and microRNAs is a novel mechanism of genetic exchange between cells. Nat. Cell Biol. 9, 654–659.

Van Roosbroeck, K., Fanini, F., Setoyama, T., Ivan, C., Rodriguez-Aguayo, C., Fuentes-Mattei, E., Xiao, L., Vannini, I., Redis, R., D'abundo, L., Zhang, X., Nicoloso, M.S., Rossi, S., Gonzalez-Villasana, V., Rupaimoole, R., Ferracin, M., Morabito, F., Neri, A., Ruvolo, P., Ruvolo, V.R., Pecot, C.V., Amadori, D., Aruzzo, L., Calin, S., Wang, X., You, M.J., Ferrajoli, A., Orlowski, R.Z., Plunkett, W., Lichtenberg, T., Davuluri, R.V., Berindan-Neagoe, I., Negrini, M., Wistuba, I., Hagop, K., Sood, A.K., Lopez-Berestein, G., Keating, M.J., Fabbri, M., Calin, G.A., 2016. Combining anti-miR-155 with chemotherapy for the treatment of lung cancers. Clin. Cancer Res. http://dx.doi.org/10.1158/1078-0432.CCR-16-1025. [Epub ahead of print].

Varjosalo, M., Taipale, J., 2008. Hedgehog: functions and mechanisms. Genes Dev. 22, 2454–2472.

Varnat, F., Duquet, A., Malerba, M., Zbinden, M., Mas, C., Gervaz, P., Ruiz I Altaba, A., 2009. Human colon cancer epithelial cells harbour active HEDGEHOG-Gli signalling that is essential for tumour growth, recurrence, metastasis and stem cell survival and expansion. EMBO Mol. Med. 1, 338–351.

Vasudevan, S., Tong, Y., Steitz, J.A., 2007. Switching from repression to activation: microRNAs can up-regulate translation. Science 318, 1931–1934.

Ventura, A., Young, A.G., Winslow, M.M., Lintault, L., Meissner, A., Erkeland, S.J., Newman, J., Bronson, R.T., Crowley, D., Stone, J.R., Jaenisch, R., Sharp, P.A., Jacks, T., 2008. Targeted deletion reveals essential and overlapping functions of the miR-17 through 92 family of miRNA clusters. Cell 132, 875–886.

Venturini, L., Battmer, K., Castoldi, M., Schultheis, B., Hochhaus, A., Muckenthaler, M.U., Ganser, A., Eder, M., Scherr, M., 2007. Expression of the miR-17-92 polycistron in chronic myeloid leukemia (CML) CD34+ cells. Blood 109, 4399–4405.

Vogelstein, B., Fearon, E.R., Hamilton, S.R., Kern, S.E., Preisinger, A.C., Leppert, M., Nakamura, Y., White, R., Smits, A.M., Bos, J.L., 1988. Genetic alterations during colorectal-tumor development. N. Engl. J. Med. 319, 525–532.

Volinia, S., Calin, G.A., Liu, C.G., Ambs, S., Cimmino, A., Petrocca, F., Visone, R., Iorio, M., Roldo, C., Ferracin, M., Prueitt, R.L., Yanaihara, N., Lanza, G., Scarpa, A., Vecchione, A., Negrini, M., Harris, C.C., Croce, C.M., 2006. A microRNA expression signature of human solid tumors defines cancer gene targets. Proc. Natl. Acad. Sci. U.S.A. 103, 2257–2261.

Vousden, K.H., Prives, C., 2009. Blinded by the light: the growing complexity of p53. Cell 137, 413–431.

Wang, J., Qiao, L., Li, Y., Yang, G., 2008. Ginsenoside Rb1 attenuates intestinal ischemia-reperfusion- induced liver injury by inhibiting NF-kappaB activation. Exp. Mol. Med. 40, 686–698.

Wang, D., Qiu, C., Zhang, H., Wang, J., Cui, Q., Yin, Y., 2010. Human microRNA oncogenes and tumor suppressors show significantly different biological patterns: from functions to targets. PLoS One 5 (9), pii: e13067. http://dx.doi.org/10.1371/journal.pone.0013067.

Wang, X.F., Shi, Z.M., Wang, X.R., Cao, L., Wang, Y.Y., Zhang, J.X., Yin, Y., Luo, H., Kang, C.S., Liu, N., Jiang, T., You, Y.P., 2012. MiR-181d acts as a tumor suppressor in glioma by targeting K-ras and Bcl-2. J. Cancer Res. Clin. Oncol. 138, 573–584.

Watkins, D.N., Berman, D.M., Burkholder, S.G., Wang, B., Beachy, P.A., Baylin, S.B., 2003. Hedgehog signalling within airway epithelial progenitors and in small-cell lung cancer. Nature 422, 313–317.

Wee, S., Wiederschain, D., Maira, S.M., Loo, A., Miller, C., Debeaumont, R., Stegmeier, F., Yao, Y.M., Lengauer, C., 2008. PTEN-deficient cancers depend on PIK3CB. Proc. Natl. Acad. Sci. U.S.A. 105, 13057–13062.

Wellbrock, C., Hurlstone, A., 2010. BRAF as therapeutic target in melanoma. Biochem. Pharmacol. 80, 561–567.

Whelan, J.T., Forbes, S.L., Bertrand, F.E., 2007. CBF-1 (RBP-J kappa) binds to the PTEN promoter and regulates PTEN gene expression. Cell Cycle 6, 80–84.

Wong, T.S., Man, O.Y., Tsang, C.M., Tsao, S.W., Tsang, R.K., Chan, J.Y., HO, W.K., Wei, W.I., To, V.S., 2011. MicroRNA let-7 suppresses nasopharyngeal carcinoma cells proliferation through downregulating c-Myc expression. J. Cancer Res. Clin. Oncol. 137, 415–422.

Worby, C.A., Dixon, J.E., 2014. Pten. Annu. Rev. Biochem. 83, 641–669.

Yauch, R.L., Gould, S.E., Scales, S.J., Tang, T., Tian, H., Ahn, C.P., Marshall, D., Fu, L., Januario, T., Kallop, D., Nannini-Pepe, M., Kotkow, K., Marsters, J.C., Rubin, L.L., de Sauvage, F.J., 2008. A paracrine requirement for hedgehog signalling in cancer. Nature 455, 406–410.

Yilmaz, O.H., Valdez, R., Theisen, B.K., Guo, W., Ferguson, D.O., Wu, H., Morrison, S.J., 2006. Pten dependence distinguishes haematopoietic stem cells from leukaemia-initiating cells. Nature 441, 475–482.

Yin, L., Velazquez, O.C., Liu, Z.J., 2010. Notch signaling: emerging molecular targets for cancer therapy. Biochem. Pharmacol. 80, 690–701.

Yun, J., Rago, C., Cheong, I., Pagliarini, R., Angenendt, P., Rajagopalan, H., Schmidt, K., Willson, J.K., Markowitz, S., Zhou, S., Diaz Jr., L.A., Velculescu, V.E., Lengauer, C., Kinzler, K.W., Vogelstein, B., Papadopoulos, N., 2009. Glucose deprivation contributes to the development of KRAS pathway mutations in tumor cells. Science 325, 1555–1559.

Zaravinos, A., 2015. The regulatory role of microRNAs in EMT and cancer. J. Oncol. 2015, 865816.

Zhang, J., Grindley, J.C., Yin, T., Jayasinghe, S., HE, X.C., Ross, J.T., Haug, J.S., Rupp, D., Porter-Westpfahl, K.S., Wiedemann, L.M., Wu, H., Li, L., 2006. PTEN maintains haematopoietic stem cells and acts in lineage choice and leukaemia prevention. Nature 441, 518–522.

Zhang, B., Pan, X., Cobb, G.P., Anderson, T.A., 2007. microRNAs as oncogenes and tumor suppressors. Dev. Biol. 302, 1–12.
Zhang, H., Wang, Y., Xu, T., Li, C., Wu, J., HE, Q., Wang, G., Ding, C., Liu, K., Tang, H., Ji, F., 2016. Increased expression of microRNA-148a in osteosarcoma promotes cancer cell growth by targeting PTEN. Oncol. Lett. 12, 3208–3214.
Zhu, H., Dougherty, U., Robinson, V., Mustafi, R., Pekow, J., Kupfer, S., LI, Y.C., Hart, J., Goss, K., Fichera, A., Joseph, L., Bissonnette, M., 2011. EGFR signals downregulate tumor suppressors miR-143 and miR-145 in Western diet-promoted murine colon cancer: role of G1 regulators. Mol. Cancer Res. 9, 960–975.

CHAPTER SEVEN

MicroRNAs as Multifaceted Players in Glioblastoma Multiforme

Neri Mercatelli*,‡, Silvia Galardi†, Silvia Anna Ciafrè†,1

*Unit of Biology, Genetics and Biochemistry, Department of Movement, Human and Health Sciences, University of Rome "Foro Italico", Rome, Italy

†Department of Biomedicine and Prevention, University of Rome "Tor Vergata", Rome, Italy

‡Laboratory of Cellular and Molecular Neurobiology, CERC Fondazione Santa Lucia, Rome, Italy

[1]Corresponding author: e-mail address: ciafre@uniroma2.it

Contents

Abstract

Glioblastoma multiforme (GBM) is the most common and inevitably lethal primary brain tumor, with a median survival rate of only 15 months from diagnosis. The current standard treatment involves maximal surgical resection flanked by radiotherapy and chemotherapy with the alkylating agent temozolomide. However, even such aggressive treatment is never curative, and recurrent tumors always arise, commonly in more aggressive, chemo- and radio-resistant forms, leading to untreatable and deadly tumors. MicroRNAs, recognized major players in cancer, are deeply involved in GBM, as shown by more than a decade of studies. In this review, we revise the main milestones of MicroRNA studies in GBM, and the latest relevant discoveries in this field. Examples are given of MicroRNAs working as "oncomiRs" or tumor suppressors, with specific connections with GBM clinical subtypes, patients' survival, and resistance to therapies. As the interaction of GBM cells with the microenvironment was proven as a key determinant of

International Review of Cell and Molecular Biology, Volume 333

ISSN 1937-6448

http://dx.doi.org/10.1016/bs.ircmb.2017.03.002

tumor growth, the role of MicroRNAs in GBM microenvironment, tumor angiogenesis, and tumor-secreted microvesicles is also reviewed. Finally, we discuss the latest findings presenting MicroRNAs as possible therapeutic targets for GBM, or their use as circulating biomarkers in diagnosis and prognosis.

1. INTRODUCTION

Gliomas are the most frequent intracranial tumors, representing more than 50% of all primary brain tumors, with glioblastoma multiforme (GBM) being by far the most common and aggressive. In the mid-19th century, Rudolph Virchow for the first time proposed a classification of brain tumors, first describing neuroglia and then relating it to brain tumors (Virchow, 1864/1865). After his work, Bailey and Cushing in 1926 set the basis of all subsequent glioma classifications, based on histogenetic characteristics, and including the concept of grading, which relates tumor histopathology with patient outcome (Bailey and Cushing, 1926). Our present classification of gliomas is guided by the World Health Organization (WHO), which started classifying brain tumors in 1979, and then published several revisions, the last of which in 2007, that has very recently undergone an important update in 2016 (Louis et al., 2007, 2016).

2. CLINICAL AND MOLECULAR SUBTYPING

Gliomas are classified into low-grade vs malignant, high-grade tumors and until 2007 their official classification was essentially based on cellular density, presence of necrosis, nuclear atypia, mitotic index, vascular proliferation, clinical course, and treatment outcome, defining pathological grades I–IV. The malignant forms of these tumors are classified by WHO as either grade III tumors such as anaplastic astrocytoma, anaplastic oligodendroglioma, anaplastic oligoastrocytoma, and anaplastic ependymomas, or grade IV tumors such as GBM. GBM is largely prevalent (80%) among malignant primary brain tumors, with an incidence of more than 3–4 cases per 100,000 per year (Dolecek et al., 2012). With a mean age at diagnosis of 64 (ranging from 50 to 70), it is clearly a tumor of the adults, and it shows a male predominance (M/F ratio around 1.5-fold) (Ostrom et al., 2015). Most GBMs are primary tumors, occurring de novo, and developing rapidly with no previous lesions of lower malignancy. However, about 10% are secondary, and arise through the progression of preexisting grade II or III gliomas, usually affecting younger patients, below the age of 45 years, and do not show a clear male predominance (Ohgaki and Kleihues, 2013). Despite their different origins, no morphological difference distinguishes primary from secondary GBMs, and their prognoses are equally dismal, ranging from 5 to 15 months after diagnosis for primary GBM to 8–31 months for secondary tumors (Dropcho and Soong, 1996; Louis et al., 2016), which makes these tumors rank among the deadliest human cancers. In fact, despite the recent advances in surgical techniques and radiotherapy, the current standard approach for GBM, consisting of maximal safe resection followed by radiotherapy and concomitant and adjuvant chemotherapy with temozolomide (TMZ), only results in a median overall survival (OS) of 12–18 months (Ohgaki and Kleihues, 2005; Wen and Kesari, 2008). An interesting exception to this rule is a small fraction (<10%) of GBM patients, referred to as long-term survivors (LTS), who survive longer than 36 months (Geisenberger et al., 2015).

From a molecular genetics perspective, some specific features are distinctive of primary vs secondary GBMs. For example, amplification of epidermal growth factor receptor (EGFR), CDK4, MDM2, and MDM4, homozygous deletion of CDKN2A and $p14^{ARF}$, together with mutation/homozygous deletion of RB1, mutations in TERT promoter, mutation of PTEN and monosomy of chromosome 10, prevalently characterize primary GBMs. On the contrary, TP53 mutations characterize more than 60% of secondary GBMs, and the whole p53 pathway is altered in secondary GBMs, which also show silencing of RB1 through promoter hypermethylation, over-expression of PDGFRA, mutations in ATRX, and allelic losses of regions in 19q and 13q.

One of the most important molecular features differentiating primary and secondary GBM is a mutation in IDH1 (Parsons et al., 2008), where a mutated state of IDH1 is a predictor of a better prognosis. On the contrary, primary GBMs show wild-type IDH1, with only a tiny fraction (4%–7%) bearing mutated forms of IDH1 (Ohgaki and Kleihues, 2013) (see Box 1). Actually, the doubt still exists whether there are true primary GBMs with IDH1 mutations: it is possible, in fact, that tumors clinically diagnosed as primary may instead be neoplasms rapidly progressed from precursor lesions that escaped clinical diagnosis and so were misclassified as primary glioblastomas.

A further relevant challenge presented by malignant brain tumors is their great inter- and intratumoral molecular heterogeneity, despite an apparent histological uniformity. Thus, while we continue to define malignant

BOX 1 IDH Gene State as a Clinical Subtype Definer for Glioblastoma

- IDH wt is typical of primary, or "de novo" glioblastoma, while IDH mut characterizes prevalently secondary GBM
- IDH wt GBM patients are diagnosed on average at 62 years, compared to IDH mut patients, diagnosed on average at 44
- IDH wt GBM patients show a median OS (after surgical resection, radio- and chemotherapy) of 15 months, compared to IDH mut patients, with a OS of 31 months
- The great majority (~92%) of IDH mut GBMs show a Proneural signature
- A large fraction (~72%) of IDH wt GBM show mutations in TERT promoter, compared to only 26% of IDH mut GBMs
- Only a minority (~27%) of IDH wt GBMs show loss or mutations of TP53, compared to ~81% of IDH mut GBMs
- ATRX mutations are extremely rare in IDH wt GBMs, while they are prevalent (~71%) in IDH mut GBMs
- EGFR amplification/mutations are more frequent (~35%) in IDH wt GBMs, while they are very rare in IDH mut GBMs
- PTEN mutations are often present (~24%) in IDH wt GBMs, while they are very rare in IDH mut GBMs
- NF1 loss/mutations characterize IDH wt GBMs vs IDH mut ones
- PIK3R1/PIK3CA mutations characterize IDH wt GBMs vs IDH mut ones

gliomas by histological criteria, several reports in the last 15 years have shown that gene expression profiles can be a much more powerful tool for the classification and the consequent prognosis of brain tumors. Gene expression studies have revealed at least two key aspects of astrocytomas and glioblastomas: one is that these tumors are usually composed of a mixture of diverse molecular genetic subtypes, pointing at the importance of unraveling the different components of each tumor, in order to develop effective therapies. The second, not by importance, observation is that, though among this great variability, some signatures exist that define glioblastoma subclasses and are useful not only for the prognosis but also for the prediction of patients response or resistance to therapy.

A significant step toward the molecular characterization of glioblastoma was taken in 2010, when the Cancer Genome Atlas (TCGA) network reported a robust gene expression profiling classification for GBM, resulting in four distinct subtypes, namely proneural (PN), neural (N), classical (CL), and mesenchymal (MES) (Verhaak et al., 2010, see Box 2). Beside deeply defining genetic and gene expression aberrations specifically tagging the four

BOX 2 Glioblastoma Classification Based on Verhaak's Molecular Signatures

	Proneural	Neural	Classical	Mesenchymal
Cell type-specific signature	Oligodendrocytic	Oligodendrocytic and astrocytic +neuronal	Astrocytic	Microglia and cultured astroglia
Core signature	FBXO3 GABRB2 SNCG MBP	DNMT1 TOP1 ABL1 BOP1	DLL3 NKX2-2 SOX2 ERBB3 OLIG2	ILR4 CHI3L1 TRADD TLR2/4 RELB CASP1/4/5/8
Age at diagnosis	Younger than all other subtypes			
OS	Trend toward longer survival than all other subtypes			
Aggressive treatment efficacy	Not effective	Trend toward reduced mortality	Significantly reduces mortality	Significantly reduces mortality
Aberrations	PDGFRA abnormalities; IDH mut; TP53 mut	EGFR amplification	Chr7 amplifications; Chr10 deletions; EGFR amplification; absence of TP53 mutations; homozygous deletion of the *Ink4a/ARF* locus	Overexpression of *CHI3L1* and *MET*; high frequency of *NF1* mutation/deletion

subtypes, this work showed that the different subtypes have signatures that are reminiscent of distinct healthy neural cell types. PN glioblastomas are highly enriched with the oligodendrocytic signature, CL tumors closely reflect an astrocytic signature, MES ones are strongly associated with the cultured astroglial signature, and the neural tumors show a strong association with oligodendrocytic and astrocytic differentiation, but they are also enriched in genes specifically expressed by neurons. Such interesting biological observations were flanked by the most important findings, from a clinical perspective, of differential features shown by the four subtypes. From several points of view, the proneural subtype is clearly different from all others: patients affected by this type of GBM were younger than the others, and, even if not statistically significant, there was a trend toward longer survival for PN GBM patients. Most importantly, the effects of intensive treatment (defined as concurrent chemo- and radiotherapy or more than three subsequent cycles of chemotherapy), on survival was found to be different in the four subtypes. While classical and mesenchymal subtypes benefited of aggressive treatment, which reduced mortality, the proneural subtype survival was not affected, indicating how therapy could be "personalized," based on the subtype definition.

To further confirm the importance of IDH1 mutation state in the classification of glioblastoma (see Box 1), the same authors also found that 92% of $IDH1^{mut}$tumors showed a proneural signature, while wt IDH1 GBMs were identified as classical, mesenchymal, neural, or even proneural (Verhaak et al., 2010). Moreover, the majority of secondary GBMs analyzed in that work was classified as proneural. These and other observations suggest that secondary GBMs are rather homogeneous, largely overlap with $IDH1^{mut}$tumors and with the proneural subtype, while primary glioblastomas are heterogeneous, with several distinct expression profiles.

3. GLIOBLASTOMA INITIATION AND PROPAGATION: THE CANCER STEM CELLS

The definition of the cells of origin of glioblastoma has been the subject of a long-lasting scientific debate which reached a peak in the early 2000, when more than one group revealed the existence of the so-called "glioblastoma stem cells" (GSCs) or "brain tumor initiating cells" (Galli et al., 2004; Hemmati et al., 2003; Ignatova et al., 2002). These cells are defined as multipotent, undifferentiated, self-sustaining, yet transformed cells that, in the specific context of glioblastoma, self-renew under clonal conditions and differentiate into neuron- and glia-like cells, with aberrant, mixed neuronal/astroglial phenotypes. In vitro, these cells grow as tumor spheres, express stem cell markers, and can be differentiated into all three neural lineages. Most importantly, a required feature of GSCs is their ability, upon intracerebral transplantation in immunosuppressed mice, to form secondary tumors that recapitulate the cellular heterogeneity present in the parental tumor, thus phenocopying it, even after multiple rounds of transplantation (Eyler et al., 2011; Flavahan et al., 2013; Lathia et al., 2010; Li et al., 2009a; Singh et al., 2004). Remarkably, these cells display resistance to radio- and chemotherapy (Bao et al., 2006a; Chen et al., 2012), promote tumor angiogenesis (Bao et al., 2006b), and are considered the cell type responsible of recurrence and consequently of the inevitable dismal prognosis of glioblastoma clinical history. Moreover, such cells are very difficult enemies to defeat, as they are not rapidly dividing, but are rather quiescent, thus escaping all antiproliferative therapeutic approaches, and they are not negatively affected by several stressful conditions that instead impair common tumor cell growth and survival, such as hypoxia, oxidative stress, inflammation, acidic stress, and low glucose (Flavahan et al., 2013; Hjelmeland et al., 2011; Li et al., 2009a; Venere et al., 2014; Ye et al., 2012). On the contrary, tumor stem cells can exploit difficult growth conditions and use them to overtake other, nonstem cells, thus rapidly repopulating the area, even after a first surgical and chemo- or radiotherapeutic intervention (for an exhaustive review about GSCs, see Lathia et al., 2015). Their ability to escape cytotoxic therapies relies on multiple mechanisms, including DNA damage control (Ahmed et al., 2015; Bao et al., 2006a), resistance to apoptosis (Venere et al., 2014), epigenetic modulation (Banelli et al., 2015; Suvà et al., 2014), and specific pathways, such as Notch (Saito et al., 2015; Wang et al., 2010), Wnt (Rheinbay et al., 2013), and NF-κB (Berger et al., 2011; Kim et al., 2016). The obvious consequence of such an intricate defense network employed by GSCs is that it would be necessary to target more than one of these aspects, and to integrate these approaches with others targeted to the microenvironment to which these cells are exposed, in order to achieve a therapeutic and stable success against glioblastoma growth and recurrence.

4. TREATMENT OPTIONS AND RESISTANCE TO THERAPY

The standard care treatment for glioblastoma has been essentially the same in the last 20 years, and involves maximal surgical resection, followed by external beam irradiation and chemotherapy with temozolomide (TMZ). This drug is an oral cytotoxic DNA-alkylating agent, approved for the first time in 1999 for its use against malignant gliomas, that promotes the methylation of the O6 in guanine, and of N7 and N3 in purines. It is used both immediately after surgery, in combination with regional radiotherapy, and then later, alone as an adjuvant. However, even this concurrent radio- and chemotherapy results in a poor median OS (14.6 months) and 2-year survival rate of only 26.5% (Stupp et al., 2005). The reason for such a scarce therapeutic efficacy can be found in the fast recurrence of TMZ-resistant tumors. A finding which confers further importance to "stem cells" in glioblastoma is that described by Chen and colleagues: they showed that these quiescent cells are at the apex of a cellular hierarchy in tumor maintenance and recurrence after TMZ chemotherapy, giving rise to a transient population of highly proliferative and TMZ-resistant relapsing cells (Chen et al., 2012).

One of the prevalent and most studied mechanisms of resistance to TMZ is the expression of MGMT (O6-methylguanine-DNA methyltransferase), a DNA repair enzyme that removes alkyl groups from the O6 position of guanine, thus impairing the effects of alkylating drugs. In fact, upon treatment with TMZ and radiotherapy, GBM patients whose tumors present a methylated promoter of MGMT gene, with the consequent repression of MGMT expression, have an improved prognosis (median OS at 2 years of 46%) (Hegi et al., 2005). Another mechanism contributing to resistance to TMZ in glioblastoma patients is via poly(ADP-ribose)polymerase-1 (PARP-1). This gene encodes for a crucial base excision repair enzyme acting on N7 and N3 lesions on purines. When PARP-1 is inactivated or inhibited, such lesions persist and contribute to TMZ cytotoxicity, even in a context where O6-methylguanine adducts are repaired by a functioning MGMT.

Very recently, the involvement of additional DNA repair systems has been shown in resistance of glioblastoma cells to temozolomide. Peng et al. demonstrated that the expression of the error-prone DNA polymerase κ was significantly upregulated in GBM cell lines and tumors upon TMZ treatment, and its overexpression correlated with poor prognosis in GBM patients treated with TMZ. Moreover, DNA polymerase κ overexpression was sufficient to confer resistance to TMZ (Peng et al., 2016). This enzyme, working in homologous recombination (HR)-mediated repair, counteracts TMZ-induced Rad17 polyubiquitination, and proteasomal degradation, thus activating ATR-Chk1 signaling, in turn allowing the restart of stalled replication forks, cell-cycle reentry and progression, and the clearance of TMZ-induced DNA breaks, ultimately leading to tolerance to temozolomide. Another recent work with strong implications from both the biological and translational perspectives uncovers a key role for retinoblastoma-binding protein 4 (RBBP4) in the resistance of glioblastoma to temozolomide. RBBP4, a component of several chromatin-modifying protein complexes, was found to negatively modulate GBM sensitivity to TMZ, by increasing MGMT expression, and also inducing several genes involved in HR, including RAD51 (Kitange et al., 2016). Thus, RBBP4, through and epigenetic mode of regulation of gene expression affecting histone acetylation, is placed upstream of at least two main mechanisms of chemoresistance, as the repair of O6-methylguanine lesions and the HR-mediated double-strand break repair.

The ability to repair DNA damage, and in particular double-strand breaks, is strictly related to sensitivity to radiotherapy. As radiation therapy is an integrating part of the standard therapy for GBM, the comprehension of radioresistance mechanisms is as important as that of chemoresistance. In fact, some pathways are shared by the two responses, such as ATM–ATR signaling, resulting in a cascade of events comprehensively known as "DNA damage response" (DDR), and involving Chk1, Chk2, Rad51, and several other enzymes playing key roles in DDR. Ionizing radiations activate DNA damage checkpoints in glioma stem cells (GSCs) and thus induce the ability of these cells to repair DNA breaks, in turn leading to radioresistance of recurrent tumors (Bao et al., 2006a). For this reason, a number of ATM inhibitors are under consideration as possible complements of radiotherapy for glioblastoma, as for example KU-60019, shown to significantly increase the survival of glioblastoma xenografts treated with radiotherapy (Biddlestone-Thorpe et al., 2013).

A recent work has provided a great contribution to the elucidation of a radioresistance mechanism in GSCs, that is upstream of ATM (De Bacco et al., 2016). After demonstrating that ionizing radiations positively select GSCs, De Bacco et al., showed that the HGF receptor MET, a functional marker sustaining self-renewal of GSCs, specifically marks radioresistant stem cells in GBM. In fact, MET-positive cells are strongly enriched in recurrent, radioresistant tumors compared to their matched primary ones. The radioresistance of MET-positive stem cells was correlated with a high basal DDR activity, represented by constitutive phosphorylation of ATM and Chk2 kinases, and, after irradiation, by a strong further induction of ATM phosphorylation and RAD51 expression. The MET-driven signaling pathway responsible for radioresistance works via the phosphorylation of AKT, in turn activating Aurora A and causing the cytoplasmic retention of p21, respectively, leading to ATM activation and DNA repair or to inhibition of apoptosis. As a consequence, MET inhibition sensitizes GSCs to ionizing radiations, by directly impinging on DDR, reducing ATM and Chk2 phosphorylation, and RAD51 expression. The ultimate message of this work is that a careful selection of glioblastoma patients on the basis of the MET positivity of their tumor cells, may help design a personalized radiotherapy associated to MET inhibitors, thus enhancing the efficacy of radiotherapy and likely counteracting recurrence of radioresistant tumors.

As for HGF, working through its receptor MET, also FGF-2 is a potent mediator of the microenvironment effects on glioblastoma cells, and specifically on their response to radiations. Upon irradiation, FGF-2, by binding to its receptor FGFR1, triggers a cascade of intracellular events involving phospholipase C γ (PLCγ) and HIF1-α, leading to impaired mitotic cell death, and consequent reduced radiosensitivity. Thus, inhibition of FGFR1 has been recently shown to increase sensitivity of GBM cells to radiation, in a manner associated with centrosome overduplication and mitotic cell death (Gouazé-Andersson et al., 2016). As for MET inhibitors, drugs targeting FGFR1 are now under evaluation as possible enhancers of radiotherapy efficacy in GBM, and to overcome radioresistance (Ader et al., 2014; Gouazé-Andersson et al., 2016).

5. MicroRNAs AND GLIOBLASTOMA

In the last years, a wide range of scientific papers have highlighted the relevance of MicroRNAs in the genetic network driving GBM oncogenesis (Sumazin et al., 2011). Among them, expression profile studies conducted in the early 2000s were pioneers in providing the first sets or MicroRNA families which would have been explored for their putative role in GBM etiology (Chan et al., 2005; Ciafrè et al., 2005). Indeed, the subsequent understanding of cancer-related molecular pathways regulated by these MicroRNAs together with the discovery of their main mRNA targets, led to the final identification of their functional role in tumorigenic processes linked to GBM pathology (Godlewski et al., 2010b).

The widespread relevance of MicroRNAs in glioblastoma pathogenesis and clinical evolution is clearly indicated by a recent work which studied the effects of knocking out DICER, the enzyme required for the cleavage of pre-miRNAs into their mature forms, in GSCs (Mansouri et al., 2016). The disruption of such a key enzyme for MiRNA biosynthesis deeply affected GSC characteristics, by of reducing the expression stemness markers and the efficiency of neurosphere formation. However, the orthotopic injection of DICER-silenced GSCs into recipient mice generated more aggressive tumors and consequent shorter survival compared to control mice. Very importantly though, these tumors were significantly more sensitive to radiation therapy and, after irradiation, grew much smaller compared to control tumors. The authors propose that such apparently contrasting results may be explained by a complex role played by MicroRNAs in GSCs, that might shift the nature of these cells toward a more differentiated and proproliferative state, producing larger, but also more radiosensitive tumors. These data in fact recapitulate previous works reporting a general reduction of MicroRNA expression in cancer, including GBM (Kumar et al., 2007), and, at the same time, a great involvement of GSCs in radiation resistance (Bao et al., 2006a).

At present, several specific MicroRNAs can be considered as master regulators of GBM and, depending on their positive or negative effect on tumor occurrence and progression, they are currently ascribed to the oncogene or tumor–suppressor gene classes (Floyd and Purow, 2014).

In this section, we will discuss in a detailed manner the most important experimental findings collected by different research groups including ours, regarding the functional role of MicroRNAs in GBM and the molecular mechanism associated to each of them, with the main purpose to provide a reasoned and accurate classification of MicroRNAs with tumorigenic or tumor–suppressor function in glioblastoma (Fig. 1).

5.1 MicroRNAs With Tumorigenic Roles in Glioblastoma

Among the MicroRNAs showing a tumorigenic effect in GBM (Table 1), a pivotal role is played by MiR-21. Initially identified by us and others as one of the most expressed MicroRNAs in human GBM tumor tissues as well as in GBM cell lines (Chan et al., 2005; Ciafrè et al., 2005), MiR-21 contributes to the malignant phenotype by promoting proliferation, survival, migration, and radio-chemoresistance of glioma cells (Chan et al., 2005; Gabriely et al., 2008; Papagiannakopoulos et al., 2008). In particular, the molecular mechanism through which MiR-21 exerts its oncogenic role is due to its pleiotropic effect acting on a network of tumor-suppressive mRNA targets (Papagiannakopoulos et al., 2008). Indeed, it has been well documented that the antiapoptotic role of MiR-21 in GBM is mediated by the repression of the apoptotic factor PDCD4 as well as of HNRPK and TAp63 (Chen et al., 2008; Papagiannakopoulos et al., 2008). The specific inhibition of MiR-21 not only is able to activate in vitro and in vivo caspase-associated apoptosis but also to decrease cell invasiveness by directly increasing the protein levels of matrix metalloproteinase inhibitors, RECK and TIMP3 (Gabriely et al., 2008).

Similarly to MiR-21, MiR-10b is one of the most upregulated MicroRNAs in GBM tumors and cell lines (Ciafrè et al., 2005; Sasayama et al., 2009). The increase of its expression has been linked to a significant reduction of survival in GBM patients (Gabriely et al., 2011). Targeting of

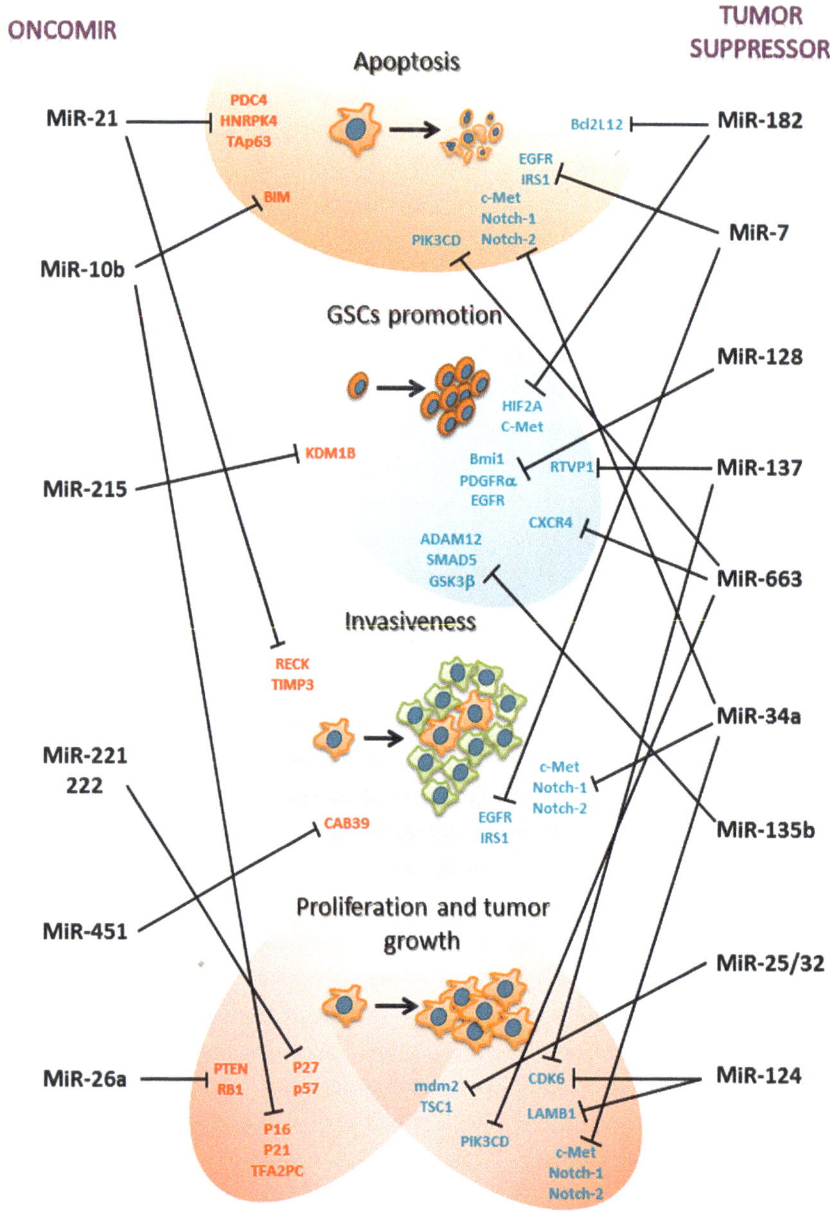

Fig. 1 An overview of relevant MiRNAs working as "oncomiRs" or tumor suppressor MiRNAs in glioblastoma. Oncogenic or tumor suppressor MiRNAs cited in this text are listed on the left or the right of the figure, respectively. The regulated targets (in *red* for targets of oncomiRs and in *blue* for targets of tumor suppressor MiRNAs) are shown for each MiRNA, together with the function(s) mainly affected.

Table 1 Examples of MicroRNAs Working as Oncogenes or Tumor Suppressors in Glioblastoma

MicroRNA	Targets	References
Oncogenic MicroRNAs		
MiR-21	PDCD4	Chen et al. (2008)
	HNRPK, TAp63	Papagiannakopoulos et al. (2008)
	RECK, TIMP3	Gabriely et al. (2008)
MiR-10b	Bim, TFAP2C, p16, and p21	Gabriely et al. (2011)
MiR-221/222	p27	Le Sage et al. (2007)
	p57	Medina et al. (2008)
	MGMT	Quintavalle et al. (2013)
MiR-26a	PTEN, RB1	Kim et al. (2010)
MiR-215	KDM1B	Hu et al. (2016)
MiR-451	CAB39	Godlewski et al. (2010a)
Tumor Suppressor MicroRNAs		
MiR-124	LAMB1	Khazenzon et al. (2003)
MiR-137	RTVP1	Rosenzweig et al. (2006)
MiR-7	EGFR, IRS1	Kefas et al. (2008)
MiR-663	PI3KCD	Shi et al. (2014)
	CXCR4	Shi et al. (2015a,b)
MiR-34a	c-Met, Notch-1, Notch-2	Li et al. (2009b)
MiR-128	Bmi-1	Godlewski et al. (2008)
	PDGFRα, EGFR	Papagiannakopoulos et al. (2012)
MiR-25/32	Mdm2, TSC1	Suh et al. (2012)
MiR-182	HIF2A, c-Met, Bcl2-like12	Kouri et al. (2015)
MiR-135b	ADAM12, SMAD5, GSK3β	Lulli et al. (2015)

The table lists some examples of MicroRNAs discussed in this review, with a recognized role as oncogenes (*upper panel*) or tumor suppressors (*lower panel*) in glioblastoma. Some relevant experimentally demonstrated targets are listed in the second column.

MiR-10b leads to a reduction of the tumorigenic potential of glioma cells by enhancing their sensitivity to apoptosis and by decreasing their growth rate. This effect is mediated by the enhanced expression of direct targets of MiR-10b such as Bim, TFAP2C, p16, and p21, as well as other indirect mRNA targets mainly involved in biological processes related to cell cycle progression and apoptosis (Gabriely et al., 2011).

MiR-221 and MiR-222 are a couple of MiRNAs located in tandem on chromosome X which are frequently overexpressed in a wide range of human malignancies, including GBM (Ciafrè et al., 2005). The oncogenic role of these MicroRNAs in GBM disease has been broadly described by our and other groups (Galardi et al., 2016; le Sage et al., 2007; Medina et al., 2008; Tomaselli et al., 2015). Through their ability to downregulate cell cycle-related tumor suppressors such as p27 and p57, MiR-221 and MiR-222 accelerate the proliferative potential of glioma cells by promoting the G1-S phase transition. In addition, MiR-221 and MiR-222 overexpression has been also associated with an increased survival and invasiveness of glioblastoma cells (Quintavalle et al., 2012; Zhang et al., 2010). The upregulation of MiR-221 and MiR-222 in GBM seems to occur through the specific activation of a transcriptional regulatory mechanism including NF-κB and c-Jun (Galardi et al., 2011). These transcription factors bind two genomic regions upstream of this MiRNA cluster and cooperatively increase the transcription of MiR-221 and MiR-222 in glioblastoma cells, by promoting the recruitment of RNA polymerase II at the transcriptional start site (Galardi et al., 2011).

Dysregulation of MiRNA expression observed in almost all kind of cancers can be also due to genomic alterations. For instance, Kim and colleagues revealed by TCGA data analysis, that MiR-26a is a component of a 12q13.3–14.1 region which is often amplified in GBM (Kim et al., 2010). This specific genomic amplification is the main determinant for the overexpression of MiR-26a observed in several GBM samples, that is associated with a shorter median survival. MiR-26a reduces the expression of the tumor suppressors PTEN and RB1, thus promoting tumor growth and resistance to JNK-dependent apoptosis by activating AKT and inhibiting MAP3K2/MEKK2 expression. Moreover, the tumorigenic effect of MiR-26a is enhanced by the overexpression of two genes, CDK4 and CENTG1, located in the same genomic locus. Interestingly, these genes display a protumorigenic role by acting on the same MiR-26a-regulated pathways, suggesting the existence of a functionally integrated oncomiR/oncogene DNA cluster cooperating in GBM progression (Kim et al., 2010). Thus, the oncogenic relevance of MicroRNAs in glioblastoma can be further extended beyond their role as single regulators of single genes/pathways, as they can be inserted in a wide network of "canonical" oncogenic pathways in a continuous crosstalk of a variety of different regulators.

In agreement with this concept Sumazin and coworkers (Sumazin et al., 2011), by mining data from TCGA, unveiled a surprisingly extensive post-transcriptional network acting in human glioma, including more than 248,000 MiR-mediated interactions. This network regulates pivotal molecular players of tumor initiation and progression, establishing a relevant crosstalk between these interactions and the canonical oncogenic pathways. In this specific context, the tumor suppressor PTEN emerges with a total of 534 interactions in the MiR program-mediated regulatory interaction network. The authors highlight the importance of this PTEN-miRNA network, which can explain the multiple levels of downregulation of such a crucial tumor suppressor gene in GBM, even in the case when PTEN locus is intact (Sumazin et al., 2011).

In our group, we have recently performed a SAGE analysis of glioblastoma tissues, that included also peritumor areas. This integrated study, addressing both "long" RNAs (mRNAs and lncRNAs) and MicroRNAs, produced a deep expression picture of tumor and peritumor tissues, highlighting relevant pathways perturbed in these tumors and governed by a crosstalk of key mRNAs and MicroRNAs (Fazi et al., 2015). One of these is tumorigenic TGFβ signaling, indicated by the dysregulation of several mRNAs and MicroRNAs both in frankly tumor areas and in the immediately surrounding regions. Included in this TGFβ distinctive MiRNA signature, we found the overexpression of the previously cited MiR-21 and MiR-10b, together with MiR-106 and MiR-183, all known to be induced by TGFβ, and the reduced expression of MiR-127, previously demonstrated to be under the negative control of TGFβ in different tumor types.

Among the MiRNAs with oncogenic role in GBM, a special mention is deserved for those which have been linked to the ability of cancer cells to adapt different stress conditions, such as hypoxia and metabolic stress (Godlewski et al., 2010a; Hu et al., 2016). In a recent paper, Hu and colleagues identified MiR-215 as a crucial MiRNA modulating the response of glioma initiating cells to hypoxia by targeting the epigenetic regulator KDM1B. Interestingly, the enhanced expression of MiR-215 observed during hypoxia in these cells is mediated by the specific interaction of HIF1α with Drosha, which in turn promotes cleavage of pri-miR-215 into its mature form. Thus, this novel molecular axis HIF1α-MiR-215-KDM1B sustains GBM oncogenesis, participating to the adaptive response of glioma cells toward hypoxic conditions occurring in tumor microenvironment (Hu et al., 2016).

MiR-451 is another relevant MicroRNA which plays a pivotal role in controlling proliferation, survival, and motility by acting as a molecular sensor of glucose levels in glioma cells. While high glucose induce MiR-451 expression, in low glucose conditions, the reduction ofmiR-451 expression stabilizes the LKB1 signaling complex by upregulating CAB39, which results in the phosphorylation of its substrates such as AMPK and MARKs. Once activated, these kinases impair cell proliferation by reducing mTOR activity but in parallel promote migration. Thus, this low MiR-451-driven mechanism results in the adaptation of glioma cells which slow down their proliferation but migrate to seek out metabolically favorable conditions and ultimately sustain tumor growth (Godlewski et al., 2010a).

5.2 MicroRNAs Acting as Tumor Suppressors in Glioblastoma

To date, a broad number of MicroRNAs with antitumorigenic role in GBM have been described and the elucidation of their mechanism of action has led them to be recognized as tumor suppressors (Table 1).

Among them, MiR-124 and MiR-137 have first emerged because of their reduced expression in GBM tumor samples and astrocytomas in comparison to normal brain tissues (Silber et al., 2008). The overexpression of these MicroRNAs decreases the proliferative rate of glioblastoma cells by endorsing G1 cell cycle arrest and is also effective in promoting neuronal differentiation of mouse oligodendroglioma and/or human GBM-derived stem cells. Finally, the identification of LAMB1 (laminin 8) and RTVP1 oncogenes (Khazenzon et al., 2003; Rosenzweig et al., 2006) as direct targets of MiR-124 and MiR-137, respectively, has provided the molecular basis of the tumor suppressor role of these MicroRNAs in GBM etiology (Bier et al., 2013; Chen et al., 2014a).

In 2008, efforts made by Kefas and colleagues contributed to the discovery of the involvement of MiR-7 in the regulation of pathways associated to the migration, invasiveness, and proliferation of GBM cells (Kefas et al., 2008). In particular, they demonstrated that MiR-7 exerts its antitumorigenic effect in a dual manner, by directly targeting the 3′ UTRs of both the EGFR and the insulin receptor substrate 1 (IRS1). The former is a well-known oncoprotein implicated in GBM pathology, and the latter is an upstream regulator of Akt pathway. Interestingly, the expression levels of MiR-7 are strongly downregulated in GBM samples vs normal brain samples and its ectopical overexpression in GBM cells reduced the levels of phosphorylated Akt, consequently impairing glioblastoma cell viability and invasiveness (Kefas et al., 2008).

Similarly to MiR-7, MiR-663 functions as a repressor of PI3K-Akt pathway through the inhibition of the phosphoinositide 3-kinase (PI3K) subunit p110δ (PIK3CD), and the inverse correlation between MiR-663 and PI3KCD in glioblastoma specimens predicts poor prognosis (Shi et al., 2014). More recently, the C–X–C motif chemokine receptor 4 (CXCR4) known for its important contribution to GBM progression (Munson et al., 2013; Ping et al., 2011) and, specifically, to the transdifferentiation of GSCs into vascular pericytes (Cheng et al., 2013), has been identified as additional target of MiR-663 (Shi et al., 2015a). From a mechanistic viewpoint, it is important to mention the intriguing result of MiR-663 regulating CXCR4 by targeting its coding sequence, instead of the more common 3′UTR. This, together with the primate specificity of MiR-663, confers a peculiar interest to it, even beyond the context of glioblastoma.

Several MicroRNAs are related to the p53-dependent tumor suppressor response and among them, MiR-34a is the major example of transcriptional target of p53 which is often downregulated in many types of cancers including glioblastoma (Hermeking, 2007; Li et al., 2009a). Indeed, MiR-34a is lower in mutant p53 as compared to wild-type p53 glioma and its expression inversely correlates to that of c-Met which, together with Notch-1 and Notch-2, are direct targets of this MicroRNA. The protein levels of such oncogenes are reduced by the forced expression of MiR-34a, which in turn decreases brain tumor cell proliferation, survival, and invasion, as well as in vivo glioma xenograft growth (Li et al., 2009b). Even the p53 accumulation itself seems to be controlled in GBM by a feedback circuitry involving specific MicroRNAs. In this specific context, MiR-25 and MiR-32 are pivotal players because they were initially identified as the most p53-repressed MiRNAs in U87 cells treated with Nutlin-3a, a known inhibitor of the Mdm2/p53 complex. Moreover, the expression of these MicroRNAs is undetectable in human GBM showing high levels of p53, whereas it is abundant in those with low p53 signal. Interestingly, the overexpression of MiR-25 and MiR-32 leads to p53 accumulation by inhibition of their direct mRNA targets mdm2 and TSC1, which are negative regulators of p53 stability, resulting in an impairment of glioma cell proliferation through cell cycle arrest (Suh et al., 2012).

Recently, an increasing number of evidences showed that MicroRNAs play critical roles in GSCs which are, as mentioned earlier, stem-like cancer cells contributing in the maintenance of tumor growth by promoting the resistance of therapeutic approach and tumor recurrence (Galli et al., 2004). Consistent with this, a series of MiRNAs specifically acting to reduce tumorigenicity of GSCs have been identified, suggesting their putative therapeutic use in the treatment of GBM tumors (Godlewski et al., 2008; Kouri et al., 2015; Lopez-Bertoni et al., 2015; Lulli et al., 2015; Papagiannakopoulos et al., 2012). Among them, MiR-128 was the first one to be characterized because of its ability to repress human glioblastoma cell growth (Godlewski et al., 2008; Papagiannakopoulos et al., 2012). Interestingly, Godlewski and colleagues demonstrated that the tumor-suppressive potential of MiR-128 in GSCs is mediated by its ability to modulate the expression of Bmi-1 oncogene which is an epigenetic silencer and a positive regulator of stem cell renewal. Thus, MiR-128 overexpression induces a series of changes in GSCs which are consistent with Bmi-1 downregulation, including the reduction in self-renewal ability of these cells (Godlewski et al., 2008). Few years later, results obtained by Papagiannakopoulos further contributed to the understanding of the molecular mechanism through which MiR-128 exerts its tumor-suppressive role in GSCs. They first identified two novel MiR-128 direct mRNA targets, PDGFRα and EGFR, which are two mitogenic receptor tyrosine kinases with a well-established oncogenic role in GBM (Brennan et al., 2013). Subsequently, they correlated the MiR-128-mediated reduction of these specific targets to a reduction in the proliferative rate of glioma-initiating stem cells, coupled to an enhancement of their neural differentiation ability (Papagiannakopoulos et al., 2012). Taking into account that restoring differentiation represents a potential therapeutic approach in gliomas, altogether these findings support the relevance of the tumor-suppressive role of MiR-128 in GBM pathology.

Similarly, other MicroRNAs showing a functional effect in the process of GSCs differentiation have been described (Kouri et al., 2015; Lopez-Bertoni et al., 2015). Their antitumorigenic activity is well documented whereas the mechanism of action of some of them remains still unknown, as in the case of MiR-148a which was found to be strongly reduced in glioma stem like cells expressing the stemness marker CD133 vs CD133 cells. The forced expression of Oct4 and Sox2, two core reprograming transcription factors able to induce self-renewal and pluripotency, in GSCs resulted in a DNA methyl transferase-mediated MiR-148a promoter repression, and in an increase of self-renewal and tumor propagating ability of GSCs, while in vitro differentiation of GSCs was accompanied by an increase of MiR-148a levels. On the contrary, its forced expression abrogated the ability of these transcription factors to induce a stem-like tumor-propagating state. According to this, the orthotopic implantation in mice of patient-derived GBM neurospheres transduced to overexpress MiR-148a shows a significant reduction in tumor development, supporting the pivotal role of MiR-148a as a suppressor of propagating capacity of GBM stem cells (Lopez-Bertoni et al., 2015).

Another example of a MicroRNA involved in glioma progression and/or initiation by its ability in modulating the process of GSCs differentiation, is MiR-182 (Kouri et al., 2015). This is a scarcely expressed MicroRNA in GBM, and a negative regulator of the transcriptome signature associated to the stemness of GSCs (Kouri et al., 2015). Its ectopical overexpression promotes the differentiation in vitro of GSCs as measured by the decrease in neurosphere size and expansion upon low-density seeding. Consistently, MiR-182 represses the expression of two mRNAs, HIF2A and c-Met, which are critical oncogenes involved in self-renewal, growth, survival, and tumor initiating potential of GSCs (Li et al., 2009a, 2011). In addition to the regulation of GSCs behavior and glioma cell plasticity, MiR-182 can also enhance the sensitivity of glioma cells to TMZ treatment by negatively regulating the expression of the antiapoptotic factor Bcl2-like12 (Bcl2L12). Thus, it appears clear that this MicroRNA exerts its tumor suppressor activity through a pleiotropic effect acting on several mRNA targets related to different aspects of GBM pathology.

In a similar manner, MiR-135b negatively affects glioma tumorigenicity through the modulation of a wide number of targets. Among them, we would like to mention ADAM12, a metalloproteinase highly expressed in GBM which is associated to cellular proliferation, SMAD 5, and the glycogen synthase kinase 3 beta (GSK3β), an enzyme known to be implicated in many signaling pathways affecting GBM (Kotliarova et al., 2008). Interestingly, the restoration of MiR-135b significantly decreases the proliferative and migration rate as well as the clonogenic ability of GSCs in vitro and in vivo (Lulli et al., 2015).

6. MicroRNAs IN GLIOBLASTOMA CLASSIFICATION

Our group was the first to perform a profiling analysis of the global MicroRNA expression in glioblastoma patient samples in 2005 (Ciafrè et al., 2005). In that work, we not only determined the MiRNAs differentially expressed in GBM samples compared to healthy brain tissues, but also we analyzed the miRNome of peritumoral regions, which are clinically relevant, as they are a very common site of tumor recurrence. The results obtained in that work were not aimed at a classification of glioblastoma, due to the small number of patients analyzed. Nevertheless, even in that restricted cohort, almost all the MicroRNAs that emerged would have been recognized, in later studies, as major ones in glioblastoma-specific signatures, and further employed to discriminate between different classes of glioblastoma patients.

One of the first studies to perform a global analysis of MicroRNA expression in GBM, and to correlate it with patients' survival was published 5 years later by Guan and coauthors, who initially analyzed the miRNomes of eight GBMs and compared them to those of four anaplastic astrocytomas, the second highest grade glioma (Guan et al., 2010). The expression of 16 MiRNAs resulted not only differential between the two tumor types, but also it was sufficient to clearly clusterize those two groups of brain tumors. When the authors analyzed a larger cohort of 74 GBMs and 28 anaplastic astrocytomas, they could confirm the statistically significant differential expression of those 16 MiRNAs, among which MiR-196a was the most relevant one. Most importantly, its expression, together with that of its cognate molecule MiR-196b, was an independent and significant predictor of short OS in glioblastoma patients (Guan et al., 2010).

A larger cohort, consisting of 222 GBM patients derived from TCGA dataset, was studied by Srinivasan and colleagues, who identified a 10-MiRNA signature classifying patients successfully into low-risk and high-risk groups with significantly different survival times (Srinivasan et al., 2011). They took a further step toward the employment of MiRNA signatures for GBM patients classification by constructing a mathematical model for survival prediction. The formula they devised assigned each patient a "risk score" that is a linear combination of the expression levels of the significant MiRNAs weighted by their respective Cox regression coefficients. Such a model is aimed at guiding clinicians in the identification of high-risk GBM patients, and possibly provides them with more aggressive adjuvant therapies in addition to standard treatments.

Some groups, after preliminary screenings of large cohorts of glioblastoma patients, focused on specific MiRNAs whose expression was revealed to neatly distinguish short-term from LTS, beside discriminating glioblastoma grade IV from other astrocytomas of lower grades. In addition to the first profiling reports, such deeper studies investigated also the mechanisms through which one or more MiRNAs may affect GBM aggressiveness and patients' survival.

Shi and colleagues established that MiR-663 expression inversely correlates with the overall (OS) and, interestingly, also the disease-free survival (DFS) in glioblastoma patients, and for this reason they proposed the measurement of MiR-663 in GBM samples as a novel and clinically feasible candidate for MiRNA-based diagnosis and subclassification for glioblastoma (Shi et al., 2014). Their observation was based on the analysis of very large number of samples, conferring a great statistical soundness to the results obtained, and was supported by the investigation of the molecular mechanism linking MiR-663 expression and glioblastoma characteristics (see above "MicroRNAs acting as tumor suppressor in glioblastoma"). By identifying the chemokine receptor CXCR4 as a target of MiR-663, they could also expand their previous results about MiR-663 significance in GBM prognosis by demonstrating that the combination of CXCR4 and MiR-663 expression levels is an independent predictor of DFS and OS of patients with GBM, as a clear advantage in terms of both DFS and OS was observed in $CXCR4^{low}$ $MiR\text{-}663^{high}$ patients.

Equally relevant, due to the large size of patient sample which was studied, is the analysis published by Kim and coauthors in 2014 (Kim et al., 2014). By mining the expression data of almost 500 glioblastoma patients deposited in the TCGA database, they uncovered that elevated MiR-148a expression is a highly significant negative risk factor for patients' survival. In addition, they showed that high levels of MiR-148a are typical of the most aggressive glioblastoma cell lines and of three patient-derived GSCs they studied. In both cell types, MiR-148 induces proliferation, migration, and invasiveness, while repressing apoptosis. These results, seemingly in contrast to those reported by Lopez-Bertoni and colleagues (Lopez-Bertoni et al., 2015, see above), who showed a repression of MiR-148a expression in GSCs, are very difficult to compare, in order to obtain an unequivocal response about the pro- vs antitumorigenic role of MiR-148a in glioblastoma. The possibility exists, in fact, that the two groups of authors have studied very different GSCs, conceivably deriving from differently classified glioblastomas, in turn resulting in such apparently divergent conclusions, which might merely be a further demonstration of the great variability characterizing these tumors and their initiating cells. However, the work by Kim et al., supported by data mining of patients' data, surely owns its clinical value. Even though their data may represent a "static" picture of a single moment of GBM evolution (when the tumor, already grown large and symptomatic, was excised), it is in fact relevant because it reflects the nature of the tumor when it is usually accessible to clinicians. Of note, at that stage, glioblastomas are probably poor of real "stem" cells, the bulk of the tumor mass being surely composed of nonstem cells. This too may account for the different conclusions drawn by the two works. Moreover, Kim and colleagues provided an explanation for MiR-148a oncogenic action in GBM by showing that it targets the proapoptotic molecule BIM, and MIG6, a critical regulator of EGFR trafficking, degradation, and activation. Thus, overexpressed MiR-148 inhibits apoptosis and impairs EGFR trafficking and degradation, in turn resulting in EGFR hyperactivation.

In 2012, Silber et al. investigated the MiRNA-related biological classification of glioblastomas, by specifically focusing on the proneural subtype of GBM (Silber et al., 2012). By data mining in TCGA, they showed the downregulation of MiR-34a in this class of glioblastomas, where they suggested that it may be due to a specific repression by PDGF. In fact, the dysregulation of PDGF signaling is tightly linked to the pathogenesis of this class of gliomas. Consistently with their observation, they also demonstrated a negative effect of MiR-34a reexpression in proneural cells with regards to their proliferation in vitro and tumorigenicity in murine xenografts. They completed their study about the functional role of MiR-34a repression in proneural GBMs by showing that this MiRNA targets PDGFRA, thus depicting a GBM subtype-specific feedforward loop. It is interesting to note that proneural GBMs, although strongly driven by PDGF signaling, rarely harbor mutations in PDGF receptor PDGFRA. Thus, MiR-34a downregulation in these GBMs might represent an alternative nongenomic mechanism boosting PDGFRA action even in the absence of canonical genomic aberrations of its gene.

Similarly to the previous work, Ma et al. reported a GBM subtype-specific MiRNA signature, with the expression of MiR-128a, -504, -124a, and -184 being inversely correlated to that of typical mesenchymal markers, such as CD44, YKL-40, and Vimentin, while the expression of MiR-21 and MiR-34a appeared positively correlated (Ma et al., 2012).

In the specific attempt of constructing respective prognostic MiRNA signatures in the five intrinsic subtypes of primary glioblastoma, Li et al. in 2014 analyzed data of 448 GBM patients deposited in the TCGA database (Li et al., 2014), classified as classical, mesenchymal, neural, proneural G-CIMP, and proneural non-G-CIMP subtypes (based on the CPG Island methylator phenotype) (Noushmehr et al., 2010). By using permutation tests and Cox regression analyses, they filtered significant MiRNAs for each GBM subtype, for which risk scores were calculated to divide patients into low- and high-risk groups. In all five subtypes, the risk scores based on MiRNA signatures could efficiently separate patients with longer OS (low-risk groups) from those with shorter OS (high-risk groups). Interestingly, only two subtypes, classical and neural, shared, in their predictive MiRNA signatures one "risky" MiRNA, MiR-222, and two closely related paralogs as "protective" MiRNAs, MiR-422a and MiR-422b (in classical and neural, respectively). MiR-378, whose mature sequence matches that of MiR-422b, was part of the protective signature in proneural non-G-CIMP subtype too. In general, apart from these MiRNAs, each subtype showed a specific risk-predicting MiRNA signature.

A few years earlier, in 2011, an innovative work was presented about MicroRNA classification of glioblastomas reflecting patients survival, age of onset, and response to temozolomide therapy (Kim et al., 2011). In that paper, Kim and colleagues proposed a novel way of classifying glioblastomas, based on MiRNA signatures, each related to a different neural precursor cell type. By analyzing the expression data of 261 glioblastomas from TCGA portal, they were able to clusterize samples into five distinct subclasses, each characterized by one of five MicroRNA expression clusters. They suggested a relationship between each subclass and a distinct stage of neural differentiation and consequently named each glioblastoma subclass according to its associated stage of neural precursor cell differentiation, that is "oligoneural," "radial glial," "neural," "neuromesenchymal," and "astrocytic." Using this classification, they compared survival of the patients belonging to the different classes and observed that patients with oligoneural glioblastomas lived significantly longer than all others and developed disease at a younger age. Earlier onset is in fact considered a favorable prognostic indicator for glioblastoma survival. Moreover, patients from the astrocytic and the radial glia subgroups showed a significantly better response to the standard temozolomide and radiation regimen in terms of survival. The importance of this work is that MicroRNAs reveal a greater diversity of glioblastoma subclasses than previously recognized. In fact, while the previous mRNA-based classifications failed to reveal significant survival differences among classes (Verhaak et al., 2010), MicroRNA-based glioblastoma subclasses showed robust survival differences among them. In particular, they demonstrated that the previously defined "proneural" subclass, associated with longer survival (Phillips et al., 2006) can be further separated into two subgroups with significant survival differences by using MicroRNA-based classification. This observation may at least in part account for the heterogeneity of "proneural" GBMs in terms of survival and is supported by the recognized existence of proneural G-CIMP and proneural non-G-CIMP subtypes, differentiated by survival (Noushmehr et al., 2010).

Another attempt to employ MicroRNA expression data as a classifier and a prognostic tool that can work per se, across already existing classes of glioblastoma was presented by Hayes and coauthors in 2015 (Hayes et al., 2015). By using a novel methodology, known as LASSO (least absolute shrinkage and selection operator) (Tibshirani, 1996), on the almost 500 GBM samples available in the TCGA dataset, they produced a nine MicroRNA prognostic signature able to stratify GBM patients, and thus predict survival, across the predefined GBM subtypes and also in treatment with the most common therapeutic agent, temozolomide. Moreover, they could validate their results in a larger set of patients, further strengthening their significance. Interestingly, most of the nine MiRNAs in the signature were already known for their role in cancer and in GBM specifically, and were already present in previously published signatures. Moreover, for many of them a biological role had been previously demonstrated in glioblastoma, supporting the use of their combination in this novel signature to predict OS and progression free survival (PFS), together with TMZ resistance in GBM patients.

Similarly, the prognostic role of the well-known oncomiR MiR-21 in GBM was further corroborated by a study by Sathyan and colleagues, who established a functional axis linking MiR-21 to the key neural stemness regulator Sox2, and used it to stratify GBM patients in two distinct survival classes (Sathyan et al., 2015). In their very comprehensive work, these authors set up a novel classification method, able to mathematically sort half of the TCGA GBM patients into either progenitor-like, high MiR-21/low Sox2 (Class A) or stem-like, low MiR-21/high Sox2 (Class B) subtypes. Patients with Class B tumors had longer OS than those with Class A tumors. In addition, the role of MiR-21 as a negative regulator of Sox2 was experimentally demonstrated in a very relevant model, by deriving cancer stem cells from tumor samples of Classes A and B patients and ectopically modulating MiR-21 levels. Sox2 expression was accordingly modulated, being suppressed in MiR-21 overexpressing cells. The inversely correlated pair of markers, MiR-21 and Sox2, showed a much greater soundness than each of the two markers alone, in defining the survival of GBM patients. Moreover, Class A vs Class B classification resulted to be a stronger predictor of patient survival than currently used parameters, such as Karnofsky performance status, age at diagnosis, and G-CIMP status. In this study, a global gene expression differential analysis was also performed, that revealed distinct gene expression clustering for Classes A and B tumors. However, compared to the existing signatures for GBM (Verhaak et al., 2010), classes A and B did not overlap with any known signatures of any specific GBM subtype; rather, they showed a mixture of all four subgroups, with a preponderance of the mesenchymal subgroup in Class A and a preponderance of the classical and proneural subgroups in Class B. A GSEA analysis predicted that Class A tumors were significantly associated with immune/inflammatory response, while Class B with chromosome organization and nervous system development, thus providing a molecular basis for the distinct nature of the two subtypes.

7. MicroRNAs AND RESISTANCE TO CHEMOTHERAPY AND RADIOTHERAPY IN GLIOBLASTOMA

Several works in the last years addressed the role of MicroRNAs in a specific clinical aspect of glioblastoma, that is resistance to chemotherapy with temozolomide. Fiore et al. showed that MiR-340 expression distinguishes LTS from short-term ones (STS) and, more widely, that patients with higher MiR-340 expression had a longer survival rate and, thus, a better prognosis compared to those with low MiR-340 levels, thus proposing MiR-340 as significant predictor of good prognosis in GBM (Fiore et al., 2016). Beside demonstrating that the tumor-suppressive action of MiR-340 can be at least in part explained by the targeting of the oncogene NRAS and the consequent blunting the downstream ERKs and AKT pathways, they reasoned that resistance to temozolomide adjuvant therapy is a relevant factor for short survival in GBM patients and demonstrated that the ectopic overexpression of MiR-340 enhanced the sensitivity of GBM cell lines to TMZ and coherently increased apoptosis. They interpreted these results by suggesting that MiR-340 higher expression in LTS GBM patients can contribute to their better response to temozolomide, thus prolonging their survival.

Given the importance of MGMT in glioblastoma resistance to temozolomide, the result found by Gao et al. about MiR-370-3p as a negative regulator of MGMT mRNA is very interesting. In fact, they showed not only that MiR-370-3p inhibits MGMT expression, but also that it is strongly underexpressed in TMZ-resistant cells and tissues, and that its ectopic expression in TMZ-resistant cell lines restores their sensitivity to temozolomide through repression of MGMT expression (Gao et al., 2016). MGMT is the final target through which another MiRNA underexpressed in TMZ-resistant glioblastoma exerts its function. MiR-101 was found to target glycogen synthase kinase 3β (GSK3β), whose inhibition in turn results in MGMT promoter methylation and consequent downregulation (Tian et al., 2016). Importantly, these authors found that MiR-101 expression is strongly reduced in TMZ-resistant cells and patients' samples, and MiR-101 ectopic expression restores TMZ sensitivity in vitro and also in vivo in GBM xenografts, thus cooperating with temozolomide in the inhibition of subcutaneous tumor growth (Tian et al., 2016).

A deep molecular analysis was performed by Haemmig and colleagues, who demonstrated that MiR-125b, a brain-enriched MiRNA shown for the first time by our laboratory to be overexpressed in GBM (Ciafrè et al., 2005), positively regulates NF-κB in glioblastoma by targeting its negative regulators TNFAIP3 and NKIRAS2 (Haemmig et al., 2014). Noticeably, NF-κB is a key factor in gliomagenesis and was previously demonstrated to drive the expression of the oncogenic MiRNAs MiR-221 and MiR-222 in GBM (Galardi et al., 2011). Functionally, MiR-125b controls proliferation and apoptosis of GBM cells and confers TMZ resistance in a way which is independent of MGMT methylation or p53 status. The clinical relevance of these data was underlined by showing that MiR-125b expression significantly correlated with nuclear localization of NF-κB in paraffin-embedded GBM samples and, most importantly, that MiR-125b expression was an unfavorable prognostic marker for patients treated with TMZ and radiotherapy. It is interesting to note that the very similar cognate MiRNA MiR-125a, although identical to MiR-125b in its seed sequence but with some differences in the rest of the sequence, and not overexpressed in GBM (Ciafrè et al., 2005), was not a predictor of patients response to temozolomide, thus underlining the GBM-specific MiRNA dysregulation and the fine mode of action of MiRNAs as regulators.

A pivotal work has been very recently published by Luo et al., who presented the results of detailed mechanistic in vitro and in vivo experiments, corroborated by the analysis of large clinical datasets, describing a new functional axis, linking c-Myc, MiR-29c, and one of the major error-prone polymerases of translesion synthesis, as a novel important regulator of TMZ resistance in GBM (Luo et al., 2016). After the preliminary identification of a limited number of MiRNAs associated with TMZ resistance between paired primary and recurrent tumors which had acquired resistance to TMZ during recurrence, they showed that MiR-29c was the most underexpressed one in recurrent and TMZ-refractory tumors. They went deep into the mechanistic basis of MiR-29c decrease in TMZ-resistant GBMs by demonstrating that it was led by the oncogene c-Myc, negatively regulating MiR-29c expression, and verified that the MiR-29c target with a key role in TMZ resistance is REVL3, whose high expression in resistant GBM cells yields increased DNA repair through HR, and subsequent reduced genomic instability and resistance to TMZ. They were able to modify the response of GBM cells and tumors to TMZ, both in vitro and in vivo, by ectopically modulating each of the three components of the regulatory axis, i.e., c-Myc, MiR-29c, and REVL3, and provided sound data in favor of a possible use of MiR-29c as a predictor of the response to temozolomide of GBM patients. In fact, by analyzing the expression pattern of MiR-29c, c-Myc, and REV3L in 138 GBM patients who had undergone TMZ-based chemotherapy, they showed an inverse correlation between MiR-29c and c-Myc and REVL3 and, most relevantly, revealed that patients whose tumors had a high MiR-29c expression responded better to TMZ treatment compared with patients with low MiR-29c expression, and survived longer, both when considering the OS and the PFS. These results led the authors to suggest that pretreatment evaluation of MiR-29c expression could be used to predict the therapeutic response to TMZ in a form of personalized prognosis for GBM patients. As for the already described MiR-125b, it is very important to underline that MiR-29c function in GBM resistance was shown to be independent on MGMT methylation status and modulation. This opens new perspectives on the mechanisms of drug resistance in GBM, and must be taken into account when performing pretreatment molecular screenings in GBM patients.

Radiotherapy is a standard treatment for GBM, often coupled with chemotherapy with temozolomide but, comparably to TMZ, GBM cells usually develop, or own an intrinsic radioresistance. The involvement of MiRNAs in this clinical aspect of glioblastomas was investigated by Moskwa and colleagues, who performed an in vitro screening of MiRNAs conferring radioresistance in an unrelated cell line, and then confirmed the differential expression of four MiRNAs identified by the screening in a set of human glioblastoma cell lines, characterized by different sensitivity to radiation (Moskwa et al., 2014). The four MiRNAs, namely MiR-1, MiR-125a, MiR-150, and MiR-425, were shown to induce radioresistance in vitro both singularly or in combination, and to work at least in part by enhancing the cellular checkpoint response after irradiation, while no impairment of double-strand break repair was found.

In conclusion, all these publications and many others we could not cite due to space limits addressed the extremely diverse nature of glioblastoma, in the attempt of classifying these tumors based on specific MiRNA expression. Even though so far no single MiRNA or MiRNA signature is definitively accepted for GBM classification, some molecules recur frequently in several signatures (Table 2), and likely represent the most robust markers to be further investigated to possibly provide a "personalized" prognostic tool in the era of "precision medicine."

8. MicroRNAs IN THE INTERACTION BETWEEN GLIOBLASTOMA AND ITS MICROENVIRONMENT

For many years, cancer was mainly considered to be a cell autonomous disease. However, studies on the tumor microenvironment have given rise to the emerging concept that tumor progression depends on an intense network of interactions among cancer cells and the surrounding stroma (Junttila and de Sauvage, 2013). Glioblastoma as well as many other tumors have a remarkable ability to affect the stromal environment to its own advantage; tumor cells influence the endogenous activity of other surrounding brain cells including endothelial cells, macrophages, astrocytes, and microglia, and thus facilitate tumor growth, invasion, chemoresistance, and immune evasion (Bonavia et al., 2011). In this scenario, the interactions between cancer cells and stroma coevolve as the tumor develops.

Recent papers have highlighted the important role played by MiRNAs in the evolution of tumor microenvironment. Alteration of MiRNA expression patterns in cancer and stromal cells can exert distal and proximal effects on the tumor stroma, respectively. Such actions include regulation of tumor angiogenesis, immune cell invasion, and tumor–stromal interactions. An interesting example that correlates MiRNA alterations in cancer cells and tumor microenvironment, is provided by a study that identified a transcriptionally regulated MicroRNA-based loop linking nutrient

Table 2 MicroRNAs as Classifiers for Glioblastoma

MicroRNA	Prognostic Value	References
MiR-21	Positively correlated to mesenchymal properties	Ma et al. (2012)
	Risky for proneural subtype	Li et al. (2014)
	High MiR-21/low Sox2 predicts shorter survival	Sathyan et al. (2015)
	Overexpressed in recurrent, TMZ-resistant GBMs	Luo et al. (2016)
	GBM class V, astrocytic precursors	Kim et al. (2011)
MiR-31	Risky signature	Srinivasan et al. (2011)
	Risky for classical subtype	Li et al. (2014)
	GBM class V, astrocytic precursors	Kim et al. (2011)
MiR-34a	Inhibits proneural features	Silber et al. (2012)
	Mesenchymal subtype marker	Ma et al. (2012)
	Risky for proneural non-G-CIMP	Li et al. (2014)
	Risky	Hayes et al. (2015)
	GBM class V, astrocytic precursors	Kim et al. (2011)
MiR-124a	Negatively correlated to mesenchymal signature	Ma et al. (2012)
	Risky	Hayes et al. (2015)
	GBM class III, neural precursors	Kim et al. (2011)
MiR-125a/b	Risky, resistance to TMZ	Haemmig et al. (2014)
	Risky, radioresistance	Moskwa et al. (2014)
	Overexpressed in recurrent, TMZ-resistant GBMs	Luo et al. (2016)
	GBM class II and III, multipotent precursors	Kim et al. (2011)
MiR-148a	Risky signature	Srinivasan et al. (2011)
	Risky	Kim et al. (2014)
	Risky	Hayes et al. (2015)
	GBM class V, astrocytic precursors	Kim et al. (2011)

Continued

Table 2 MicroRNAs as Classifiers for Glioblastoma—cont'd

MicroRNA	Prognostic Value	References
MiR-222	Risky signature	Srinivasan et al. (2011)
	Risky for classical and neural subtypes	Li et al. (2014)
	Risky	Hayes et al. (2015)
	Overexpressed in recurrent, TMZ-resistant GBMs	Luo et al. (2016)
	GBM class V, astrocytic precursors	Kim et al. (2011)
MiR-17-5p	Protective signature	Srinivasan et al. (2011)
	Lost in recurrent, TMZ-resistant GBMs	Luo et al. (2016)
	GBM class I, oligoneural precursors	Kim et al. (2011)
MiR-20a	Protective signature	Srinivasan et al. (2011)
	Lost in recurrent, TMZ-resistant GBMs	Luo et al. (2016)
	GBM class I, oligoneural precursors	Kim et al. (2011)
MiR-106a/b	Protective signature Lost in recurrent, TMZ-resistant GBMs GBM class I, oligoneural precursors	Srinivasan et al. (2011) Luo et al. (2016) Kim et al. (2011)
MiR-145	Protective in proneural non-G-CIMP	Li et al. (2014)
	Protective	Hayes et al. (2015)
	GBM class IV, neuromesenchymal precursors	Kim et al. (2011)
mir-182	Protective in proneural non-G-CIMP	Li et al. (2014)
	Protective	Hayes et al. (2015)
	GBM class I, oligoneural precursors	Kim et al. (2011)
MiR-340	Protective	Fiore et al. (2016)
	Lost in recurrent, TMZ-resistant GBMs	Luo et al. (2016)
	GBM class I, oligoneural precursors	Kim et al. (2011)
MiR-370	Protective in neural subtype	Li et al. (2014)
	Protective	Hayes et al. (2015)
	GBM class IV, neuromesenchymal precursors	Kim et al. (2011)

The table lists MicroRNAs which were reproducibly reported by at least three independent groups to own a prognostic/diagnostic value in glioblastoma. The *thicker horizontal line* separates MicroRNAs included in "risky" signatures from those comprised in "protective" signatures.

availability to activation of AMPK, the major energy sensor of the cell (Ansari et al., 2015). Cancer cells face frequent dynamic changes in their microenvironment including the availability of glucose and other nutrients. As mentioned in a previous paragraph, MiR-451 is regulated by glucose availability in GBM cells: MiR-451 levels are high in a glucose-rich environment and are low in glucose depletion (Godlewski et al., 2010a). High glucose leads to the induction of MiR-451 expression by the DNA-binding activity of transcription factor OCT1; in this context, MiR-451 suppresses AMPK pathway and promotes cell proliferation. Conversely, in low glucose environment, the AMPK complex becomes functional and prevents Oct1 transcriptional activity, resulting in rapidly declining levels of MiR-451 and turning on the respective phenotypic behaviors of the cell from proliferative to migratory (Ansari et al., 2015). This, in addition to execution of an energy saving and less glucose-dependent metabolic shift, allows effective implementation of promigratory behavior of the cells, seeking for more favorable environments (Ansari et al., 2015; Godlewski et al., 2010a). Thus, MiR-451 results finely regulated in glioblastoma and can be considered the major effector of glucose-regulated AMPK signaling, allowing tumor cell adaptation to variations in nutrient availability in the tumor microenvironment.

An important aspect of the evolution of tumor microenvironment is the evasion of antitumor immune responses (Junttila and de Sauvage, 2013). Recent analyses have shed light on the involvement of multiple MiRNAs in immune responses. For instance, MiR-124 has been identified as a lead candidate for modulating *STAT3* signaling, a key pathway mediating immunosuppression in the tumor microenvironment (Wei et al., 2013). Wei and coauthors showed that MiR-124, by interacting with the STAT3 pathway, regulates the immunosuppressive properties of GSCs and that MiR-124 upregulation or administration in vivo exerted potent antitumor immune effects.

It was recently found that MiR-142-3p is the most downregulated MiRNA in monocyte-derived glioma-associated macrophages (M2), which are the largest infiltrating immune cell population in glioblastomas (Xu et al., 2014). MiR-142-3p overexpression induced selective modulation of TGFBR1, which led to subsequent preferential apoptosis in the M2 subset. In vivo MiR-142-3p administration resulted in glioma growth inhibition and extended median survival, with an associated decrease in infiltrating macrophages. Moreover, a MiR-142-3p-driven autocrine and paracrine-positive loop was shown that epigenetically regulates the progression and cancer stem-like property of GBM by targeting the secretion of the proinflammatory cytokine interleukin-6 (Chiou et al., 2013).

While the majority of MicroRNAs is found intracellularly, a significant number of MicroRNAs have been observed outside cells, often encapsulated in secreted extracellular vesicles (EVs) through which they are transferred to recipient cells and mediate gene regulation (see below; Skog et al., 2008). In tumor microenvironments, extracellular MiRNAs have been suggested to influence tumor progression via bidirectional tumor-to-stromal and stromal-to-tumor communication. In a recent study, Van der Vos and coauthors visualized, by intravital microscopy, the transfer of EVs and the effects of tumor-released extracellular MiRNAs on the phenotype of microglia. They found that microglia cells took up GBM vesicles and were thus induced to increased proliferation and to shifting of their cytokine profile toward immune suppression. These functional effects are associated in part with increased MiRNA levels, decreased target mRNAs, and encoded proteins (van der Vos et al., 2016). Li and coauthors showed that the exposure of brain microvascular endothelial cells to GBM EVs resulted in significant changes in endothelial cell gene expression (Li et al., 2013). Many transcripts downregulated in endothelial cells exposed to GBM vesicles were targets of MiRNAs that were found to be either abundant or specifically enriched in such vesicles. Notably, among these MiRNAs, Li et al. identified several oncogenic MicroRNAs as MiR-21, MiR-23a, MiR-30a, MiR-221, and MiR-451.

Brosniz et al. demonstrated that MiR-1 is downregulated in human glioblastoma, and its ectopic expression blocked in vivo growth, neovascularization, and invasiveness of glioblastoma cells. Interestingly, these authors showed that MiR-1 mediates EV function in glioblastoma (Bronisz et al., 2014). MiR-1 effects as a tumor suppressor in fact were associated with changes in the EV proteome, linked to cancer-related signaling networks. The ectopic expression of MiR-1 resulted in the coordinated targeting of key oncogenic factors as ANXA2, MET, JNK, and members of the Polycomb repressor complex. Analysis of the composition of microvesicles secreted by MiR-1 overexpressing GBM cells also revealed that the most downregulated protein was Annexin A2 (ANXA2). This protein, a member of calcium-dependent phospholipid-binding protein family, is highly expressed in GBM (Zhai et al., 2011) where it plays a critical role in driving invasion and progression and is one of the most abundant proteins in GBM EVs. In vivo, MiR-1 overexpressing glioblastoma cells produced tumors that were dramatically less vascularized than control ones, suggesting a strong effect of the microvesicle-contained MiR-1 on new blood vessel formation and more widely as an inducer of rearrangements of tumor microenvironment. MiR-1 overexpressed in donor cells was in fact loaded into EVs and transferred to recipient cells to functionally target its effectors there.

All the above-mentioned papers and many others underline a key role for MicroRNAs in the regulation of intercellular communication in the glioblastoma microenvironment. Further studies will likely add a needed additional layer of complexity to the glioblastoma network, consisting of intricate intercellular communications that result in a potent boosting of the tumorigenic potential of glioblastoma cells.

9. MicroRNAs AS NOVEL THERAPEUTIC AGENTS AND TARGETS IN GLIOBLASTOMA

Molecular therapy is evolving quickly because of its lower toxicity compared to radiotherapy and chemotherapy, and adaptability to personalization. In this perspective, MiRNAs represent a novel and attractive target, and their involvement as wide range regulators of gene expression makes them ideal entities for molecular therapeutic applications. We can consider two approaches for the development of MiRNA-based therapeutics: MiRNA mimics and MiRNA antagonists. MiRNA mimics should restore a loss of function; this "MiRNA replacement" strategy aims to reintroduce into cancer cells MiRNAs that are normally expressed in healthy cells but lost or reduced in tumoral ones. MiRNA antagonists should inhibit endogenous MiRNAs that show a gain of function in diseased cells; this methodology is comparable to other inhibitory therapeutics targeting a single gene product such as small molecule inhibitors and short interfering RNAs (siRNAs). The inhibition of endogenous MiRNAs can be achieved by the employment of the so called "MiRNA sponges," longer RNA molecules designed to contain repeated MiRNA-binding sequences that can thus sequester MiRNAs from their endogenous target mRNAs (Ebert et al., 2007). A stably expressed MiRNA sponge is a particularly valuable tool for long-term loss-of-function studies (Chen et al., 2014b).

Recent data demonstrate that the selective modulation of MiRNA expression through antisense inhibition or replacement can be achieved in many different ways. DNA sequences encoding MiRNAs or MiRNA antagonists can be introduced into vectors such as adenoviruses, adenoassociated viruses, or retroviruses, that can deliver them to the tissue of interest in a long-lasting way (Chen et al., 2014b; Kota et al., 2009). An alternative is the direct use of synthetic RNAs that mimic MicroRNAs or antisense RNAs complementary to their cognate MicroRNAs (Gebert et al., 2014; Krützfeldt et al., 2005; Liu et al., 2011). These synthetic RNAs are usually chemically modified and associated to carrier systems that increase RNAs half-life in vivo and allow a major efficiency of delivery (Costa et al., 2015; Di Martino et al., 2014; Neuberg and Kichler, 2014; Semple et al., 2010).

Delivery of RNA-based therapeutics to the central nervous system is particularly challenging due to a multitude of factors such as the blood–brain or blood–tumor barriers, blood nucleases that easily degrade naked synthetic RNAs, extracellular matrix, and the difficult cellular uptake by target tissues. In spite of these problems, many improved cellular delivery strategies originally developed for therapeutic siRNAs and based on the use of nanoparticles (i.e., stable nucleic acid lipid particles, gold particles) or cationic polymers (i.e., polyethylenimine, PEI) are now used for therapeutic MiRNAs (Costa et al., 2015; Di Martino et al., 2014; Neuberg and Kichler, 2014; Semple et al., 2010). For instance, Jensen and coauthors initially developed gold nanoparticles covalently functionalized with densely packed, highly oriented siRNA duplexes (spherical nucleic acids, SNAs) designed to target the oncoprotein Bcl2L12. They showed that, in vivo, the SNAs targeting Bcl2L12, penetrated the blood–brain and blood–tumor barrier to disseminate throughout xenogeneic glioma explants, and were effective in knocking down endogenous Bcl2L12 (Jensen et al., 2013). Subsequently, these authors applied the same SNA technology to harness the tumor-suppressive activities of MiR-182 (see a previous paragraph for its role as s tumor suppressor). Using MiR-182-functionalized SNAs, they demonstrated that MiR-182 sequences effectively reduce tumor burden, thereby demonstrating that systemically administered MiRNAs can cross the blood–brain/blood–tumor barriers, infiltrate glioma parenchyma, inhibit GBM progression, and prolong animal subject survival (Kouri et al., 2015). Together with toxicity studies confirming the absence of significant adverse side effects and inflammatory responses associated with systemic introduction of MiR-182 SNAs, these authors proposed that MiRNA–SNAs can represent a novel therapeutic paradigm for the treatment of GBM.

The great interest regarding MiRNA use as therapeutic molecules has also led to a considerable increase in the number of patent applications filed over the decade. With regard to glioblastoma, a clinical trial is currently ongoing on MiR-10b, which is targeted as an onco-miRNA in GBM (ClinicalTrials.gov Identifier: NCT01849952). The work from several laboratories has revealed in fact that inhibition of MiR-10b reduces the growth of glioma lines and GSCs, while not affecting normal neural cells, and therefore MiR-10b possesses strong therapeutic potential (Gabriely et al., 2011b; Guessous et al., 2013; Lin et al., 2012). In a recent research, Teplyuk and coauthors explored various strategies for MiR-10b inhibition in orthotopic GBM xenografts (Teplyuk et al., 2016). They delivered chemically modified MiR-10b antisense oligonucleotide inhibitors (ASO), conjugated with cationic lipid nanoparticles, into two distinct orthotopic GBM models, one based on human GSC xenografts that form diffusely invasive brain tumors in nude mice (Wakimoto et al., 2012), while another based on mouse GL261 glioma allografts in syngeneic immune-competent mice. Three delivery systems were employed: direct intratumoral injections, continuous osmotic delivery, and systemic intravenous injections. In all cases, the treatment with MiR-10b ASO led to targets' derepression, and attenuated growth and progression of established intracranial GBM. One of the most significant aspects of this study is associated with promising data provided by the systemic administration of MiR-10b ASO. Authors found that even though only a small portion of the ASO was delivered to GBM, it resulted in twofold downregulation of MiR-10b in the tumor tissues, significant derepression of its mRNA targets, sufficient delay of GBM progression, and absence of toxicity. Furthermore, longer administration of high-dose MiR-10b ASOs proved safe in mice, suggesting that systemic MiR-10b inhibition can be tolerated well clinically, despite high MiR-10b levels in various normal extracranial tissues and cells.

Similar experiments were conducted by Wei and coauthors, who highlighted the advantage of the intravenous administration of an RNA duplex that mimics the precursor form of the oncosuppressor MiR-124 (Wei et al., 2013). They observed prolonged survival and no evidence of CNS toxicity or induced autoimmunity in treated mice, and suggested the upgrading of delivery strategies by the complexation of MiRNA mimics with specialized vectors, as nanoparticles, liposomes, or exosomes, to improve MiR-124 immunotherapeutic potential.

Studies as those cited, provide rationale for increasing research and development on MiRNAs as therapeutics in glioblastoma. More efficacy data in human subjects over the next few years are granted to offer more insight into MiRNA mimics and antagonist biology, strengthening the possibility to employ them as new class of anticancer therapeutic agents.

10. MicroRNAs AS CIRCULATING BIOMARKERS FOR GLIOBLASTOMA

As previously introduced, increasing evidence shows that the communication between tumor and microenvironment cells is mediated by the secretion and uptake of exosomes loaded with proteins and RNAs. Exosomes are small membrane-bound vesicles (50–200 nm in diameter) that contain unique proteins and nucleic acids reflecting their cells of origin (Graner et al., 2009; Shao et al., 2012), show specific integrin expression patterns directing them to specific recipient cells (Hoshino et al., 2015), and are actively secreted by mammalian cells, in particular by dividing tumor cells (Lötvall et al., 2014). Their molecular cargoes are transferred to other tumor cells and to stromal cells, affecting their phenotype in a way that favors tumor growth and spreading. The great importance of MicroRNAs in cell communication mediated by exosomes is confirmed by the observation that cancer cell-released exosomes, unlike those released by normal cells, are endowed with complete pre-miRNA processing machinery, resulting in a prompt release of mature MiRNAs into recipient cells, in turn inducing rapid changes in their gene expression profiles (Melo et al., 2014). The finding that glioblastoma cells release microvesicles containing proteins, mRNAs, and MicroRNAs is relatively recent: in 2008, Skog and colleagues showed for the first time that such microvesicles transport proteins and RNAs and deliver them to recipient cells, including both other glioblastoma cells and cells of the microenvironment, as endothelial cells (Skog et al., 2008). By finding tumor exosomes in the sera of glioblastoma patients, that work was also the first to propose that microvesicle cargoes, including specific MicroRNAs, can provide easily accessible diagnostic biomarkers for glioblastoma.

After that publication, a few years passed before a study specifically aimed at finding a blood-derived specific MiRNA fingerprint was performed for the first time (Roth et al., 2011). Roth and colleagues compared the MiRNA profiles from the blood of 20 glioblastoma patients to those of 20 age- and sex-matched healthy controls, and found that the upregulation of MiR-128 and the downregulation of MiR-342-3p reproducibly distinguished glioblastoma patients' blood from that of healthy controls. However, that work employed whole blood samples, with the major drawback that MiRNAs deriving from blood cells, rather than circulating ones, might interfere with the accuracy of the results, by masking MiRNAs contained in exosomes and microvesicles released by tumor cells. All the following studies in fact, analyzed MicroRNAs in plasma or serum, thus excluding the background deriving from MiRNAs in blood cells.

The expression of select MiRNAs was studied by Wang et al. in the plasma of 10 glioblastoma patients (Wang et al., 2012). By comparing GBM plasma levels with those of healthy controls, the authors found increased levels of MiR-21 and reduction of MiR-128 and MiR-342-3p. The same three plasmatic MiRNAs specifically discriminated gliomas from different brain tumors such as pituitary adenomas or meningiomas. In addition, while MiR-21 expression could not discriminate among glioma grades (grade II vs grade III vs grade IV, cohorts of 10 patients each), MiR-342-3p levels were inversely correlated with glioma grade, being notably decreased in glioma with ascending tumor grades. Very importantly from a diagnostic point of view, after chemo-radiotherapy the expression of MiR-21 and MiR-342-3p was restored to the levels found in healthy controls, strongly suggesting their possible employment as suitable diagnostic biomarkers. While MiR-21 and MiR-342-3p results are in agreement with established data about a general MiR-21 overexpression in glioblastoma (Papagiannakopoulos et al., 2008) and the specific MiR-342-3p reduction in glioblastoma patients' blood (Roth et al., 2011), MiR-128 reported decrease in GBM plasma apparently is in contrast with previous whole-blood data (Roth et al., 2011). However, as already noted, whole-blood results may be confused by the prevalence of cellular MiRNAs; this, together with the recognized MiR-128 downregulation in glioblastoma tissues (Ciafrè et al., 2005), indicates that MiR-128 decrease in GBM plasma, rather than increase in whole blood, is more reliable as a biomarker.

A further step in the investigation of circulating MicroRNAs in GBM was taken by refining the samples where MicroRNAs were measured: Manterola and colleagues performed an accurate purification and characterization of exosomes from GBM patient sera, where they detected the differential expression of three small noncoding RNAs compared to exosomes purified from healthy control sera. Their characterization, based on transmission electron microscopy (TEM) and dynamic light scattering for morphology and size, and on the detection of specific exosomal markers by Western blot and FACS analyses, confirmed the pureness of their exosomal preparations, thus allowing for a more specific study than those performed on total plasma or sera. Moreover, a special attention was paid to the homogeneity of sample collection, as all sera were collected at initial diagnosis, immediately before surgery and without any prior treatment. The small noncoding RNAs whose levels were increased in GBM exosomes were two MicroRNAs—MiR-320 and MiR-574-3p—and one small nuclear RNA, RNU6-1. After a preliminary analysis in 25 GBM patients, they validated their results in a validation cohort of 50 GBM patient sera, thus reaching the conclusion that RNU6-1 alone or the MiR-320/MiR-574-3p/RNU6-1 combined signature could have diagnostic value in GBM patients (Manterola et al., 2014).

A further effort was made by two groups in 2015 to investigate if the measurement of specific circulating MiRNAs might be used not only for diagnosis but also for prognosis of glioblastoma patients. Herman et al. analyzed a small cohort of 16 glioblastoma patients by measuring their MicroRNA plasma levels compared to 16 healthy controls, matched by age range and sex. Besides finding 19 MicroRNAs differentially expressed in GBM patient plasma, they also revealed 24 MiRNAs whose plasma expression levels significantly correlated with the survival of GBM patients, among which 20 were directly correlated with survival and 4 were inversely correlated (Herman et al., 2015). Interestingly, a subset of 10 survival-correlated MiRNAs (among which MiR-302c-3p, MiR-592, MiR-484, MiR-1260a, MiR-493-3p, MiR-145-5p, MiR-30a-5p, MiR-483-5p inversely correlated, while MiR-514a-3p and MiR-124-3p directly correlated) overlapped with plasmatic MiRNAs differentiating GBM patient samples from healthy controls, making them appealing candidates as both diagnostic and prognostic molecules.

Lai and colleagues, starting with a training group of 10 glioblastoma patient sera, compared to an equal number of healthy controls, did a focused search specifically for the expression of MiR-210, whose increased expression had been previously shown to be associated with worse prognosis in glioma biopsies (Wang et al., 2014). They were able to detect MiR-210 in small volumes of serum, and to reveal that it was increased approximately sevenfold compared to healthy controls (Lai et al., 2015). When assaying a validation cohort, composed of sera from 40 healthy controls and 126 gliomas, among which 32 glioblastomas, they confirmed a general fourfold increase of MiR-210 expression in glioma patient sera compared to controls, but the results were statistically significant only for high grade glioma patients (WHO grades III and IV). In agreement with this, when patient samples were compared based on histological grade, a statistically significant increase in serum MiR-210 expression was evident in high-grade relative to low-grade samples. Of great interest from a prognostic perspective is the observation that high levels of expression of serum MiR-210 were found to be associated with worse OS in the whole cohort of glioma patients. Even though this work did not specifically focus on glioblastoma patients, but was performed on a wider group of different kinds of gliomas, ranging from pilocytic astrocytomas to glioblastomas, it was the first one to demonstrate that the measurement of serum MicroRNAs can serve as a noninvasive biomarker for glioma diagnosis and prognosis. In a similar work, though involving a smaller cohort of patients, D'Urso et al. showed that the plasma levels of MiR-16, MiR-15b, and MiR-21 efficiently discriminate between glioma patients and a control group of patients suffering from various, nontumoral, and neurological disorders (D'Urso et al., 2015). While MiR-15b and MiR-21 were increased in glioma patients, MiR-16 was significantly decreased and, most importantly, only MiR-16 expression in plasma was able to discriminate among different grades of glioma, its levels being significantly decreased in glioblastoma compared to low grade and anaplastic gliomas.

From this review of the available literature, one issue mostly emerges, that is the very limited overlap of the results obtained by the different groups. This is surely due to the fact that the methods for sample collection, MiRNA extraction, and measurement in biological liquids are not completely standardized yet. For example, samples range from serum to plasma, in this latter case treated with different anticoagulants, or even conserved in different conditions; measurement techniques can be RT-qPCR or microarrays, with different sensitivity and specificity, and while in some cases microarrays may not be able to clearly distinguish between precursor and mature MiRNAs, on the other hand for RT-qPCR in extracellular samples no universally recognized "housekeeping" normalizer has been established, making assays performed in different labs hard to compare. In summary, for these reasons and also because of the still restricted number of samples analyzed so far, this approach, especially for glioblastoma, is still in its infancy. Nonetheless, at least one MicroRNA, MiR-21, holds a great promise as a circulating biomarker for glioblastoma. This is in agreement with its recognized role as "oncomiR" in GBM (Papagiannakopoulos et al., 2008), actually the first one to be studied in this tumor (Chan et al., 2005).

In agreement and in strict correlation with these results, MiR-21 arises as a relevant candidate also in a different body fluid, cerebrospinal fluid (CSF), tightly connected to glioblastoma and recently investigated by some groups in search of alternative biomarkers. In fact, one of the limiting factors for the use of plasma or serum in search of biomarkers for CNS diseases is the blood–brain barrier, thought to impair the release of tumor-derived molecules into blood. On the contrary, CSF is in close contact with brain, is the medium through which nutrients are delivered and waste is drained, and is well known to contain several types of molecules, as proteins, glucose, chemokines, and many others.

A candidate approach, based on published MiRNA expression data in glioma tissues, was used by Baraniskin et al., who compared the expression of nine candidate MiRNA biomarkers in CSF samples from 10 patients diagnosed with glioma (among which five with glioblastoma WHO IV) to that of 10 CSF samples from patients with miscellaneous neurologic disorders (Baraniskin et al., 2012). They found that MiR-21 and MiR-15b were significantly increased in glioma CSF samples. This result interestingly resembled that found by D'Urso et al. (2015) in plasma. Their most relevant result, however, was that deriving from the comparison of glioma CSFs with CSFs from primary diffuse large B-cell lymphoma of the CNS (PCNSL), which represents an important differential diagnosis of glioma. This comparison revealed that MiR-15b was significantly overexpressed while MiR-21 was decreased in glioma patients' CSFs compared to PCNSL. Moreover, by combining the relative expression levels of MiR-15b and MiR-21, they developed a "diagnostic tree" able to correctly discriminate glioma patients by considering CSF samples with MiR-15b above the cutoff and MiR-21 under the cutoff. One limit of this work is evidently the small number of patients analyzed. A relatively larger cohort was studied by Teplyuk and collaborators, who enrolled 19 GBM patients, and compared their CSFs to those of 15 patients affected by nonneoplastic neurological conditions, and to those of the CSFs of a total of 64 patients with metastases from either breast or lung cancer (Teplyuk et al., 2012). As in Baraniskin's approach, they a priori selected specific MicroRNAs to measure by RT-qPCR. Among these, and in agreement with previous results, they found that MiR-21 levels were significantly increased in the CSF of most GBM patients compared to nonneoplastic samples, but this overexpression was not specific for GBM exclusively, as it did not discriminate between the CSF of GBM and metastatic patients. The same findings of increased expression applied to MiR-10b, making it a highly indicative marker of high-grade primary and metastatic brain cancers, but again in a way which is not exclusive for GBM. However, besides building a simple "diagnostic tree" employing their results in MiRNA differential expression in CSF, they successfully employed the RBF kernel "machine-learning technique" based on the SVM concept (Cortes and Vapnik, 1995) to more accurately classify and separate nonneoplastic controls vs GBMs vs metastases. The best classification accuracy was achieved using the levels of seven MiRNAs—MiR-10b,

MiR-21, MiR-125b, MiR-141, MiR-200a, MiR-200b, and MiR-200c—as independent variables. This method allowed to distinguish among the different types of tumoral patients as well as nonneoplastic controls with the average cross-validation accuracy of about 90%.

As for serum and plasma, also in CSF a more sophisticated approach can envision the purification of EVs, followed by the specific measurement of their MicroRNA load. Akers and colleagues used this approach to determine the absolute levels of MiR-21 in EVs from CSF of 13 glioblastoma and 14 nononcologic patients (Akers et al., 2013). Their work differed from all previous ones because they purified and measured the number of EVs produced by all samples they tested, and used these numbers to normalize EV MiR-21 levels. In GBM patients, they found EV MiR-21 ranging from 0.14 to 1.04 copies/EV, compared with nononcologic patients' CSF, where MiR-21 ranged from 5.26×10^{-4} to 1.48×10^{-1} copies/EV. This data, in agreement with previous ones found in GBM tumor, serum/plasma, and CSF samples, were further refined by distinguishing between MiR-21 located inside EVs and freely floating MiR-21, and revealed that less than 10% of total CSF MiR-21 was detectable in the EV-depleted CSF fractions, while the bulk was within EVs. Moreover, they were able to set the threshold value of CSF EV MiR-21 at 0.25 copy/EV, and used it to correctly classify an independent cohort of patients, composed of 15 GBMs and 14 nononcologic CSF EV samples.

In agreement with this work, Shi et al. too purified and characterized exosomes from CSF and sera of 70 glioma patients, among which 25 diagnosed with I or II WHO grade tumors, and 45 with III or IV WHO grade (Shi et al., 2015b). They employed a traditional ultracentrifugation protocol followed by TEM and Western blotting, through which they determined the spherical nature and the diameters of about 100 nm and the expression of the typical exosomal marker CD63 for both serum- and CSF-purified EVs. The EVs from serum and CSF differed in the total amount of RNA, which was much less in CSF EVs, but much more enriched in the small molecular ranges, corresponding to MiRNA sizes. These authors too chose MiR-21 as the most promising candidate MiRNA biomarker, and measured it in CSF and serum EVs, thus finding that MiR-21 expression in CSF EVs was significantly increased in both high- and low-grade gliomas, compared to a set of nontumor controls. On the contrary, serum EV levels of MiR-21 did not discriminate between glioma and nonglioma patients. This observation indicates that CSF might be a better source than serum for the diagnostic evaluation of glioma MiRNA biomarkers, in part because CSF, compared to serum or plasma, has a lower risk of being contaminated by MiRNAs released by blood cells that escaped the purification. Interestingly, they also reported that the high and low MiR-21 groups showed a significant difference in the development of diffuse intracranial and spinal dissemination, as the CSF derive dexosomal MiR-21 levels correlated with tumor recurrence at those sites.

The use of a high-throughput microarray-like technology, called NanoString, to measure MicroRNAs in CSF samples of 34 patients with CNS benign and malignant tumors (among which four GBMs) and in 14 healthy controls, allowed Drusco and colleagues to identify a MicroRNA profile that could differentiate between and among some classes of CNS primary tumors or metastases (Drusco et al., 2015). This signature, composed bymir-451, MiR-223, MiR-125b, MiR-711, and MiR-935, was validated by RT-qPCR, and was employed to design a diagnostic flowchart for CNS tumors. In particular regarding GBMs, the signature requires the absence of MiR-935 in CSF, flanked by high levels of MiR-711 and MiR-451, and a strong expression of MiR-223. Based on this analysis, the CSF MiRNA profiles most similar to the GBM ones were those of medulloblastoma. However, the concomitant absence of MiR-935 and the overexpression of MiR-711 and MiR-223 could discriminate between the two groups, while medulloblastoma CSF showed the highest expression of MiR-125b compared to any other sample.

As for, and even more than for serum/plasma MicroRNAs, one issue is clear in this field of MiRNA analysis in CSF for diagnostic or prognostic use in GBM: the numbers analyzed so far are extremely low, the sample types and collection methods, and the measurement techniques are variable. All this lead to results that sometimes may appear contradictory, but that probably only need some refinement and a significant amplification of the cohorts analyzed. Nonetheless, this approach has already led to some shared findings, as MiR-21 overexpression in GBM CSF, and is surely a promising one that deserves more studies. These will not only enroll larger groups of patients, but also employ more sophisticated and sensitive methods for MiRNA detection, such as next-generation sequencing.

11. CONCLUDING REMARKS

Since their first discovery in 2001, MicroRNAs have been acquiring more and more importance for their role as regulators of gene expression in physiology and pathology. The number of annotated MicroRNAs in the human genome is today larger than 2500, and these molecules, as tiny as powerful, have been shown to regulate essentially all aspects of cell biology in eukaryotes in a pleiotropic and versatile mode. Specifically, the study of their involvement in glioblastoma pathogenesis and clinical history is a field of great importance, as it deals with a tumor which is still untreatable, despite decades of research, and efforts made to reveal the molecular basis of glioblastoma, to be exploited for therapy. A peculiarity of glioblastoma, rendering its comprehension and therapy very difficult, is its heterogeneity, in particular at the molecular level, rarely reflected in macroscopic or even histologic features, so that the entities hidden behind a single term represent in fact a variegated population composed of tissues as different among them as they are from a healthy tissue. The finding that specific MicroRNAs can induce or repress GBM growth, can classify glioblastomas as well or even better than mRNAs, can mediate the interaction with tumor microenvironment, and can even be used as prognostic circulating biomarkers (Fig. 2), is an exciting and stimulating one. From a clinical perspective, MicroRNAs

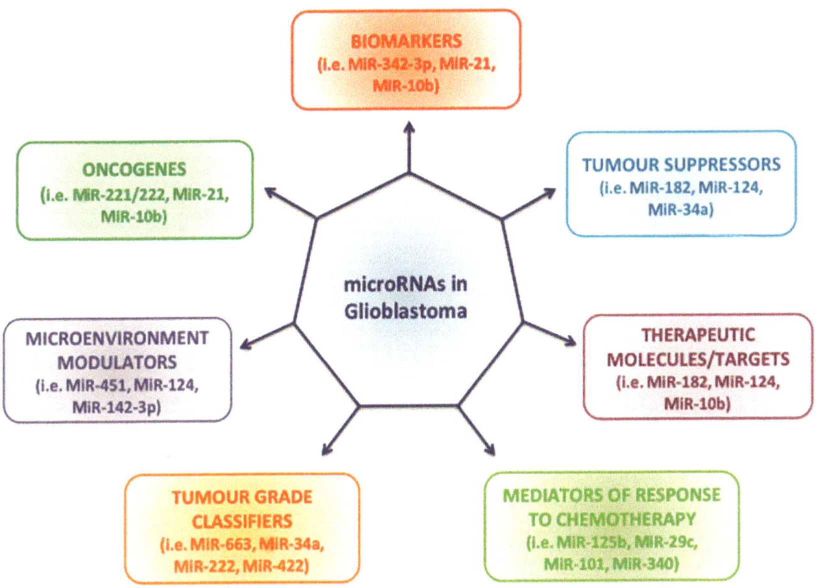

Fig. 2 Main roles of MicroRNAs in glioblastoma. Studies conducted in the last decades highlight the multiple roles of different MicroRNAs working as oncogenes or tumor suppressors and as mediators of GBM crosstalk with the microenvironment, with specific connections with GBM clinical subtypes and resistance to therapies. Moreover, latest findings present the use of MicroRNAs as possible therapeutic molecules/targets or as circulating biomarkers for diagnosis and prognosis.

own several added values, such as their high stability, which can be exploited in the search for biomarkers, their pleiotropic mode of action, implying that the targeting of a single MicroRNA can be translated in the modulation of several mRNAs, all targets of the same MiRNA, and many others. The combination of sophisticated technologies for the detection of MicroRNA expression, with the development of novel delivery methods are now paving the way for a realistic employment of such tiny molecules as a weapon against glioblastoma, and we can reasonably expect that knowledge about MicroRNAs will contribute to the fight to this tumor, and ultimately to its defeat.

ACKNOWLEDGMENTS

This work was partially supported by a Fondazione Umberto Veronesi grant to S.A.C. We thank Banca Fideuram S.p.A. for their contribution to SAC work on MicroRNAs in glioblastoma. N.M. was supported by a fellowship from Fondazione Umberto Veronesi.

REFERENCES

Ader, I., Delmas, C., Skuli, N., Bonnet, J., Schaeffer, P., Bono, F., Cohen-Jonathan-Moyal, E., Toulas, C., 2014. Preclinical evidence that SSR128129—a novel small-molecule multi-FGF receptor blocker—radiosensitizes human glioblastoma. Eur. J. Cancer 50, 2351–2359.

Ahmed, S.U., Carruthers, R., Gilmour, L., Yildirim, S., Watts, C., Chalmers, A.J., 2015. Selective inhibition of parallel DNA damage response pathways optimizes radiosensitization of glioblastoma stem-like cells. Cancer Res. 75 (20), 4416–4428.

Akers, J.C., Ramakrishnan, V., Kim, R., Skog, J., Nakano, I., Pingle, S., Kalinina, J., Hua, W., Kesari, S., Mao, Y., Breakefield, X.O., Hochberg, F.H., Van Meir, E.G., Carter, B.S., Chen, C.C., 2013. MiR-21 in the extracellular vesicles (EVs) of cerebrospinal fluid (CSF): a platform for glioblastoma biomarker development. PLoS One 8 (10), e78115.

Ansari, K.I., Ogawa, D., Rooj, A.K., Lawler, S.E., Krichevsky, A.M., Johnson, M.D., Chiocca, E.A., Bronisz, A., Godlewski, J., 2015. Glucose-based regulation of miR-451/AMPK signaling depends on the OCT1 transcription factor. Cell Rep. 11 (6), 902–909.

Bailey, P., Cushing, H., 1926. A Classification of the Tumours of the Glioma Group on a Histogenetic Basis With a Correlated Study of Prognosis. Lippincott, Philadelphia.

Banelli, B., Carra, E., Barbieri, F., Würth, R., Parodi, F., Pattarozzi, A., Carosio, R., Forlani, A., Allemanni, G., Marubbi, D., Florio, T., Daga, A., Romani, M., 2015. The histone demethylase KDM5A is a key factor for the resistance to temozolomide in glioblastoma. Cell Cycle 14 (21), 3418–3429.

Bao, S., Wu, Q., McLendon, R.E., Hao, Y., Shi, Q., Hjelmeland, A.B., Dewhirst, M.W., Bigner, D.D., Rich, J.N., 2006a. Glioma stem cells promote radioresistance by preferential activation of the DNA damage response. Nature 444, 756–760.

Bao, S., Wu, Q., Sathornsumetee, S., Hao, Y., Li, Z., Hjelmeland, A.B., Shi, Q., McLendon, R.E., Bigner, D.D., Rich, J.N., 2006b. Stem cell-like glioma cells promote tumour angiogenesis through vascular endothelial growth factor. Cancer Res. 66, 7843–7848.

Baraniskin, A., Kuhnhenn, J., Schlegel, U., Maghnouj, A., Zöllner, H., Schmiegel, W., Hahn, S., Schroers, R., 2012. Identification of micro RNAs in the cerebrospinal fluid as biomarker for the diagnosis of glioma. Neuro Oncol. 14 (1), 29–33.

Berger, R., Jennewein, C., Marschall, V., Karl, S., Cristofanon, S., Wagner, L., Vellanki, S.H., Hehlgans, S., Rödel, F., Debatin, K.M., Ludolph, A.C., Fulda, S., 2011. NF-κB is required for Smac mimetic-mediated sensitization of glioblastoma cells for γ-irradiation-induced apoptosis. Mol. Cancer Ther. 10 (10), 1867–1875.
Biddlestone-Thorpe, L., Sajjad, M., Rosenberg, E., Beckta, J.M., Valerie, N.C., Tokarz, M., Adams, B.R., Wagner, A.F., Khalil, A., Gilfor, D., Golding, S.E., Deb, S., Temesi, D.G., Lau, A., O'Connor, M.J., Choe, K.S., Parada, L.F., Lim, S.K., Mukhopadhyay, N.D., Valerie, K., 2013. ATM kinase inhibition preferentially sensitizes p53-mutant glioma to ionizing radiation. Clin. Cancer Res. 19 (12), 3189–3200.
Bier, A., Giladi, N., Kronfeld, N., Lee, H.K., Cazacu, S., Finniss, S., Xiang, C., Poisson, L., deCarvalho, A.C., Slavin, S., Jacoby, E., Yalon, M., Toren, A., Mikkelsen, T., Brodie, C., 2013. MicroRNA-137 is downregulated in glioblastoma and inhibits the stemness of glioma stem cells by targeting RTVP-1. Oncotarget 4 (5), 665–676.
Bonavia, R., Inda, M.M., Cavenee, W.K., Furnari, F.B., 2011. Heterogeneity maintenance in glioblastoma: a social network. Cancer Res. 71, 4055–4060.
Brennan, C.W., Verhaak, R.G., McKenna, A., Campos, B., Noushmehr, H., Salama, S.R., Zheng, S., Chakravarty, D., Sanborn, J.Z., Berman, S.H., Beroukhim, R., Bernard, B., Wu, C.J., Genovese, G., Shmulevich, I., Barnholtz-Sloan, J., Zou, L., Vegesna, R., Shukla, S.A., Ciriello, G., Yung, W.K., Zhang, W., Sougnez, C., Mikkelsen, T., Aldape, K., Bigner, D.D., Van Meir, E.G., Prados, M., Sloan, A., Black, K.L., Eschbacher, J., Finocchiaro, G., Friedman, W., Andrews, D.W., Guha, A., Iacocca, M., O'Neill, B.P., Foltz, G., Myers, J., Weisenberger, D.J., Penny, R., Kucherlapati, R., Perou, C.M., Hayes, D.N., Gibbs, R., Marra, M., Mills, G.B., Lander, E., Spellman, P., Wilson, R., Sander, C., Weinstein, J., Meyerson, M., Gabriel, S., Laird, P.W., Haussler, D., Getz, G., Chin, L., Research Network, T.C.G.A., 2013. The somatic genomic landscape of glioblastoma. Cell 155 (2), 462–477.
Bronisz, A., Wang, Y., Nowicki, M.O., Peruzzi, P., Ansari, K.I., Ogawa, D., Balaj, L., De Rienzo, G., Mineo, M., Nakano, I., Ostrowski, M.C., Hochberg, F., Weissleder, R., Lawler, S.E., Chiocca, E.A., Godlewski, J., 2014. Extracellular vesicles modulate the glioblastoma microenvironment via a tumor suppression signaling network directed by miR-1. Cancer Res. 74 (3), 738–750.
Chan, J.A., Krichevsky, A.M., Kosik, K.S., 2005. MicroRNA-21 is an antiapoptotic factor in human glioblastoma cells. Cancer Res. 65 (14), 6029–6033.
Chen, Y., Liu, W., Chao, T., Zhang, Y., Yan, X., Gong, Y., Qiang, B., Yuan, J., Sun, M., Peng, X., 2008. MicroRNA-21 down-regulates the expression of tumor suppressor PDCD4 in human glioblastoma cell T98G. Cancer Lett. 272 (2), 197–205.
Chen, J., Li, Y., Yu, T.-S., McKay, R.M., Burns, D.K., Kernie, S.G., Parada, L.F., 2012. A restricted cell population propagates glioblastoma growth after chemotherapy. Nature 488, 522–526.
Chen, Q., Lu, G., Cai, Y., Li, Y., Xu, R., Ke, Y., Zhang, S., 2014a. MiR-124-5p inhibits the growth of high-grade gliomas through posttranscriptional regulation of LAMB1. Neuro Oncol. 16 (5), 637–651.
Chen, L., Zhang, K., Shi, Z., Zhang, A., Jia, Z., Wang, G., Pu, P., Kang, C., Han, L., 2014b. Lentivirus-mediated Mir-23b sponge diminishes the malignant phenotype of glioma cells *in vitro* and *in vivo*. Oncol. Rep. 31, 1573–1580.
Cheng, L., Huang, Z., Zhou, W., Wu, Q., Donnola, S., Liu, J.K., Fang, X., Sloan, A.E., Mao, Y., Lathia, J.D., Min, W., McLendon, R.E., Rich, J.N., Bao, S., 2013. Glioblastoma stem cells generate vascular pericytes to support vessel function and tumor growth. Cell 153, 139–152.
Chiou, G.Y., Chien, C.S., Wang, M.L., Chen, M.T., Yang, Y.P., Yu, Y.L., Chien, Y., Chang, Y., Shen, C.C., Chio, C.C., Lu, K.H., Ma, H.I., Chen, K.H., Liu, D.M.,

Miller, S.A., Chen, Y.W., Huang, P.I., Shih, Y.H., Hung, M.C., Chiou, S.H., 2013. Epigenetic regulation of the miR142-3p/interleukin-6 circuit in glioblastoma. Mol. Cell 52 (5), 693–706.

Ciafrè, S.A., Galardi, S., Mangiola, A., Ferracin, M., Liu, C.G., Sabatino, G., Negrini, M., Maira, G., Croce, C.M., Farace, M.G., 2005. Extensive modulation of a set of microRNAs in primary glioblastoma. Biochem. Biophys. Res. Commun. 334 (4), 1351–1358.

Cortes, C., Vapnik, V., 1995. Support-vector networks. Mach. Learn. 20 (3), 273–297.

Costa, P.M., Cardoso, A.L., Custódia, C., Cunha, P., Pereira de Almeida, L., Pedroso de Lima, M.C., 2015. MiRNA-21 silencing mediated by tumor-targeted nanoparticles combined with sunitinib: a new multimodal gene therapy approach for glioblastoma. J. Control. Release 207, 31–39.

De Bacco, F., D'Ambrosio, A., Casanova, E., Orzan, F., Neggia, R., Albano, R., Verginelli, F., Cominelli, M., Poliani, P.L., Luraghi, P., Reato, G., Pellegatta, S., Finocchiaro, G., Perera, T., Garibaldi, E., Gabriele, P., Comoglio, P.M., Boccaccio, C., 2016. MET inhibition overcomes radiation resistance of glioblastoma stem-like cells. EMBO Mol. Med. 8 (5), 550–568.

Di Martino, M.T., Campani, V., Misso, G., Cantafio, M.E., Gullà, A., Foresta, U., Guzzi, P.H., Castellano, M., Grimaldi, A., Gigantino, V., Franco, R., Lusa, S., Cannataro, M., Tagliaferri, P., De Rosa, G., Tassone, P., Caraglia, M., 2014. In vivo activity of miR-34a mimics delivered by stablenucleic acidlipidparticles (SNALPs) against multiple myeloma. PLoS One 9, e90005.

Dolecek, T.A., Propp, J.M., Stroup, N.E., Kruchko, C., 2012. CBTRUS statistical report: primary brain and central nervous system tumours diagnosed in the United States in 2005–2009. Neuro Oncol. 14 (Suppl. 5), v1–49.

Dropcho, E.J., Soong, S.J., 1996. The prognostic impact of prior low grade histology in patients with anaplastic gliomas: a case-control study. Neurology 47, 684–690.

Drusco, A., Bottoni, A., Laganà, A., Acunzo, M., Fassan, M., Cascione, L., Antenucci, A., Kumchala, P., Vicentini, C., Gardiman, M.P., Alder, H., Carosi, M.A., Ammirati, M., Gherardi, S., Luscrì, M., Carapella, C., Zanesi, N., Croce, C.M., 2015. A differentially expressed set of microRNAs in cerebro-spinal fluid (CSF) can diagnose CNS malignancies. Oncotarget 6 (25), 20829–20839.

D'Urso, P.I., D'Urso, O.F., Gianfreda, C.D., Mezzolla, V., Storelli, C., Marsigliante, S., 2015. miR-15b and miR-21 as circulating biomarkers for diagnosis of glioma. Curr. Genomics 16 (5), 304–311.

Ebert, M.S., Neilson, J.R., Sharp, P.A., 2007. MicroRNA sponges: competitive inhibitors of small RNAs in mammalian cells. Nat. Methods 9, 721–726.

Eyler, C.E., Wu, Q., Yan, K., MacSwords, J.M., Chandler-Militello, D., Misuraca, K.L., Lathia, J.D., Forrester, M.T., Lee, J., Stamler, J.S., Goldman, S.A., Bredel, M., McLendon, R.E., Sloan, A.E., Hjelmeland, A.B., Rich, J.N., 2011. Glioma stem cell proliferation and tumour growth are promoted by nitric oxide synthase-2. Cell 146, 53–66.

Fazi, B., Felsani, A., Grassi, L., Moles, A., D'Andrea, D., Toschi, N., Sicari, D., De Bonis, P., Anile, C., Guerrisi, M.G., Luca, E., Farace, M.G., Maira, G., Ciafrè, S.A., Mangiola, A., 2015. The transcriptome and miRNome profiling of glioblastoma tissues and peritumoral regions highlights molecular pathways shared by tumors and surrounding areas and reveals differencesbetween short-term and long-term survivors. Oncotarget 6 (26), 22526–22552.

Fiore, D., Donnarumma, E., Roscigno, G., Iaboni, M., Russo, V., Affinito, A., Adamo, A., De Martino, F., Quintavalle, C., Romano, G., Greco, A., Soini, Y., Brunetti, A., Croce, C.M., Condorelli, G., 2016. MiR-340 predicts glioblastoma survival and modulates key cancer hallmarks through down-regulation of NRAS. Oncotarget 7 (15), 19531–19547.

Flavahan, W.A., Wu, Q., Hitomi, M., Rahim, N., Kim, Y., Sloan, A.E., Weil, R.J., Nakano, I., Sarkaria, J.N., Stringer, B.W., Day, B.W., Li, M., Lathia, J.D., Rich, J.N., Hjelmeland, A.B., 2013. Brain tumour initiating cells adapt to restricted nutrition through preferential glucose uptake. Nat. Neurosci. 16, 1373–1382.

Floyd, D., Purow, B., 2014. Micro-masters of glioblastoma biology and therapy: increasingly recognized roles for microRNAs. Neuro Oncol. 16 (5), 622–627.

Gabriely, G., Wurdinger, T., Kesari, S., Esau, C.C., Burchard, J., Linsley, P.S., Krichevsky, A.M., 2008. MicroRNA 21 promotes glioma invasion by targeting matrix metalloproteinase regulators. Mol. Cell. Biol. 28 (17), 5369–5380.

Gabriely, G., Yi, M., Narayan, R.S., Niers, J.M., Wurdinger, T., Imitola, J., Ligon, K.L., Kesari, S., Esau, C., Stephens, R.M., Tannous, B.A., Krichevsky, A.M., 2011. Human glioma growth is controlled by microRNA-10b. Cancer Res. 71, 3563–3572.

Galardi, S., Mercatelli, N., Farace, M.G., Ciafrè, S.A., 2011. NF-kB and c-Jun induce the expression of the oncogenic miR-221 and miR-222 in prostate carcinoma and glioblastoma cells. Nucleic Acids Res. 39 (9), 3892–3902.

Galardi, S., Petretich, M., Pinna, G., D'Amico, S., Loreni, F., Michienzi, A., Groisman, I., Ciafrè, S.A., 2016. CPEB1 restrains proliferation of glioblastoma cells through the regulation of p27(Kip1) mRNA translation. Sci. Rep. 6, 25219.

Galli, R., Binda, E., Orfanelli, U., Cipelletti, B., Gritti, A., De Vitis, S., Fiocco, R., Foroni, C., Dimeco, F., Vescovi, A., 2004. Isolation and characterization of tumourigenic, stem-like neural precursors from human glioblastoma. Cancer Res. 64, 7011–7021.

Gao, Y.T., Chen, X.B., Liu, H.L., 2016. Up-regulation of miR-370-3p restores glioblastoma multiforme sensitivity to temozolomide by influencing MGMT expression. Sci. Rep. 6, 32972.

Gebert, L.F., Rebhan, M.A., Crivelli, S.E., Denzler, R., Stoffel, M., Hall, J., 2014. Miravirsen (SPC3649) can inhibit the biogenesis of miR-122. Nucleic Acids Res. 42, 609–621.

Geisenberger, C., Mock, A., Warta, R., Rapp, C., Schwager, C., Korshunov, A., Nied, A.K., Capper, D., Brors, B., Jungk, C., Jones, D., Collins, V.P., Ichimura, K., Bäcklund, L.M., Schnabel, E., Mittelbron, M., Lahrmann, B., Zheng, S., Verhaak, R.G., Grabe, N., Pfister, S.M., Hartmann, C., von Deimling, A., Debus, J., Unterberg, A., Abdollahi, A., Herold-Mende, C., 2015. Molecular profiling of long-term survivors identifies a subgroup of glioblastoma characterized by chromosome 19/20 co-gain. Acta Neuropathol. 130 (3), 419–434.

Godlewski, J., Bronisż, A., Nowicki, M.O., Chiocca, E.A., Lawler, S., 2010a. microRNA-451: a conditional switch controlling glioma cell proliferation and migration. Cell Cycle 9 (14), 2742–2748.

Godlewski, J., Newton, H.B., Chiocca, E.A., Lawler, S.E., 2010b. MicroRNAs and glioblastoma; the stem cell connection. Cell Death Differ. 17 (2), 221–228.

Godlewski, J., Nowicki, M.O., Bronisz, A., Williams, S., Otsuki, A., Nuovo, G., Raychaudhury, A., Newton, H.B., Chiocca, E.A., Lawler, S., 2008. Targeting of the Bmi-1 oncogene/stem cell renewal factor by microRNA-128 inhibits glioma proliferation and self-renewal. Cancer Res. 68 (22), 9125–9130.

Gouazé-Andersson, V., Delmas, C., Taurand, M., Martinez-Gala, J., Evrard, S., Mazoyer, S., Toulas, C., Cohen-Jonathan-Moyal, E., 2016. FGFR1 induces glioblastoma radioresistance through the PLCγ/Hif1α pathway. Cancer Res. 76 (10), 3036–3044.

Graner, M.W., Alzate, O., Dechkovskaia, A.M., Keene, J.D., Sampson, J.H., Mitchell, D.A., Bigner, D.D., 2009. Proteomic and immunologic analyses of brain tumor exosomes. FASEB J. 23 (5), 1541–1557.

Guan, Y., Mizoguchi, M., Yoshimoto, K., Hata, N., Shono, T., Suzuki, S.O., Araki, Y., Kuga, D., Nakamizo, A., Amano, T., Ma, X., Hayashi, K., Sasaki, T., 2010.

MiRNA-196 is upregulated in glioblastoma but not in anaplastic astrocytoma and has prognostic significance. Clin. Cancer Res. 16 (16), 4289–4297.
Guessous, F., Alvarado-Velez, M., Marcinkiewicz, L., Zhang, Y., Kim, J., Heister, S., Kefas, B., Godlewski, J., Schiff, D., Purow, B., Abounader, R., 2013. Oncogenic effects of miR-10b in glioblastoma stem cells. J. Neurooncol. 112, 153–163.
Haemmig, S., Baumgartner, U., Gluck, A., Zbinden, S., Tschan, M.P., Kappeler, A., Mariani, L., Vajtai, I., Vassella, E., 2014. MiR-125b controls apoptosis and temozolomide resistance by targeting TNFAIP3 and NKIRAS2 in glioblastomas. Cell Death Dis. 5, e1279.
Hayes, J., Thygesen, H., Tumilson, C., Droop, A., Boissinot, M., Hughes, T.A., Westhead, D., Alder, J.E., Shaw, L., Short, S.C., Lawler, S.E., 2015. Prediction of clinical outcome in glioblastoma using a biologically relevant nine-microRNA signature. Mol. Oncol. 9, 704–714.
Hegi, M.E., Diserens, A.-C., Gorlia, T., Hamou, M.-F., de Tribolet, N., Weller, M., Kros, J.M., Hainfellner, J.A., Mason, W., Mariani, L., Bromberg, J.E.C., Hau, P., Mirimanoff, R.O., Cairncross, J.G., Janzer, R.C., Stupp, R., 2005. MGMT gene silencing and benefit from temozolomide in glioblastoma. N. Engl. J. Med. 352, 997–1003.
Hemmati, H.D., Nakano, I., Lazareff, J.A., Masterman-Smith, M., Geschwind, D.H., Bronner-Fraser, M., Kornblum, H.I., 2003. Cancerous stem cells can arise from pediatric brain tumours. Proc. Natl. Acad. Sci. 100, 15178–15183.
Herman, A., Gruden, K., Blejec, A., Podpečan, V., Motaln, H., Rožman, P., Hren, M., Zupančič, K., Veber, M., Verbovšek, U., Lah Turnšek, T., Porčnik, A., Koršič, M., Knežević, M., Jeras, M., 2015. Analysis of glioblastoma patients' plasma revealed the presence of MicroRNAs with a prognostic impact on survival and those of viral origin. PLoS One 10 (5), e0125791.
Hermeking, H., 2007. p53 enters the microRNA world. Cancer Cell 12 (5), 414–418.
Hjelmeland, A.B., Wu, Q., Heddleston, J.M., Choudhary, G.S., MacSwords, J., Lathia, J.D., McLendon, R., Lindner, D., Sloan, A., Rich, J.N., 2011. Acidic stress promotes a glioma stem cell phenotype. Cell Death Differ. 18, 829–840.
Hoshino, A., Costa-Silva, B., Shen, T.L., Rodrigues, G., Hashimoto, A., Tesic Mark, M., Molina, H., Kohsaka, S., Di Giannatale, A., Ceder, S., Singh, S., Williams, C., Soplop, N., Uryu, K., Pharmer, L., King, T., Bojmar, L., Davies, A.E., Ararso, Y., Zhang, T., Zhang, H., Hernandez, J., Weiss, J.M., Dumont-Cole, V.D., Kramer, K., Wexler, L.H., Narendran, A., Schwartz, G.K., Healey, J.H., Sandstrom, P., Labori, K.J., Kure, E.H., Grandgenett, P.M., Hollingsworth, M.A., de Sousa, M., Kaur, S., Jain, M., Mallya, K., Batra, S.K., Jarnagin, W.R., Brady, M.S., Fodstad, O., Muller, V., Pantel, K., Minn, A.J., Bissell, M.J., Garcia, B.A., Kang, Y., Rajasekhar, V.K., Ghajar, C.M., Matei, I., Peinado, H., Bromberg, J., Lyden, D., 2015. Tumour exosome integrins determine organotropic metastasis. Nature 527 (7578), 329–335.
Hu, J., Sun, T., Wang, H., Chen, Z., Wang, S., Yuan, L., Liu, T., Li, H.R., Wang, P., Feng, Y., Wang, Q., McLendon, R.E., Friedman, A.H., Keir, S.T., Bigner, D.D., Rathmell, J., Fu, X.D., Li, Q.J., Wang, H., Wang, X.F., 2016. MiR-215 is induced post-transcriptionally via hif-drosha complex and mediates glioma-initiating cell adaptation to hypoxia by targeting KDM1B. Cancer Cell 29 (1), 49–60.
Ignatova, T.N., Kukekov, V.G., Laywell, E.D., Suslov, O.N., Vrionis, F.D., Steindler, D.A., 2002. Human cortical glial tumours contain neural stem-like cells expressing astroglial and neuronal markers in vitro. Glia 39, 193–206.
Jensen, S.A., Day, E.S., Ko, C.H., Hurley, L.A., Luciano, J.P., Kouri, F.M., Merkel, T.J., Luthi, A.J., Patel, P.C., Cutler, J.I., Daniel, W.L., Scott, A.W., Rotz, M.W., Meade, T.J., Giljohann, D.A., Mirkin, C.A., Stegh, A.H., 2013. Spherical nucleic acid nanoparticle conjugates as an RNAi-based therapy for glioblastoma. Sci. Transl. Med 5, 209ra152.

Junttila, M.R., de Sauvage, F.J., 2013. Influence of tumour micro-environment heterogeneity on therapeutic response. Nature 501, 346–354.

Kefas, B., Godlewski, J., Comeau, L., Li, Y., Abounader, R., Hawkinson, M., Lee, J., Fine, H., Chiocca, E.A., Lawler, S., Purow, B., 2008. microRNA-7 inhibits the epidermal growth factor receptor and the Akt pathway and is down-regulated in glioblastoma. Cancer Res. 68 (10), 3566–3572.

Khazenzon, N.M., Ljubimov, A.V., Lakhter, A.J., Fujita, M., Fujiwara, H., Sekiguchi, K., Sorokin, L.M., Petäjäniemi, N., Virtanen, I., Black, K.L., Ljubimova, J.Y., 2003. Antisense inhibition of laminin-8 expression reduces invasion of human gliomas in vitro. Mol. Cancer Ther. 2 (10), 985–994.

Kim, S.H., Ezhilarasan, R., Phillips, E., Gallego-Perez, D., Sparks, A., Taylor, D., Ladner, K., Furuta, T., Sabit, H., Chhipa, R., Cho, J.H., Mohyeldin, A., Beck, S., Kurozumi, K., Kuroiwa, T., Iwata, R., Asai, A., Kim, J., Sulman, E.P., Cheng, S.Y., Lee, L.J., Nakada, M., Guttridge, D., DasGupta, B., Goidts, V., Bhat, K.P., Nakano, I., 2016. Serine/threonine kinase MLK4 determines mesenchymal identity in glioma stem cells in an NF-κB-dependent manner. Cancer Cell 29 (2), 201–213.

Kim, H., Huang, W., Jiang, X., Pennicooke, B., Park, P.J., Johnson, M.D., 2010. Integrative genome analysis reveals an oncomir/oncogene cluster regulating glioblastoma survivorship. Proc. Natl. Acad. Sci. USA 107 (5), 2183–2188.

Kim, T.M., Huang, W., Park, R., Park, P.J., Johnson, M.D., 2011. A developmental taxonomy of glioblastoma defined and maintained by MicroRNAs. Cancer Res. 71 (9), 3387–3399.

Kim, J., Zhang, Y., Skalski, M., Hayes, J., Kefas, B., Schiff, D., Purow, B., Parsons, S., Lawler, S., Abounader, R., 2014. microRNA-148a is a prognostic oncomir that targets MIG6 and BIM to regulate EGFR and apoptosis in glioblastoma. Cancer Res. 74 (5), 1541–1553.

Kitange, G.J., Mladek, A.C., Schroeder, M.A., Pokorny, J.C., Carlson, B.L., Zhang, Y., Nair, A.A., Lee, J.H., Yan, H., Decker, P.A., Zhang, Z., Sarkaria, J.N., 2016. Retinoblastoma binding protein 4 modulates temozolomide sensitivity in glioblastoma by regulating DNA repair proteins. Cell Rep. 14 (11), 2587–2598.

Kota, J., Chivukula, R.R., O'Donnell, K.A., Wentzel, E.A., Montgomery, C.L., Hwang, H.W., Chang, T.C., Vivekanandan, P., Torbenson, M., Clark, K.R., Mendell, J.R., Mendell, J.T., 2009. Therapeutic microRNA delivery suppresses tumorigenesis in a murine liver cancer model. Cell 137 (6), 1005–1017.

Kotliarova, S., Pastorino, S., Kovell, L.C., Kotliarov, Y., Song, H., Zhang, W., Bailey, R., Maric, D., Zenklusen, J.C., Lee, J., Fine, H.A., 2008. Glycogen synthase kinase-3 inhibition induces glioma cell death through c-MYC, nuclear factor-kappaB, and glucose regulation. Cancer Res. 68 (16), 6643–6651.

Kouri, F.M., Hurley, L.A., Daniel, W.L., Day, E.S., Hua, Y., Hao, L., Peng, C.Y., Merkel, T.J., Queisser, M.A., Ritner, C., Zhang, H., James, C.D., Sznajder, J.I., Chin, L., Giljohann, D.A., Kessler, J.A., Peter, M.E., Mirkin, C.A., Stegh, A.H., 2015. miR-182 integrates apoptosis, growth, and differentiation programs in glioblastoma. GenesDev. 29 (7), 732–745.

Krützfeldt, J., Rajewsky, N., Braich, R., Rajeev, K.G., Tuschl, T., Manoharan, M., Stoffel, M., 2005. Silencing of microRNAs in vivo with 'antagomirs'. Nature 438, 685–689.

Kumar, M.S., Lu, J., Mercer, K.L., Golub, T.R., Jacks, T., 2007. Impaired microRNA processing enhances cellular transformation andtumorigenesis. Nat. Genet. 39, 673–677.

Lai, N.S., Wu, D.G., Fang, X.G., Lin, Y.C., Chen, S.S., Li, Z.B., Xu, S.S., 2015. Serum microRNA-210 as a potential noninvasive biomarker for the diagnosis and prognosis of glioma. Br. J. Cancer 112 (7), 1241–1246.

Lathia, J.D., Gallagher, J., Heddleston, J.M., Wang, J., Eyler, C.E., Macswords, J., Wu, Q., Vasanji, A., McLendon, R.E., Hjelmeland, A.B., Rich, J.N., 2010. Integrin alpha 6 regulates glioblastoma stem cells. Cell Stem Cell 6, 421–432.
Lathia, J.D., Mack, S.C., Mulkearns-Hubert, E.E., Valentim, C.L., Rich, J.N., 2015. Cancer stem cells in glioblastoma. Genes Dev. 29 (12), 1203–1217.
le Sage, C., Nagel, R., Egan, D.A., Schrier, M., Mesman, E., Mangiola, A., Anile, C., Maira, G., Mercatelli, N., Ciafrè, S.A., Farace, M.G., Agami, R., 2007. Regulation of the p27(Kip1) tumor suppressor by miR-221 and miR-222 promotes cancer cell proliferation. EMBO J. 26 (15), 3699–3708.
Li, Z., Bao, S., Wu, Q., Wang, H., Eyler, C., Sathornsumetee, S., Shi, Q., Cao, Y., Lathia, J., McLendon, R.E., Hjelmeland, A.B., Rich, J.N., 2009a. Hypoxia-inducible factors regulate tumourigenic capacity of glioma stem cells. Cancer Cell 15, 501–513.
Li, C.C., Eaton, S.A., Young, P.E., Lee, M., Shuttleworth, R., Humphreys, D.T., Grau, G.E., Combes, V., Bebawy, M., Gong, J., Brammah, S., Buckland, M.E., Suter, C.M., 2013. Glioma microvesicles carry selectively packaged coding and non-coding RNAs which alter gene expression in recipient cells. RNA Biol. 10 (8), 1333–1344.
Li, R., Gao, K., Luo, H., Wang, X., Shi, Y., Dong, Q., Luan, W., You, Y., 2014. Identification of intrinsic subtype-specific prognostic microRNAs in primary glioblastoma. J. Exp. Clin. Cancer Res. 33, 9.
Li, Y., Guessous, F., Zhang, Y., Dipierro, C., Kefas, B., Johnson, E., Marcinkiewicz, L., Jiang, J., Yang, Y., Schmittgen, T.D., Lopes, B., Schiff, D., Purow, B., Abounader, R., 2009b. MicroRNA-34a inhibits glioblastoma growth by targeting multiple oncogenes. Cancer Res. 69 (19), 7569–7576.
Li, Y., Li, A., Glas, M., Lal, B., Ying, M., Sang, Y., Xia, S., Trageser, D., Guerrero-Cázares, H., Eberhart, C.G., Quiñones-Hinojosa, A., Scheffler, B., Laterra, J., 2011. c-Met signaling induces a reprogramming network and supports the glioblastoma stem-like phenotype. Proc. Natl. Acad. Sci. USA. 108 (24), 9951–9956.
Lin, J., Teo, S., Lam, D.H., Jeyaseelan, K., Wang, S., 2012. MicroRNA-10b pleiotropically regulates invasion, angiogenicity and apoptosis of tumor cells resembling mesenchymal subtype of glioblastoma multiforme. Cell Death Dis. 3, e398.
Liu, C., Kelnar, K., Liu, B., Chen, X., Calhoun-Davis, T., Li, H., Patrawala, L., Yan, H., Jeter, C., Honorio, S., Wiggins, J.F., Bader, A.G., Fagin, R., Brown, D., Tang, D.G., 2011. The microRNA miR-34a inhibits prostate cancer stem cells and metastasis by directly repressing CD44. Nat. Med. 17, 211–215.
Lopez-Bertoni, H., Lal, B., Li, A., Caplan, M., Guerrero-Cázares, H., Eberhart, C.G., Quiñones-Hinojosa, A., Glas, M., Scheffler, B., Laterra, J., Li, Y., 2015. DNMT-dependent suppression of microRNA regulates the induction of GBM tumor-propagating phenotype by Oct4 and Sox2. Oncogene 34 (30), 3994–4004.
Lötvall, J., Hill, A.F., Hochberg, F., Buzás, E.I., Di Vizio, D., Gardiner, C., Gho, Y.S., Kurochkin, I.V., Mathivanan, S., Quesenberry, P., Sahoo, S., Tahara, H., Wauben, M.H., Witwer, K.W., Théry, C., 2014. Minimal experimental requirements for definition of extracellular vesicles and their functions: a position statement from the International Society for Extracellular Vesicles. J. Extracell Vesicles 3, 26913.
Louis, D.N., Ohgaki, H., Wiestler, O.D., Cavenee, W.K., Burger, P.C., Jouvet, A., Scheithauer, B.W., Kleihues, P., 2007. The 2007 WHO classification of tumours of the central nervous system. Acta Neuropathol. (Berl.) 114, 97–109.
Louis, D.N., Perry, A., Reifenberger, G., von Deimling, A., Figarella-Branger, D., Cavenee, W.K., Ohgaki, H., Wiestler, O.D., Kleihues, P., Ellison, D.W., 2016. The 2016 World Health Organization classification of tumours of the central nervous system: a summary. Acta Neuropathol. 131 (6), 803–820.

Lulli, V., Buccarelli, M., Martini, M., Signore, M., Biffoni, M., Giannetti, S., Morgante, L., Marziali, G., Ilari, R., Pagliuca, A., Larocca, L.M., De Maria, R., Pallini, R., Ricci-Vitiani, L., 2015. miR-135b suppresses tumorigenesis in glioblastoma stem-like cells impairing proliferation, migration and self renewal. Oncotarget 6 (35), 37241–37256.

Luo, H., Chen, Z., Wang, S., Zhang, R., Qiu, W., Zhao, L., Peng, C., Xu, R., Chen, W., Wang, H.-W., Chen, Y., Yang, J., Zhang, X., Zhang, S., Chen, D., Wu, W., Zhao, C., Cheng, G., Jiang, T., Lu, D., You, Y., Liu, N., Wang, H., 2016. c-Myc–miR-29c–REV3L signalling pathway drives the acquisition of temozolomide resistance in glioblastoma. Brain 138, 3654–3672.

Ma, X., Yoshimoto, K., Guan, Y., Hata, N., Mizoguchi, M., Sagata, N., Murata, H., Kuga, D., Amano, T., Nakamizo, A., Sasaki, T., 2012. Associations between microRNA expression and mesenchymal marker gene expression in glioblastoma. Neuro Oncol. 14 (9), 1153–1162.

Mansouri, S., Singh, S., Alamsahebpur, A., Burrell, K., Li, M., Karabork, M., Ekinci, C., Koch, E., Solaroglu, I., Chang, J.T., Wouters, B., Aldape, K., Zadeh, G., 2016. DICER governs characteristics of glioma stem cells and theresulting tumors in xenograft mouse models of glioblastoma. Oncotarget 7 (35), 56431–56446. http://dx.doi.org/10.18632/oncotarget.

Manterola, L., Guruceaga, E., Gállego Pérez-Larraya, J., González-Huarriz, M., Jauregui, P., Tejada, S., Diez-Valle, R., Segura, V., Samprón, N., Barrena, C., Ruiz, I., Agirre, A., Ayuso, A., Rodríguez, J., González, A., Xipell, E., Matheu, A., López de Munain, A., Tuñón, T., Zazpe, I., García-Foncillas, J., Paris, S., Delattre, J.Y., Alonso, M.M., 2014. A small noncoding RNA signature found in exosomes of GBM patient serum as a diagnostic tool. Neuro Oncol. 16 (4), 520–527.

Medina, R., Zaidi, S.K., Liu, C.G., Stein, J.L., van Wijnen, A.J., Croce, C.M., Stein, G.S., 2008. MicroRNAs 221 and 222 bypass quiescence and compromise cell survival. Cancer Res. 68 (8), 2773–2780.

Melo, S.A., Sugimoto, H., O'Connell, J.T., Kato, N., Villanueva, A., Vidal, A., Qiu, L., Vitkin, E., Perelman, L.T., Melo, C.A., Lucci, A., Ivan, C., Calin, G.A., Kalluri, R., 2014. Cancer exosomes perform cell-independent microRNA biogenesis and promote tumorigenesis. Cancer Cell 26 (5), 707–721.

Moskwa, P., Zinn, P.O., Choi, Y.E., Shukla, S.A., Fendler, W., Chen, C.C., Lu, J., Golub, T.R., Hjelmeland, A., Chowdhury, D., 2014. A functional screen identifies mirs that induce radioresistance in glioblastomas. Mol. Cancer Res. 12 (12), 1767–1778.

Munson, J.M., Bellamkonda, R.V., Swartz, M.A., 2013. Interstitial flow in a 3D microenvironment increases glioma invasion by a CXCR4-dependent mechanism. Cancer Res. 73, 1536–1546.

Neuberg, P., Kichler, A., 2014. Recent developments in nucleic acid delivery with polyethylenimines. Adv. Genet. 88, 263–288.

Noushmehr, H., Weisenberger, D.J., Diefes, K., Phillips, H.S., Pujara, K., Berman, B.P., Pan, F., Pelloski, C.E., Sulman, E.P., Bhat, K.P., Verhaak, R.G., Hoadley, K.A., Hayes, D.N., Perou, C.M., Schmidt, H.K., Ding, L., Wilson, R.K., Van Den Berg, D., Shen, H., Bengtsson, H., Neuvial, P., Cope, L.M., Buckley, J., Herman, J.G., Baylin, S.B., Laird, P.W., Aldape, K., Network, C.G.A.R., 2010. Identification of a CpG island methylator phenotype that defines a distinct subgroup of glioma. Cancer Cell 17 (5), 510–522.

Ohgaki, H., Kleihues, P., 2005. Epidemiology and etiology of gliomas. Acta Neuropathol. 109, 93–108.

Ohgaki, H., Kleihues, P., 2013. The definition of primary and secondary glioblastoma. Clin. Cancer Res. 19, 764–772.

Ostrom, Q.T., Gittleman, H., Stetson, L., Virk, S.M., Barnholtz-Sloan, J.S., 2015. Epidemiology of gliomas. Cancer Treat. Res. 163, 1–14.

Papagiannakopoulos, T., Friedmann-Morvinski, D., Neveu, P., Dugas, J.C., Gill, R.M., Huillard, E., Liu, C., Zong, H., Rowitch, D.H., Barres, B.A., Verma, I.M., Kosik, K.S., 2012. Pro-neural miR-128 is a glioma tumor suppressor that targets mitogenic kinases. Oncogene 31 (15), 1884–1895.

Papagiannakopoulos, T., Shapiro, A., Kosik, K.S., 2008. MicroRNA-21 targets a network of key tumor-suppressive pathways in glioblastoma cells. Cancer Res. 68 (19), 8164–8172.

Parsons, D.W., Jones, S., Zhang, X., Lin, J.C.-H., Leary, R.J., Angenendt, P., Mankoo, P., Carter, H., Siu, I.-M., Gallia, G.L., Olivi, A., McLendon, R., Rasheed, B.A., Keir, S., Nikolskaya, T., Nikolsky, Y., Busam, D.A., Tekleab, H., Diaz, L.A., Hartigan, J., Smith, D.R., Strausberg, R.L., Marie, S.K.N., Shinjo, S.M.O., Yan, H., Riggins, G.J., Bigner, D.D., Karchin, R., Papadopoulos, N., Parmigiani, G., Vogelstein, B., Velculescu, V.E., Kinzler, K.W., 2008. An integrated genomic analysis of human glioblastoma multiforme. Science 321, 1807–1812.

Peng, C., Chen, Z., Wang, S., Wang, H.W., Qiu, W., Zhao, L., Xu, R., Luo, H., Chen, Y., Chen, D., You, Y., Liu, N., Wang, H., 2016. The error-prone DNA polymerase κ promotes temozolomide resistance in glioblastoma through rad17-dependent activation of ATR-Chk1 signaling. Cancer Res. 76 (8), 2340–2353.

Phillips, H.S., Kharbanda, S., Chen, R., Forrest, W.F., Soriano, R.H., Wu, T.D., Misra, A., Nigro, J.M., Colman, H., Soroceanu, L., Williams, P.M., Modrusan, Z., Feuerstein, B.G., Aldape, K., 2006. Molecular subclasses of high-grade glioma predict prognosis, delineate a pattern of disease progression, and resemble stages in neurogenesis. Cancer Cell 9, 157–173.

Ping, Y.F., Yao, X.H., Jiang, J.Y., Zhao, L.T., Yu, S.C., Jiang, T., Lin, M.C., Chen, J.H., Wang, B., Zhang, R., Cui, Y.H., Qian, C., Wang, J., Bian, X.W., 2011. The chemokine CXCL12 and its receptor CXCR4 promote glioma stem cell-mediated VEGF production and tumour angiogenesis via PI3K/AKT signalling. J. Pathol. 224, 344–354.

Quintavalle, C., Garofalo, M., Zanca, C., Romano, G., Iaboni, M., del Basso, De Caro, M., Martinez-Montero, J.C., Incoronato, M., Nuovo, G., Croce, C.M., Condorelli, G., 2012. miR-221/222overexpession in human glioblastoma increases invasiveness by targeting the protein phosphate PTPμ. Oncogene 31 (7), 858–868.

Quintavalle, C., Mangani, D., Roscigno, G., Romano, G., Diaz-Lagares, A., Iaboni, M., Donnarumma, E., Fiore, D., De Marinis, P., Soini, Y., Esteller, M., Condorelli, G., 2013. MiR-221/222 target the DNA methyltransferase MGMT in glioma cells. PLoS One 8 (9), e74466.

Rheinbay, E., Suvà, M.L., Gillespie, S.M., Wakimoto, H., Patel, A.P., Shahid, M., Oksuz, O., Rabkin, S.D., Martuza, R.L., Rivera, M.N., Louis, D.N., Kasif, S., Chi, A.S., Bernstein, B.E., 2013. An aberrant transcription factor network essential for Wnt signaling and stem cell maintenance in glioblastoma. Cell Rep. 3 (5), 1567–1579.

Rosenzweig, T., Ziv-Av, A., Xiang, C., Lu, W., Cazacu, S., Taler, D., Miller, C.G., Reich, R., Shoshan, Y., Anikster, Y., Kazimirsky, G., Sarid, R., Brodie, C., 2006. Related to testes-specific, vespid, and pathogenesis protein-1 (RTVP-1) is overexpressed in gliomas and regulates the growth, survival, and invasion of glioma cells. Cancer Res. 66 (8), 4139–4148.

Roth, P., Wischhusen, J., Happold, C., Chandran, P.A., Hofer, S., Eisele, G., Weller, M., Keller, A., 2011. A specific miRNA signature in the peripheral blood of glioblastoma patients. J. Neurochem. 118 (3), 449–457.

Saito, N., Aoki, K., Hirai, N., Fujita, S., Iwama, J., Hiramoto, Y., Ishii, M., Sato, K., Nakayama, H., Harashina, J., Hayashi, M., Izukura, H., Kimura, H., Ito, K., Sakurai, T., Yokouchi, Y., Oharazeki, T., Takahashi, K., Iwabuchi, S., 2015. Effect

of Notch expression in glioma stem cells on therapeutic response to chemo-radiotherapy in recurrent glioblastoma. Brain Tumor Pathol. 32 (3), 176–183.
Sasayama, T., Nishihara, M., Kondoh, T., Hosoda, K., Kohmura, E., 2009. MicroRNA-10b is overexpressed in malignant glioma and associated with tumor invasive factors, uPAR and RhoC. Int. J. Cancer 125 (6), 1407–1413.
Sathyan, P., Zinn, P.O., Marisetty, A.L., Liu, B., Kamal, M.M., Singh, S.K., Bady, P., Lu, L., Wani, K.M., Veo, B.L., Gumin, J., Kassem, D.H., Robinson, F., Weng, C., Baladandayuthapani, V., Suki, D., Colman, H., Bhat, K.P., Sulman, E.P., Aldape, K., Colen, R.R., Verhaak, R.G.W., Lu, Z., Fuller, G.N., Huang, S., Lang, F.F., Sawaya, R., Hegi, M., Majumder, M., 2015. Mir-21–Sox2 axis delineates glioblastoma subtypes with prognostic impact. J. Neurosci. 35 (45), 15097–15112.
Semple, S.C., Akinc, A., Chen, J., Sandhu, A.P., Mui, B.L., Cho, C.K., Sah, D.W., Stebbing, D., Crosley, E.J., Yaworski, E., Hafez, I.M., Dorkin, J.R., Qin, J., Lam, K., Rajeev, K.G., Wong, K.F., Jeffs, L.B., Nechev, L., Eisenhardt, M.L., Jayaraman, M., Kazem, M., Maier, M.A., Srinivasulu, M., Weinstein, M.J., Chen, Q., Alvarez, R., Barros, S.A., De, S., Klimuk, S.K., Borland, T., Kosovrasti, V., Cantley, W.L., Tam, Y.K., Manoharan, M., Ciufolini, M.A., Tracy, M.A., de Fougerolles, A., MacLachlan, I., Cullis, P.R., Madden, T.D., Hope, M.J., 2010. Rational design of cationic lipids for siRNA delivery. Nat. Biotechnol. 28, 172–176.
Shao, H., Chung, J., Balaj, L., Charest, A., Bigner, D.D., Carter, B.S., Hochberg, F.H., Breakefield, X.O., Weissleder, R., Lee, H., 2012. Protein typing of circulating microvesicles allows real-time monitoring of glioblastoma therapy. Nat. Med. 18 (12), 1835–1840.
Shi, Y., Chen, C., Yu, S.-Z., Liu, Q., Rao, J., Zhang, H.-R., Xiao, H.-L., Fu, T.-W., Long, H., He, Z.-C., Zhou, K., Yao, X.-H., Cui, Y.-H., Zhang, X., Ping, Y.-F., Bian, X.-W., 2015a. miR-663 suppresses oncogenic function of CXCR4 in glioblastoma. Clin. Cancer Res. 21 (17), 4004–4013.
Shi, Y., Chen, C., Zhang, X., Liu, Q., Xu, J.-L., Zhang, H.-R., Yao, X.H., Jiang, T., He, Z.-C., Ren, Y., Cui, W., Xu, C., Liu, L., Cui, Y.-H., Yu, S.-Z., Ping, Y.-F., Bian, X.W., 2014. Primate-specific mir-663 functions as a tumor suppressor by targeting PIK3CD and predicts the prognosis of human glioblastoma. Clin. Cancer Res. 20 (7), 1803–1813.
Shi, R., Wang, P.Y., Li, X.Y., Chen, J.X., Li, Y., Zhang, X.Z., Zhang, C.G., Jiang, T., Li, W.B., Ding, W., Cheng, S.J., 2015b. Exosomal levels of miRNA-21 from cerebrospinal fluids associated with poor prognosis and tumor recurrence of glioma patients. Oncotarget 6 (29), 26971–26981.
Silber, J., Jacobsen, A., Ozawa, T., Harinath, G., Pedraza, A., Sander, C., Holland, E.C., Huse, J.T., 2012. MiR-34a repression in proneural malignant gliomas upregulates expression of its target PDGFRA and promotes tumorigenesis. PLoS One 7 (3), e33844.
Silber, J., Lim, D.A., Petritsch, C., Persson, A.I., Maunakea, A.K., Yu, M., Vandenberg, S.R., Ginzinger, D.G., James, C.D., Costello, J.F., Bergers, G., Weiss, W.A., Alvarez-Buylla, A., Hodgson, J.G., 2008. miR-124 and miR-137 inhibit proliferation of glioblastoma multiforme cells and induce differentiation of brain tumor stem cells. BMC Med. 24, 6–14.
Singh, S.K., Hawkins, C., Clarke, I.D., Squire, J.A., Bayani, J., Hide, T., Henkelman, R.M., Cusimano, M.D., Dirks, P.B., 2004. Identification of human brain tumour initiating cells. Nature 432, 396–401.
Skog, J., Würdinger, T., van Rijn, S., Meijer, D.H., Gainche, L., Sena-Esteves, M., Curry Jr., W.T., Carter, B.S., Krichevsky, A.M., Breakefield, X.O., 2008. Glioblastoma microvesicles transport RNA and proteins that promote tumour growth and provide diagnostic biomarkers. Nat. Cell Biol. 10 (12), 1470–1476.

Srinivasan, S., Patric, I.R., Somasundaram, K.A., 2011. A ten-microRNA expression signature predicts survival in glioblastoma. PLoS One 6 (3), e17438.

Stupp, R., Mason, W.P., van den Bent, M.J., Weller, M., Fisher, B., Taphoorn, M.J.B., Belanger, K., Brandes, A.A., Marosi, C., Bogdahn, U., Curschmann, J., Janzer, R.C., Ludwin, S.K., Gorlia, T., Allgeier, A., Lacombe, D., Cairncross, J.G., Eisenhauer, E., Mirimanoff, R.O., European Organisation for Research and Treatment of Cancer Brain Tumour and Radiotherapy Groups, National Cancer Institute of Canada Clinical Trials Group, 2005. Radiotherapy plus concomitant and adjuvant temozolomide for glioblastoma. N. Engl. J. Med. 352, 987–996.

Suh, S.S., Yoo, J.Y., Nuovo, G.J., Jeon, Y.J., Kim, S., Lee, T.J., Kim, T., Bakàcs, A., Alder, H., Kaur, B., Aqeilan, R.I., Pichiorri, F., Croce, C.M., 2012. MicroRNAs/TP53 feedback circuitry in glioblastoma multiforme. Proc. Natl. Acad. Sci. USA. 109 (14), 5316–5321.

Sumazin, P., Yang, X., Chiu, H.S., Chung, W.J., Iyer, A., Llobet-Navas, D., Rajbhandari, P., Bansal, M., Guarnieri, P., Silva, J., Califano, A., 2011. An extensive microRNA-mediated network of RNA-RNA interactions regulates established oncogenic pathways in glioblastoma. Cell 147 (2), 370–381.

Suvà, M.L., Rheinbay, E., Gillespie, S.M., Patel, A.P., Wakimoto, H., Rabkin, S.D., Riggi, N., Chi, A.S., Cahill, D.P., Nahed, B.V., Curry, W.T., Martuza, R.L., Rivera, M.N., Rossetti, N., Kasif, S., Beik, S., Kadri, S., Tirosh, I., Wortman, I., Shalek, A.K., Rozenblatt-Rosen, O., Regev, A., Louis, D.N., Bernstein, B.E., 2014. Reconstructing and reprogramming the tumor-propagating potential of glioblastoma stem-like cells. Cell 157 (3), 580–594.

Teplyuk, N.M., Mollenhauer, B., Gabriely, G., Giese, A., Kim, E., Smolsky, M., Kim, R.Y., Saria, M.G., Pastorino, S., Kesari, S., Krichevsky, A.M., 2012. MicroRNAs in cerebrospinal fluid identify glioblastoma and metastatic brain cancers and reflect disease activity. Neuro Oncol. 14 (6), 689–700.

Teplyuk, N.M., Uhlmann, E.J., Gabriely, G., Volfovsky, N., Wang, Y., Teng, J., Karmali, P., Marcusson, E., Peter, M., Mohan, A., Kraytsberg, Y., Cialic, R., Chiocca, E.A., Godlewski, J., Tannous, B., Krichevsky, A.M., 2016. Therapeutic potential of targeting microRNA-10b in established intracranial glioblastoma: first steps toward the clinic. EMBO Mol. Med. 8 (3), 268–287.

Tian, T., Mingyi, M., Qiu, X., Qiu, Y., 2016. MicroRNA-101 reverses temozolomide resistance by inhibition of GSK3β in glioblastoma. Oncotarget 7, 79584–79595. http://dx.doi.org/10.18632/oncotarget.

Tibshirani, R., 1996. Regression shrinkage and selection via the Lasso. J. R. Stat. Soc. B. Methodol. 58, 267–288.

Tomaselli, S., Galeano, F., Alon, S., Raho, S., Galardi, S., Polito, V.A., Presutti, C., Vincenti, S., Eisenberg, E., Locatelli, F., Gallo, A., 2015. Modulation of microRNA editing, expression and processing by ADAR2 deaminase in glioblastoma. Genome Biol. 16, 5.

van der Vos, K.E., Abels, E.R., Zhang, X., Lai, C., Carrizosa, E., Oakley, D., Prabhakar, S., Mardini, O., Crommentuijn, M.H., Skog, J., Krichevsky, A.M., Stemmer-Rachamimov, A., Mempel, T.R., El Khoury, J., Hickman, S.E., Breakefield, X.O., 2016. Directly visualized glioblastoma-derived extracellular vesicles transferRNA to microglia/macrophages in the brain. Neuro Oncol. 18 (1), 58–69.

Venere, M., Hamerlik, P., Wu, Q., Rasmussen, R.D., Song, L.A., Vasanji, A., Tenley, N., Flavahan, W.A., Hjelmeland, A.B., Bartek, J., Rich, J.N., 2014. Therapeutic targeting of constitutive PARP activation compromises stem cell phenotype and survival of glioblastoma-initiating cells. Cell Death Differ. 21, 258–269.

Verhaak, R.G., Hoadley, K.A., Purdom, E., Wang, V., Qi, Y., Wilkerson, M.D., Miller, C.R., Ding, L., Golub, T., Mesirov, J.P., Alexe, G., Lawrence, M.,

O'Kelly, M., Tamayo, P., Weir, B.A., Gabriel, S., Winckler, W., Gupta, S., Jakkula, L., Feiler, H.S., Hodgson, J.G., James, C.D., Sarkaria, J.N., Brennan, C., Kahn, A., Spellman, P.T., Wilson, R.K., Speed, T.P., Gray, J.W., Meyerson, M., Getz, G., Perou, C.M., Hayes, D.N., Network, C.G.A.R., 2010. Integrated genomic analysis identifies clinically relevant subtypes of glioblastoma characterized by abnormalities in PDGFRA, IDH1, EGFR, and NF1. Cancer Cell 17 (1), 98–110.

Virchow, R., 1864/1865. Die krankhaften Geschwulste. Achtzehnte Vorlesung: Psammome, Melanome, Gliome. Hirschwald, Berlin, pp. 106–169.

Wakimoto, H., Mohapatra, G., Kanai, R., Curry Jr., W.T., Yip, S., Nitta, M., Patel, A.P., Barnard, Z.R., Stemmer-Rachamimov, A.O., Louis, D.N., Martuza, R.L., Rabkin, S.D., 2012. Maintenance of primary tumor phenotype and genotype in glioblastoma stem cells. Neuro Oncol. 14, 132–144.

Wang, Q., Li, P., Li, A., Jiang, W., Wang, H., Wang, J., Xie, K., 2012. Plasma specific miRNAs as predictive biomarkers for diagnosis and prognosis of glioma. J. Exp. Clin. Cancer Res. 31, 97.

Wang, J., Wakeman, T.P., Lathia, J.D., Hjelmeland, A.B., Wang, X.F., White, R.R., Rich, J.N., Sullenger, B.A., 2010. Notch promotes radioresistance of glioma stem cells. Stem Cells 28 (1), 17–28.

Wang, J., Zhao, J., Shi, M., Ding, Y., Sun, H., Yuan, F., Zou, Z., 2014. Elevated expression of miR-210 predicts poor survival of cancer patients: a systematic review and meta-analysis. PLoS One 9 (2), e89223.

Wei, J., Wang, F., Kong, L.Y., Xu, S., Doucette, T., Ferguson, S.D., Yang, Y., McEnery, K., Jethwa, K., Gjyshi, O., Qiao, W., Levine, N.B., Lang, F.F., Rao, G., Fuller, G.N., Calin, G.A., Heimberger, A.B., 2013. MiR-124 inhibits STAT3 signaling to enhance T cell-mediated immune clearance of glioma. Cancer Res. 73 (13), 3913–3926.

Wen, P.Y., Kesari, S., 2008. Malignant gliomas in adults. N. Engl. J. Med. 359, 492–507.

Xu, S., Wei, J., Wang, F., Kong, L.Y., Ling, X.Y., Nduom, E., Gabrusiewicz, K., Doucette, T., Yang, Y., Yaghi, N.K., Fajt, V., Levine, J.M., Qiao, W., Li, X.G., Lang, F.F., Rao, G., Fuller, G.N., Calin, G.A., Heimberger, A.B., 2014. Effect of miR-142-3p on the M2 macrophage and therapeutic efficacy against murine glioblastoma. J. Natl. Cancer Inst. 106 (8). http://dx.doi.org/10.1093/jnci/dju162.

Ye, X.Z., Xu, S.L., Xin, Y.H., Yu, S.C., Ping, Y.F., Chen, L., Xiao, H.L., Wang, B., Yi, L., Wang, Q.L., Jiang, X.F., Yang, L., Zhang, P., Qian, C., Cui, Y.H., Zhang, X., Bian, X.W., 2012. Tumour-associated microglia/macrophages enhance the invasion of glioma stem-like cells via TGF-β1 signaling pathway. J. Immunol. 189, 444–453.

Zhai, H., Acharya, S., Gravanis, I., Mehmood, S., Seidman, R.J., Shroyer, K.R., Hajjar, K.A., Tsirka, S.E., 2011. Annexin A2 promotes glioma cell invasion and tumor progression. J. Neurosci. 31, 14346–14360.

Zhang, C.Z., Zhang, J.X., Zhang, A.L., Shi, Z.D., Han, L., Jia, Z.F., Yang, W.D., Wang, G.X., Jiang, T., You, Y.P., Pu, P.Y., Cheng, J.Q., Kang, C.S., 2010. MiR-221 and miR-222 target PUMA to induce cell survival in glioblastoma. Mol. Cancer 9, 229.

FSC
www.fsc.org
MIX
Papier | Fördert
gute Waldnutzung
FSC® C083411

Kolibri 360 GmbH
Ferdinand-Jühlke-Straße 7
99095 Erfurt, Deutschland
produktsicherheit@kolibri360.de